Women Warriors and National Heroes

Women Warriors and National Heroes

Global Histories

Edited by
Boyd Cothran, Joan Judge, and
Adrian Shubert

BLOOMSBURY ACADEMIC
LONDON • NEW YORK • OXFORD • NEW DELHI • SYDNEY

BLOOMSBURY ACADEMIC
Bloomsbury Publishing Plc
50 Bedford Square, London, WC1B 3DP, UK
1385 Broadway, New York, NY 10018, USA
29 Earlsfort Terrace, Dublin 2, Ireland

BLOOMSBURY, BLOOMSBURY ACADEMIC and the Diana logo are
trademarks of Bloomsbury Publishing Plc

First published in Great Britain 2020
Paperback edition published 2021

Copyright © Boyd Cothran, Joan Judge, Adrian Shubert and contributors, 2020

Boyd Cothran, Joan Judge, and Adrian Shubert have asserted their rights under the
Copyright, Designs and Patents Act, 1988, to be identified as Editors of this work.

Cover design: Tjaša Krivec
Cover image: The Rani of Jhansi on horseback kills an Englishman
with her sword, 1179 (© Art Collection 2/Alamy Stock Photo)

This work is published subject to a Creative Commons Attribution Non-commercial
No Derivatives Licence. You may share this work for non-commercial purposes only,
provided you give attribution to the copyright holder and the publisher.

Bloomsbury Publishing Plc does not have any control over, or responsibility for,
any third-party websites referred to or in this book. All internet addresses given
in this book were correct at the time of going to press. The author and publisher
regret any inconvenience caused if addresses have changed or sites have
ceased to exist, but can accept no responsibility for any such changes.

A catalogue record for this book is available from the British Library.

A catalog record for this book is available from the Library of Congress.

ISBN:	HB:	978-1-3501-2113-3
	PB:	978-1-3502-4041-4
	ePDF:	978-1-3501-2114-0
	eBook:	978-1-3501-2115-7

Typeset by Integra Software Services Pvt. Ltd.

To find out more about our authors and books visit www.bloomsbury.com
and sign up for our newsletters.

To Emiko Cothran, my sister and the bravest of women warriors, and to every woman who has had to fight for a place in this world.

Contents

List of Figures	ix
Acknowledgments	xi
Introduction *Boyd Cothran, Joan Judge, and Adrian Shubert*	1

Part 1 Process

1. India's Rebel Queen: Rani Lakshmi Bai and the 1857 Uprising
   Harleen Singh — 23
2. Historians and Nehanda of Zimbabwe in History and Memory *Ruramisai Charumbira* — 39
3. Women Warriors or Mothers of the Fatherland: Hero Cults and Gender in Basque Nationalism *Nerea Aresti* — 55

Part 2 Violence

4. Murderous Daughters as "Exemplary Women": Filial Piety, Revenge, and Heroism in Early Modern and Modern Japan *Marcia Yonemoto* — 75
5. Women Warriors and the Mobilization of Colonial Memory in the Nineteenth-Century United States *Gina M. Martino* — 93
6. From the Nation to Emancipation: Greek Women Warriors from the Revolution (1820s) to the Civil War (1940s) *Sakis Gekas* — 113

Part 3 Gender Fluidity

7. Madeleine de Verchères (1678–1747): Woman Warrior of French Canada *Colin M. Coates* — 131
8. Jeanne d'Arc, Arab Hero: Warrior Women, Gender Confusion, and Feminine Political Authority in the Arab-Ottoman *Fin de Siècle*
   Marilyn Booth — 149
9. Gender and Transgender in the Mexican Revolution: The Shifting Memory of Amelio Robles *Gabriela Cano* — 179

Part 4 Survivors

10 "Amazons" in the Pantheon? Women Warriors, Nationalism, and Hero Cults in Nineteenth- and Twentieth-Century Chile and Peru *Gabriel Cid* 199

11 Commemorating China's Wartime Spies: Red Agents Guan Lu and Jiang Zhuyun, and the Problem of Female Fidelity *Louise Edwards* 217

12 Vietnam's Martial Women: The Costs of Transgressing Boundaries *Karen Gottschang Turner* 233

Index 252

Figures

1.1	Popular Congress Party Election Posters depicting Sonia Gandhi as the Rani of Jhansi	33
3.1	Polixene Trabudua and Haydée Aguirre, both EAB leaders, leaving Larrinaga Prison, Bilbao, 1933	62
3.2	Cover of Manuel de la Sota's *Libe: Melodrama histórico* (1903, Bilbao: E. Verdes Achirica, 1934)	63
4.1	Utagawa Hiroshige, "Shiroishi banashi (The Tale of Shiroishi)," from Chūkō adauchi zue (Illustrations of Loyalty and Vengeance), 1844–45. Photograph © 2019 Museum of Fine Arts, Boston, William Sturgis Bigelow Collection	79
4.2	Tsukioka Yoshitoshi, "The Courtesan Miyagino and Her Younger Sister Shinobu," from *Kōkoku nijūshikō*, 1881. Photograph © 2019 Museum of Fine Arts, Boston, William Sturgis Bigelow Collection	80
5.1	"New England Kitchen Scene" in Joel Dorman Steele, *A Brief History of the United States* (New York: A. S. Barnes & Company, 1885), 94	94
5.2	"Weetamoo Swimming the Matapoisett" by C.S. Reinhart and S.A. Schoff in William Cullen Bryant and Sydney Howard Gay, *A Popular History of the United States*, vol. 2 (New York: Charles Scribner's Sons, 1883), 405	95
6.1	Georgios Miniatis. *Souliotisses*. Second half of the nineteenth century, Public Gallery, Corfu	116
6.2	Spiros Meletzis, "Determined for everything," featuring Eleni Panagiotidou and an unknown soldier from the collection *With the Partisans in the Mountains*, Agrafa: 1944, 242	121
7.1	Gerald Sinclair Hayward, "Madeleine de Verchères," Library and Archives Canada, Acc. No. 1989-497-1, C-083513	137
7.2	Statue of Madeleine de Verchères, Verchères, QC, about 1925. Archives de la Ville de Montréal, BM42-G0464	139
7.3	Still photograph from scene of motion picture *Madeleine de Verchères*, Library and Archives Canada, PA-028626	142
8.1	Frontispiece to *al-Hilal* 4: 4 (October 15, 1895). Private collection of M. Booth	158

8.2	Jeanne D'Arc, from the statue by Princess Marie of Orléans in the gallery at Versailles, from Mark Twain, *Personal Recollections of Joan of Arc*. Courtesy of Wikimedia Commons	159
9.1	Colonel Robles School, Xochipala, Guerrero in 2012. Photo credit: Gabriela Cano	180
9.2	Community Museum Coronela Amelia Robles in 2012. Photo credit: Gabriela Cano	181
10.1	Irene Morales, *c.* 1881 (National History Museum)	207
10.2	Leonor Ordoñez Monument, Huancaní, Perú	212
11.1	A glamorous Guan Lu as a celebrity author. Photo credit: Unknown	219
11.2	Jiang Zhuyun, her husband, and their son. Photo credit: Unknown	227
12.1	Young People on the Trail, date unknown. Courtesy of Colonel Le Trung Tam	235
12.2	Veterans of Company 814, Hanoi, 1996. Courtesy of Karen Turner	242

Acknowledgments

We offer our thanks to the many people who made this book possible. Our colleague Marcel Martel was the first and greatest champion of this project. We thank him for his support, both financial and logistical, in organizing the Women Warriors and National Heroes conference at York University in September 2018 as part of the Annual Avie Bennett Historica Canada Conference from which this book was born. As is always the case with international conferences, it took an entire village or university to pull this one off. For their support and co-sponsorship, we thank the following units at York University: the Department of History, the Department of Humanities, the Office of the Vice-President Research & Innovation, Founders College, the Faculty of Liberal Arts & Professional Studies, the Robarts Centre for Canadian Studies, the York Centre for Asian Research, and the Centre for Research on Latin America and the Caribbean. We owe a particular debt of gratitude to several individuals in the History Department: Thabit Abdullah, the Chair, and the department's highly competent and supportive staff, including Lisa Hoffmann, Patricia Di Benigno, Anita Szucsko, and Jeannine Flint. Thank you to Bàrbara Molas, who ably swooped in at the last second to prepare the index. And we also extend thanks to Spain Arts and Culture and the Cultural Office of the Embassy of Spain in Canada for their support.

This book has benefited greatly from the wisdom, generosity, and patience of our editors and the production staff at Bloomsbury Academic. We would like to thank especially Maddie Holder for enthusiastically supporting this project from the beginning, and Dan Hutchins who has cheerfully and effectively shepherded the manuscript through the design and production process.

Finally, perhaps our greatest debt is to the contributors to this book. They have all put an immense amount of time and labor into writing and rewriting their own fine essays, and have collectively, generously, and creatively thought globally about the woman warriors' problematics in ways that have shaped the volume.

Introduction

Boyd Cothran, Joan Judge, and Adrian Shubert

In February 2018, the government of Saudi Arabia announced that women between the ages of twenty-five and thirty-five would now be permitted to join the armed forces, although they would continue to be barred from combat roles. It was the latest, and most surprising, addition to a still-small group of countries whose armed forces are open to women. Eight months later, the UK made all parts of its armed forces, including frontline infantry and the Royal Marines, available to women.[1] Twenty years earlier, in 1988, Norway and Israel had become the first countries in the world "to completely abolish gender barriers in the armed forces."[2] This recent, and belated, acceptance by the modern nation-state of women's war-making potential reflects the reality that at almost all times and in almost all places fighting wars, and being recognized for doing so, has officially been men's business.[3]

In spite of this, stories of women taking up arms and receiving national recognition for having done so are a global phenomenon—and an enduring one. Not only do the ongoing emergence of new political movements and the creation of new polities generate the creation of new heroes, but changing political and social currents within existing states promote the rediscovery, re-appropriation, and redefinition of old ones. The twelve cases discussed in this volume represent only the tip of the iceberg. The very long list of others, touching almost all parts of the world, would include, to name a few: Njinga (Angola), Juana de Azurduy (Bolivia), Mulan (China), Cut Nyak Dhien (Indonesia), Queen Nanny (Jamaica), Mekatilili wa Menza (Kenya), Emilia Plater (Poland), Agustina de Aragón (Spain), Gabriela Silang (Philippines), Molly Pitcher (United States), and Manuela Saenz (Venezuela).

Turning women warriors into national heroes is not a straightforward process, and even once a woman—or group of women—has been recognized, their heroic status remains both contingent and contested. Some figures have remarkable staying power. The most famous, as well as possibly the most contentious, is undoubtedly Jeanne d'Arc, the peasant girl who led French armies during the Hundred Years War in the early-fifteenth century, and who became a global icon of female heroism and patriotism from the late-nineteenth century. She has served as "a magic mirror of personal and political idealism and, in particular, of changing ideas about women's heroism," according to historian and mythographer Marina Warner.[4] Jeanne d'Arc has been claimed by clericals and republicans in the nineteenth century, by Vichy and La Résistance during the Second World War, and, most recently, by both the far-right Front National and gay rights activists. She is "the heroine every movement has wanted as its figurehead."[5] She has also become the standard reference for women warrior figures worldwide, as Marilyn Booth's chapter in this volume demonstrates.

While having less global reach, the stories of countless women warriors across the continents have been appropriated to a range of political ends. Agustina of Aragón, who was immortalized for firing a canon at the French army besieging Zaragoza during Napoleon's invasion of Spain in 1808, has been a hero for all seasons ever since, being mobilized by both sides during the Spanish Civil War and made the subject of motion pictures in 1929 and 1950.[6] Boudica, or Boadicea, the warrior queen who led a failed revolt against the Romans in the first century CE, has been conscripted into a number of roles over the last two hundred years: "harbinger of the British Empire, a figure of nationalism, a symbol for suffragists or a supporter of Brexit," according to historian Caitlin Gillespie.[7] The Chinese Hua Mulan, who was celebrated in a sixth-century ballad for disguising herself as a man and fighting for twelve years in her infirm father's place in the ultimately victorious Toba Wei army, was long heralded as a paragon of filiality first and bravery second. From the turn of the twentieth century, she was repeatedly recast, however, as a militant female citizen and defender of her race in the context of the 1911 Republican Revolution, and as a "new woman" fighting for equal rights in the 1920s and 1930s.[8] The heroine of a number of China's earliest feature films, her story of singular bravery is so universally compelling—and fungible—that it was made into a Disney feature film in 1998.

Mulan, Boudica, Agustina, and Jeanne d'Arc have become timeless heroines; other women warriors lose their initial luster over time or are quickly consigned to oblivion. Such is the case with long-forgotten Spanish women engaged in the same struggle against Napoleon as the enduring Agustina de Aragón: among them were members of the all-woman Santa Barbara Company who fought to defend the city of Gerona against a French siege, and Casta Alvarez and María Agustín, who also took part in the defense of Zaragoza.[9] The Chilean heroes of the Confederation War described by Gabriel Cid in his chapter met a similar fate. As did the Modoc woman Toby "Winema" Riddle in the United States. Widely celebrated on the stage and in the dime novels of the late-nineteenth century as a "Woman Chief" and the "Pocahontas of the Lava Beds" for helping her people navigate their conflicts with Americans, and one of the few Native American women to be awarded a military pension by the United States Congress for her heroic actions during the Modoc War of 1872–73, her memory has been consigned to a national forest named in her honor in southern Oregon.[10]

Still others have been revived after being long ignored. The "Heroines of Cochabamba" (Bolivia), a group of mixed-race women who died defending the town against the Spanish in 1812, were recognized only in the 1920s. The monument, the work of a group of elite women from the city and intended to inspire patriotism, proved controversial precisely because the choice of mestizo women to represent the struggle for independence "ran the risk of suggesting other visions of that history and of the nation."[11] Juana de Azurduy, a mixed-race woman from Upper Peru, what is now Bolivia, who fought against Spanish rule together with her husband, came to command an army of 6,000 men, reaching the rank of lieutenant colonel. After dying in poverty and being forgotten for a century, she was declared a national hero by the Bolivian government in 1962, "heroine of the Americas" by an international Latin American commission in 1980, and since 2006 her name has been frequently invoked by President Evo Morales's pro-indigenous government. She was also the subject of "exaltation" by Argentina's president Cristina Kirschner: a 2007 law declared July 12, her birthday, the "Day of the Heroines and Martyrs of American Independence;" in 2009 she was posthumously promoted to the rank of general; and she has been directly connected to contemporary struggles, such as that of the Mothers of Plaza de Mayo. On the other hand, her presentation as an indigenous hero has been challenged by a number of indigenous groups in Argentina.[12] When Kenya's new 2010 constitution enshrined an official Heroes Day and the subsequent Kenyan Heroes Act set out plans for an official pantheon, one frequently invoked name was Mekatilili wa Menza, a woman from the Giriama people who was central to their resistance to the British in 1913. In the search for national unity after the election violence of 2008, this woman from a marginal people was seen as "an unproblematic symbol of national reconciliation, unlike the fraught memorialization of the Mau Mau."[13]

Clearly, stories of women warriors who became national—or transnational—heroes are a nearly ubiquitous phenomenon. While many of these stories have been rehearsed by amateur historians,[14] and numerous scholarly studies have focused on individuals or groups of women warriors within their specific national context,[15] the phenomenon of women warriors has not been broadly or systematically explored. This is in contrast to their male counterparts. The topic of male warriors and hero cults has, in recent years, received considerable attention. This has been especially true in Latin American[16] and European[17] history but also in African[18] and even Australian history.[19] Overwhelmingly, these studies have emphasized the centrality of male virility in the development of national hero cults. Some have also observed the power of war and violence to destabilize established gender norms, creating the space for women warriors to enter the canon.[20]

The stories of women warriors and national heroes, thus, call out for sustained examination across space and time. In order to interrogate the enduring qualities and inherent tensions in these stories, to understand their evolution over time, and to problematize the construction and deployment of real or imagined figures in different national and transnational contexts, it is necessary to widen the lens of analysis beyond the specific nation-states from within which these stories have typically flourished. To that end, this volume seeks to understand the myth of the women warrior as a transnational and global phenomenon.[21] It highlights both common patterns that have emerged across cultures through the telling and retelling of these stories, and points of divergence that reveal the specificities of each local context. In telling these global histories, we ask: what kinds of cultural work did these figures do at particular historical moments, and when and why do their stories endure? Which heroines traveled in space and time and which did not? What forms of media have been used to circulate these stories and how have these forms influenced the stories they tell? Exploring these questions is critical to understanding the ways tales of woman warriors have operated globally throughout history.

Networks of Circulation: Toward Global Histories of Women Warriors and National Heroes

One of the key goals of this volume is to emphasize that stories of women warriors were not only a global phenomenon insofar as they appeared in many different regions and countries of the world; rather, they were global because they traveled. We emphasize four routes of travel or patterns of circulation of women warrior stories: (1) The circulation of Western heroines to corners of the non-Western world particularly from the mid-nineteenth century, (2) the circulation of stories of women warriors from the colonies in the metropole, (3) the circulation of heroines from one non-Western location to another non-Western location, and (4) the circulation of tales of heroines within the Western world.

Existing research on the prominence of Western heroines in non-Western women's periodicals, textbooks, and biographical collections—in Egypt, India, Japan, China, and elsewhere—suggests that a stock pool of Western female biographies—and most likely a stock set of texts—circulated in the non-Western world from the late-nineteenth century. We catch a glimpse of this mode of global circulation in Marilyn Booth's chapter that demonstrates how Jeanne d'Arc featured in the Arab world from the late-nineteenth century. Jeanne's stature as a patriot, a warrior, and a martyr who was on her way to sainthood around the time her story began to circulate globally, made her the woman warrior par excellence. As with all women warriors, her story was tweaked and reshaped in order to serve different ends in its various retellings. Some versions emphasized her religiosity while others elided it, some skimmed over her gender inversion while others celebrated or occluded it, but all versions underscored Jeanne's bravery and patriotism.[22]

In her chapter, Booth describes Jeanne's trajectory from fifteenth-century France to nineteen- and twentieth-century Egypt. But Jeanne underwent myriad transformations and followed many routes of travel. She was exceedingly popular in India and East Asia as well as in the Middle East.[23] Translations of her story began to appear in Japanese, for example, from the 1880s. Some of these were renderings of English works, including Janet Tuckey's *Joan of Arc: "The Maid"* which was published in New York in 1886 and translated into Japanese by Awaya Kan'ichi 粟屋関一 in 1887.[24] Others came predictably from a French hand, including one written in Japanese by the Christian missionary François-Alfred Désiré Ligneul (1847–1922) and published in 1910.[25]

Jeanne d'Arc's tale was also included in at least one of the many volumes of biographies of heroic Western women published in Japan at the turn of the twentieth century: a 1902 collection of *Twelve World Heroines*. Here she appeared alongside a number of other, later European women of valor who traveled extensively in the non-Western world at the turn of the twentieth century. They include Madame Roland (1754–93), who bravely faced the guillotine on November 8, 1793, in the midst of the terror that followed the French Revolution; Charlotte Corday (1768–93), who stabbed the Jacobin leader Jean-Paul Marat (1743–93) in his bath in 1793; Anita Garibaldi (1821–49), the heroic Brazilian-born wife and comrade-in-arms of the Italian revolutionary Giuseppe Garibaldi (1807–82); and Louise Michel (1830–1905), the irrepressible anarchist who was deported to New Caledonia in 1873 for her involvement in the 1871 Paris Commune.[26] Several of these heroines made their way to China—another stop in the global circulation of Western heroines—via Japanese collections such as the *Twelve World Heroines*.[27] As in all instances of translation, the Chinese rendering of the Japanese accounts were reinterpretations that aligned the stories of intrepid Western women warriors with the vision and political agenda of the Chinese translator. In the case of Jeanne d'Arc, for example, the secular reformist translator attempted to dispel any elements of religiosity from her tale as it appeared in the Japanese collection.[28]

While the stories of Western women warriors were inevitably reshaped when they were appropriated into non-Western contexts, the process of appropriation was more violent when tales of women warriors flowed from the non-West to the West, particularly, but not exclusively, at moments of colonization and military resistance. The re-inscription of these women warriors often took one of two forms. The colonial woman warrior was presented either as a radical and hyper-sexualized or even deviant figure, or as an unnatural and inhuman genderless aberration.

Queen Njinga Ana de Sousa of Angola, the seventeenth-century Angolan ruler first introduced to European audiences through the writings of Capuchin missionaries, is an example of the former. By the eighteenth century, she had become a sensation amongst European audiences, especially in the salons of French avant-guard intellectuals and artists. According to historian Linda Heywood, the image of Njinga that they constructed stripped the Angolan queen of her historical and religious complexity, and occluded her statesmanlike prowess. Between the 1780s and 1850s writers such as Marquis de Sade in *La Philosophie dans le Boudoir* (*Philosophy in the Bedroom*), Laure Junot, Duchess of Abrantés, in *Les femmes célébres de tous le pays* (*Celebrated Women of All Countries*), as well as German philosopher G.W.F. Hegel, and erotic lithographer Achille Deveira, all contributed to a new legend of Njinga as a gendered, sexualized, savage, and blood-thirsty other, a female fiend meant to serve as a warning to elite women who yearned for power. For Europeans, Njinga became a "stand-in for the African other."[29]

Rani Lakshmi Bai, the Queen of Jhansi who fought against the British in the 1857 First War of Independence, was somewhat better treated in subsequent metropolitan representations, as Harleen Singh explains in her chapter. On the one hand, she was a "rapacious whore, effectively 'raping' the white British male and a murderous despot responsible for the massacre of some sixty British men, women and children." On the other hand, she was "an Aryan model of heroic womanhood comparable to Joan of Arc." And as Singh points out, the Rani did not sit easily with Indians, whose representations struggled with both ambiguities and paradox.

Escalating tensions between Japan and the United States during the Second World War served as another context in which non-Western women stood in for the abhorrent features, in this case, of a military enemy of the West. Between 1942 and 1944, American media projected frightening images of dangerously armed Japanese women in uniform. This distortion—Japanese women were not mobilized in the war effort until 1945—served two local purposes. The first was to legitimize the extreme measures the United States was to take in neutralizing the Japanese enemy by emphasizing that America was at war with an unnatural foe that blurred both gendered and domestic/battlefield boundaries. The second purpose was to galvanize American women: if their Japanese counterparts were so valiantly serving their country, American women had to do the same.[30]

The global circulation of women warriors was not only a story of colonial entanglements between the West and the rest, however. The circulation of these stories amongst and between non-European societies was equally important, revealing the extent to which non-Western figures could mutually inspire and bolster south-south solidarity. Queen Njinga, for instance, traveled not only to Europe and North America but to Latin America and the Caribbean. Long a part of Afro-Brazilian culture in northeastern Brazil, by the early-twenty-first century, big-city samba poets were writing lyrics "praising her bravery and what they interpreted as her promotion of women's rights and black power." She has also been the heroic subject of an "outpouring" of plays, poems and other works in such varied places as Cuba, Jamaica, and the United States.[31]

Non-Western women warriors also circulate in more concrete forms in other non-Western contexts as a marker of solidarity. The recently renamed Plurinational State of Bolivia presented Argentina with a statue of the Bolivian woman warrior and national hero Juana de Azurduy as a gift at the time of the bicentennial of Argentina's independence. The massive monument: a nine-meter-high statue standing atop a seven-meter plinth, which President Cristina Kirschner of Argentina dedicated on July 15, 2015, served as a mechanism for erasing the imperial past. Juana de Azurduy's massive form replaced a statue of Christopher Columbus.[32]

Women warriors also circulated extensively within the Western world. Jeanne d'Arc is, of course, a standard reference across Europe. Agustina of Aragón is another example. She achieved a certain measure of fame in the English-speaking world after appearing in Byron's famous poem *Childe Harold's Pilgrimage* (1812–14). She appeared in more than 900 collections of female biographies published in the UK and the United States between 1830 and 1940.[33] A very feminine Agustina also formed part of a group of warriors that included Jeanne d'Arc, Boadicea, Agnes of Dunbar, Emilie Plater, the Rani of Jhansi, Christian Davies, Hannah Snell, and Mary Ann Talbot, in suffragette Cicely Hamilton's stage spectacle, *A Pageant of Great Women*, which was performed across Britain in 1909.[34]

In addition to inspiring poets and politicians, stories of women warriors also inspired other women warriors, gesturing to a connected history of women's active defense of nation or homeland. In one Arab retelling, Jeanne d'Arc was, anachronistically, inspired by the patriotic women of Bohemia. The author of this biography of Jeanne had most likely been aware of the Bohemian Amazons who were ensconced in Czech nationalist and feminist imaginaries by the 1880s, and drew fascinated attention from writers across Europe. Libuse and Vlasta were mythical warrior heroines who were integrated into the Czech national culture that emerged during the nineteenth century and, in the process, were converted by "the mainstream—male—national discourse from 'marginal ... even negative' figures into 'new patriotic ones.'" Even in their new guise, these women warriors had to be managed, their potentially feminist implications obscured. But the creators of the national discourse could not fully control how they were understood, and starting in the 1860s "women cited them when asking Czech men to support their emancipation."[35] However, as contexts changed, existing women warriors such as Jeanne d'Arc and Agustina of Aragón could go from being embraced as exemplars to rejected as irrelevant, as Nerea Aresti demonstrates.

Creating Heroes, Creating Nations: Women Warriors and Nationalism in Global History

Women warrior stories circulated and flourished during distinct global historical moments. Two periods emerge from this volume as important. The first is the final decades of the nineteenth century and first decades of the twentieth century, a period of nation-building through the invention of traditions in many parts of the world. Heroic victories and sacrificial defeats, foundational events, and sacred places were all part of these invented traditions. So too were heroic figures who defended their nation or community against an invading force and thus came to personify the nation-state. Although generally men, storytellers also turned to tales of women warriors at the turn of the twentieth century. These stories often appealed to a broad array of readers and spoke particularly to women, who sought to define their place in the emergent nation-state. The end result was that the late-nineteenth and early-twentieth centuries were a particularly prolific moment in the development of both nationalism and the canonization of women warriors.

This process of national heroic mythmaking was especially germane to countries or sub-state communities struggling to establish themselves at the turn of the twentieth century. As Nerea Aresti and Colin Coates demonstrate in their chapters, emerging sub-state nationalisms, in the Basque Country and Quebec respectively, included women warriors from the distant past in their nascent pantheons.[36] In both of these intensely Catholic contexts, however, their promoters struggled with how to represent them. Both Quebecois and Basque manufacturers of heroes, nonetheless, opted for warriors who, as Coates says, had to "emphatically, remain a woman," although in the Basque case these figures were subject to two distinct gender ideologies that ascribed very different feminine qualities. Ultimately, a woman warrior's womanhood was an aspect of this invented tradition that had to be managed.

This was true beyond the Catholic context: storytellers had to contain the more radical or unruly aspects of the tales of women warriors in all of the grand national narratives of the late-nineteenth and early-twentieth centuries. As several authors in this collection demonstrate, many of the purveyors of these stories of women warriors espoused fairly conservative social values. For every Zaynab Fawwaz who subtly used the story of Jeanne d'Arc to champion women's rights, there was a conservative Lebanese writer like Hanna Kurani or the anonymous commentators in *Lisan al-hal*, the Beirut-based newspaper, who penned articles that used the image of the Amazons to mock British suffragists and their calls for greater political rights for women. In an age of anxiety over increasing political rights for women across the globe, successive retellings of these stories throughout the late-nineteenth and early-twentieth centuries often downplayed the women warrior's more radical potential, and even used their stories to promote conservative agendas.

If the women warrior could serve as both a national and conservative figure during the late-nineteenth and early-twentieth centuries, she embodied equally complex and contradictory roles in the tumultuous period during and after the Second World War. In a period of widespread total war in which the boundaries between civilian and combatant were shattered, opportunities for connecting contemporary stories of women combatants with historical examples from the national past abounded. These stories served to address questions of occupation, resistance, and civil war as in the example of Greece in the 1940s documented by Sakis Gekas. They were also deployed in the context of anti-colonial and national liberation struggles against, successively, the Japanese, the French, and the Americans in Vietnam, as demonstrated by Karen Turner, and between the Nationalists and Communists in China as discussed by Louise Edwards. These stories often connected the experiences of women warriors engaged in these struggles with exemplary female figures from the past such as the ancient Vietnamese poetess Ho Xuan Huong and the "female admiral" of the Greek War of Independence, Bouboulina. But as with the nationalist movements of the late-nineteenth and early-twentieth centuries, these narratives had to navigate traditional gender hierarchies and ideals, which modern warfare threatened to invert or upend.

The often-violent process of decolonization in the decades after the Second World War saw the most extensive creation of new nation-states in history: thirty-six by 1960 alone, and many more in the 1960s and 1970s. These new nations needed new pantheons—and heroes to fill them. Some, like the Rani of Jhansi discussed by Harleen Singh in her chapter in this volume, were women warriors who could be connected in some way to a tradition of struggle against the imperial power. Indonesia's official pantheon of National Heroes, which was initiated in 1958, includes two women warriors from the nineteenth-century struggles against the Dutch. Cut Nyak Dhien (1848–1908), who succeeded her husband as military leader of the Aceh in the lengthy Aceh War (1873–1904), was made a National Hero in 1964, and appears on stamps and paper money. She was also the subject of the 1988 Academy Award-nominated film, *Tjoet Nja' Dhien*. Martha Christina Tiahahu (1800–18), who participated in a guerrilla struggle against the Dutch in Molucca, was captured twice, and died en route to serving a sentence as a slave laborer, was honored in 1969. January 2, the day she died, was proclaimed Martha Christina Tiahahu Day.[37] In Jamaica, Queen Nanny was the first Maroon, as well as the first woman, to be named to the Order of the National Hero in 1976. She had led Maroons, slaves who had escaped from plantations and established independent communities, in the First Maroon War (1720–39), ultimately winning British governmental recognition of their land rights. She appears on the $500 bill; was the subject of a 2015 movie, *Queen Nanny: Legendary Maroon Chieftainess*; and has been claimed as an ancestor by Portia Simpson-Miller, Jamaica's first female prime minister (2006–07, 2012–16).[38]

Perhaps the most striking phenomenon of this kind is the canonization of the seventeenth-century Queen Njinga of Angola, whom we have already introduced as the hyper-sexualized sensation amongst European audiences. In the context of the anti-colonial struggle of the 1960s and 1970s, she was kept alive in oral culture as a "proud ruler" and "a revolutionary hero who united her people in an epic struggle against Portuguese aggression." The post-independence government made her a centerpiece of national history taught in the schools.[39] Following the long civil war that ended in 2002, she was mobilized as a symbol of national unity and "mother of the Angolan nation."[40]

The array of global woman warriors is so vast, not all fit into the two time frames highlighted here, nor necessarily into the context of the nation-state. Guadeloupe Solitude (c. 1772–1802) presents a distinctive case in that she was not a heroine who became an icon of a newly liberated state, rather she fought on behalf of a *département* of France and struggled against the actions of the state. Sometimes known as "Mulatto Solitude," Guadeloupe Solitude fought, while pregnant, with a band of insurgent ex-slaves and *gens de couleur* in 1802 against the army Napoleon had sent to the island colony to re-impose slavery that had been abolished in 1794. She was captured and executed the day after giving birth. Solitude remained alive almost exclusively in oral tradition in Guadelupe, but began to appear in novels from 1972. In 1999, a statue of a heavily pregnant Solitude was erected in Point-à-Pitre, the island's largest city, and soon became a "site for a variety of commemorations and political gatherings."[41] She has also been recently commemorated in the metropole. In 2007, the French town of Bagneux inaugurated a statue in her honor intended as a "homage to and recognition of the victims and resistors of the slave trade and slavery."[42] There have also been demands from both feminists and black organizations that Solitude, along with Olympe de Gouges, be admitted to the Panthéon, the holy of holies of French national memory which at the time only included two women among the seventy-three honorees.[43]

The two Latin American cases Gabriel Cid discusses in his chapter are distinctive in terms of chronology too. The women warriors he describes fought in wars between the independent states of Chile and Peru, first in the 1830s and then in the 1880s rather than in the period when these former Spanish colonies struggled for their independence (1810 to 1825). The Chilean warrior hero of the first war, Candelaria Pérez, had been acclaimed, forgotten, and revived by the early 1870s while the second conflict, the War of the Pacific (1879–84), was followed by four decades of generating a series of war-related hero cults. The zenith of hero-making in the Peruvian case, in contrast, came in the context of a decolonization of official national narratives in the 1960s and 1970s under a left-wing military regime that gave women, together with indigenous and mixed-race people, a central role. Gabriela Cano's astonishing story of Amelia/o Robles, the transgendered Zapatista fighter of the Mexican Revolution, also follows a different chronology.

Whether in service of new political regimes, decolonization, ascendant nationalism, or the struggle for sub-state identities, stories of women warriors have played a distinct role in global history. In some instances, these stories have drawn on real historical figures with some moderate embellishment or propagandistic reconfiguration such as the narratives of Jiang Zhuyun and Guan Lu analyzed in Louise Edwards's chapter; the melding of an individual woman with a political-religious title described by Ruramisai Charumbira; or the stories of the colonial woman warrior Hannah Duston and Indigenous female warrior and sachem Weetamoo discussed in Gina Martino's chapter; at other times, the stories are complete fabrications as in the case of the intrepid heroines Miyagino and Shinobu featured in Marcia Yonemoto's chapter and Libe, the champion of the Battle of Munguía described by Nerea Aresti. In either case, these stories of women warriors have been repeatedly reproduced in a range of sources, largely because they have played such powerful social roles. These widespread productions and reproductions have given rise to a series of shared problematics that weave through these stories and have given shape and form to this volume.

Problematizing Women Warriorhood: Themes and Methodology

The phenomenon of the women warrior has multiple sources and manifestations. When considered in a global and comparative perspective, as in the case of this volume, however, significant problematics surrounding the ways these real or imagined figures were constructed and deployed come into focus. This volume is organized around four such problematics which emerged out of and are explored through specific case studies in individual chapters.

The three chapters in **Part One** focus on the processes of heroine-making—and unmaking over time. They examine questions of power and authority as well as the role of professional and institutional history in the production, circulation, and strategic redeployment of stories of women warriors. Harleen Singh examines the process through which the radical, seditious, and even anti-nationalist Queen of Jhansi, Rani Lakshmi Bai who led troops against the British during the Uprising of 1857, was domesticated to become a celebrated icon of traditional forms of kinship, femininity, and Indian nationalism. Ruramisai Charumbira analyzes how the title of "Nehanda" a politico-religious position, was attached in the mid-1890s to a woman commanding a rag-tag army of truculent "Natives" resisting British colonialism in what is today Zimbabwe. After her execution, this Nehanda underwent multiple transformations from maligned witchdoctor fighting the march of European civilization in Africa to a symbol of national resistance and a beacon of powerful womanhood. Nerea Aresti highlights the tensions that drove the multiple iterations of women warriors in the Basque nationalist movement over time: they could embody gender inferiority and racial virility, the weaknesses of their sex and an exultant exceptionalism, and they could assume the role of "mothers of the nation" without escaping their subordination.

Part Two examines the critical disjuncture in the very term "woman warrior": a juxtaposition of the feminine qualities attendant in the word "woman," and the requisite violence that is fundamental to the role of "warrior." The chapters in this section highlight multiple representations of violent women. Both Gina Martino and Marcia Yonemoto analyze case studies in which righteous revenge justifies female violence. Their two cases, set in colonial America and early modern Japan respectively, nonetheless highlight widely divergent degrees of cultural acceptance of women's violent acts. In the American context, female violence is a redemptive, ultimately civilizing component of colonialism. It is undertaken in the wilderness, far away from the domestic, private sphere. In the early modern Japanese case, violence is clearly situated within the domestic realm: filial devotion—in this case the extreme loyalty of daughters to father—transformed murderous violence into upright devotion. Finally, Sakis Gekas offers a third justification for female violence in his case study exploring representations of women warriors in modern Greece: drawing upon the memory of legendary women warriors from the country's past, women fighters on the Resistance side of the civil war of the 1940s donned uniforms and rifles to fight for both political rights and national liberation; their inherent vulnerability rendered their violence acceptable.

The gender subversiveness of women warriors—women who took to the battlefield alone, alongside or in the place of men—was as problematic as their violence. The standard mechanism for dealing with this subversiveness was to have these intrepid heroines revert to their feminine roles shortly after performing their feat of bravery.[44] The chapters in **Part Three**, "Gender Fluidity," probe both the question of how women warriors strategically used gender ambiguity and how those around them attempted to manage it. In Gabriela Cano's remarkable case study, Amelia Robles took part in the 1910 Mexican Revolution as a combatant in Emiliano Zapata's popular agrarian movement as Amelio Robles, a transgendered male. While he successfully upheld this identity until his death in 1984, the state would attempt to overturn it thereafter. In Colin Coates's chapter, in contrast, Madeleine de Verchères (1678–1747) plays on gender ambiguity to highlight her heroism in French Canada—she outran the Indigenous attackers of her settlement, ignored crying women, donned a soldier's hat, shot a cannon, and admitted to being as drawn to glory as men—without, however, abandoning her feminine identity. In her chapter on Jeanne d'Arc in Egypt, Marilyn Booth underlines the extent to which retellers of tales of women warriors shape the gendered valence of those stories. She delineates the fleeting gender fluidity inherent in representations of Jeanne in the 1890s, a fluidity that would be absorbed into a rigid role-separation in both tales of Jeanne and Egyptian society more broadly in the successive decades.

Jeanne d'Arc's gender fluidity could be differently emphasized by different authors in different periods because she lived centuries before these turn-of-the-twentieth-century retellings of her story. Narratives of women warriors are significantly less malleable when the women live on. While martyrdom for a righteous cause unproblematically confers heroic status, women who outlive their deeds pose significant challenges. The three chapters in **Part Four** grapple with this question of the treatment of women warriors who lived on. Gabriel Cid contrasts the situation in Chile and Peru: in the former case women soldiers were recognized during their lifetime but swiftly forgotten, whereas in Peru, living combatants were not officially acknowledged until after their deaths. The first woman warrior to be recognized officially, "Sergeant" Candelaria, was hailed as Chile's Joan of Arc at the time but was soon forgotten, surviving for decades in official oblivion—if not popular memory—and increasing penury, before a brief return to the limelight after her death. Louise Edwards compares the treatment of two female spies involved in the Chinese Communist Party's struggle to gain control of China, one who is martyred and another who lives on. She demonstrates that the female spy who survives is never cleansed of the suspicion that she has insufficient fidelity, sexual or political, to warrant commemoration. Finally, Karen Turner poignantly outlines the fate of young women who compromised their health and fertility in answering the Vietnamese government's call to war, only to be confronted with ostracism and judgments of impurity in the war's aftermath.

The Enduring Woman Warrior

This volume is principally a historical examination of the global phenomenon of the woman warrior and how her stories circulated over time, across borders, and in various textual and visual media. Another important feature of these global histories is the ways their stories continue to circulate through new forms of media such as film, television, graphic histories and novels, and the internet. Similar to the retellings and translations of these stories in the past, these new forms of media often reshape the narratives to suit particular ends. When Hua Mulan's story was retold by Disney, for example, it required a Hollywood ending: while in the original ballad she hands in her camel and returns, the ever-filial daughter, to hearth and home; in the movie, she marries.

These media retellings predate the early-twenty-first century. Mulan had already been the subject of a number of films in China from the 1920s. Almost all of the woman warriors discussed in this volume had at least one, and often several, transmedial afterlives. A statue of Madeleine de Verchères (1678–1747), the figure at the center of Coates's chapter, was erected in 1913 and she was the subject of the first French-Canadian feature-length film, released in 1922. Amelia Robles (1885–1984), the subject of Cano's chapter, was honored in a modern dance performance called *La Coronela*, which premiered in Mexico in 1940. Local stone monuments to Shinobu and Miyagino, the avenging early modern Japanese sisters whose story Yonemato analyzes, were erected in the countryside in the 1960s. Duc Hoan presented unsanitized film versions of the lives of female combatants in the Vietnam War in the post-war era, and Nehanda of Zimbawe was eulogized in a 1993 novel. As Martino discusses in her chapter, Hannah Duston, the colonial Massachusetts Puritan mother who killed and scalped her Indigenous captors, is believed to be the first American woman honored with a statue in 1874.

The pace has only picked up in recent decades. The Rani of Jhansi, who Singh discusses in her chapter, has been the subject of a television series and a number of films. The most recent film *Manikarnika*, made in 2019, generated controversy even before its release because of rumors it portrayed this national hero having a romantic relationship with an Englishman. Guan Lu, the Communist spy whose story of survival is recounted by Edwards, has been featured in television programs in China only in the current century, and in a less than heroic light. The recently rediscovered Alava Network, a clandestine intelligence network organized by four Basque women during the Spanish Civil War mentioned by Aresti, was the subject of a comic book and a television documentary in 2018.[45]

Examples beyond those immediately treated in the volume abound. The Dahomey Amazons, a west African all-female group of Fon warriors from the Kingdom of Dahomey in present-day Benin, underwent a transformation when they served as the inspiration for the Dora Milaje, the special forces unit of the fictional kingdom of Wakanda in Marvel's *Black Panther* franchise.[46] Their portrayal in the 2018 movie made such an impact that Marvel announced they would get their own series.[47] The Dahomey Amazons have also appeared in Werner Herzog's film *Cobra Verde*, the video game *Empire: Total War* and even Stieg Larsson's novel *The Girl Who Kicked the Hornets' Nest*.[48]

Njinga is not only omnipresent in Angola itself, including in such banal ways as the face of Ginga Café; her face has started to appear on consumer goods in other African markets, Europe, and, very recently, on Alibaba, China's Amazon.[49] Through Angola's lobbying efforts at the United Nations, she has been included in UNESCO's "Women in African History" e-learning tool comprising "digital comic strips, rap/slam soundtracks" and other educational resources which was launched in 2013.[50] This internet tool, which is intended "to expand and disseminate knowledge of the role of women in African history to counter prejudices and stereotypes," also includes the Mulatto Solitude and the women soldiers of Dahomey.[51]

Indeed, new technologies and alternative forms of media have proliferated stories of women warriors in recent years. For instance, the story of Mai Bhago, an eighteenth-century Punjabi woman who, dressed as a man, led Sikh soldiers against the Mughals at the battle of Muktsar in 1705, has recently become an internet meme and an inspirational figure for young Sikh women around the world.[52] Two centuries after the events that made her famous, Agustina de Aragón was represented in a graphic novel as a Lara Croft precursor.[53] New scientific advances such as DNA analysis can also play a role, as when scientists announced that Casimir Pulaski, the Polish nobleman who fought with George Washington and who in 2018 was named "Father of the American Cavalry," may have been a woman or intersex. The story was the subject of a PBS television documentary broadcast in April 2019.[54] And, at the time of writing, global streaming service Netflix was carrying *Warrior Women*, a five-part series hosted by Lucy Lawless (aka Xena the Warrior Princess), with episodes devoted to Joan of Arc, Mulan, Boudica, Irish pirate Grace O'Malley, and Apache warrior Lozen.

Whether on the internet, in video games, or through comic books and even packages of coffee, women warriors are only increasing in visibility in the early-twenty-first century. In many cases, latter-day women warriors are harnessing these new forms of media to their own ends. Consider the military and political career of junior US Senator from Arizona, Martha McSally. A retired colonel in the United States Air Force (USAF), McSally was one of the highest-ranking female pilots in American history, the first female commander of a USAF fighter squadron, and the first American woman to fly in combat. She was also an advocate for equality. In 2001, she sued the US Defense Department, challenging their policy requiring servicewomen, but not service men, stationed in Saudi Arabia to wear a head scarf when traveling off base. McSally retired from the armed forces in 2010 and soon parlayed her military experience into a political career. Frequently referring to herself as a "woman warrior" and posing in campaign ads and YouTube videos in front of fighter jets, she ran first for the US House of Representatives in 2012 and then for the US Senate 2018. On the political stump, she made her groundbreaking service in the Air Force a cornerstone of her campaign message and social media brand.[55] A Republican, McSally often used her military experience and her reputation as a woman warrior to advocate for positions other GOP candidates largely shied away from, such as gender equality.[56] McSally made headlines in March 2019 when she announced during a Senate hearing on Sexual Assault and Misconduct in the Military that she had been raped by a superior officer but did not report the assault because she, "didn't trust the system," blamed herself, and felt "ashamed and confused ... powerless."[57] Across the internet and social media platforms, McSally was widely praised for publicly discussing her personal experience and breaking the silence on the issue of sexual assault in the military, sometimes under the banner of #WhyIdidntReport #MeToo.[58]

What effect will the rise of social movements such as #MeToo, #BalanceTonPorc, and the global LGBTQ2 movement have on the stories of women warriors in the future? Will such movements add to the prevalence and potency of these narratives, or will the age-old stories lose some of their appeal as the gender landscape shifts in what some consider to be a post-feminist age? What new uses of the figure of the woman warrior will emerge as the already established global phenomenon morphs to accommodate future forms of media and storytelling?

Such questions are beyond the scope of this volume. But it seems evident that the problems and themes explored in the following chapters will continue to haunt accounts of women who assert themselves in what are still considered to be non-normative ways. As increasing numbers of women are admitted into the armed forces across the globe, we will have to remain attentive to the ways their stories are framed in different contexts and scrutinize the motives behind their telling. Those who wish to extoll these women's martial virtues will have to confront their violence, and concerns with gender fluidity will inevitably arise despite increasing, if tentative, openness within certain military contexts.[59] For good or bad, women warriors with all of their complexity will continue to be an enduring part of our global present and our global future as well as a prominent part of our global past.

Notes

1. BBC News, "Saudi Arabia allows women to join military," February 26, 2018, https://www.bbc.com/news/world-middle-east-43197048; http://saudigazette.com.sa/article/529295/SAUDI-ARABIA/12-criteria-for-Saudi-women-to-join-army-as-soldiers (accessed January 19, 2019); Steven Morris, "All roles in UK military to be open to women, Williamson announces," October 25, 2018, https://www.theguardian.com/uk-news/2018/oct/25/all-roles-in-uk-military-to-be-open-to-women-williamson-announces (accessed January 19, 2019).
2. Elizabeth Braw, "Norway's Radical Military Experiment," *Foreign Affairs*, January 19, 2017, https://www.foreignaffairs.com/articles/norway/2017-01-19/norways-radical-military-experiment (accessed January 19, 2019). Even so, in Israel allowing women into combat roles remains contested, especially as the numbers skyrocketed after 2013. Lizi Hameiri, "Women in the Israeli Military Just Aren't Cut Out for Combat Roles," *Haaretz*, January 10, 2018, https://www.haaretz.com/opinion/.premium-women-just-aren-t-cut-out-for-combat-roles-1.5730157 (accessed April 7, 2019).
3. This does not mean, however, that women had no role in wartime. Before the creation of modern quartermaster systems, they frequently followed men to the front

to provide laundry, cooking, and other domestic services, a topic that Gabriel Cid explores in his chapter. As late as the Crimean War, the British army allowed four men in each company to take their wives "on the strength," at the army's expense. In return, the women cooked and did laundry. Even as the *levee en masse* of 1792 created the (male) citizen-soldier, the French state called on women "to act upon their patriotic feelings"; women responded by creating "new practices to support the troops, creating new relationships with national and imperial, as well as more local, communities that fostered a sense of common cause, solidarity, and cultural affinity," Karen Hagemann, Karen Aalestad and Judith Miller, "Introduction," *European Historical Quarterly* 37, no. 4 (October 2007): 503–504. As discussed in Gina Martino's and Colin Coates's chapters, women in societies that lacked formal military institutions such as North American frontiers of the seventeenth and eighteenth centuries often picked up arms and engaged in combat. Similar in form though larger in scale, women serving in military insurgencies such as the Vietcong and the Communists in the Greek Civil War also participated in war as discussed by Karen Turner and Sakis Gekas.

4 Marina Warner, *Joan of Arc: The Image of Female Heroism* (Oxford: Oxford University Press, 2013), xv–xvi.
5 Warner, *Joan of Arc*, xv–xvi. The selection of a mixed-race girl to portray Joan in the annual Joan of Arc festival in Orléans in February 2018 generated a torrent of racist abuse. "French Far Right Attack Choice of Mixed-Race Girl for Joan of Arc Role," *The Guardian*, February 23, 2018, https://www.theguardian.com/world/2018/feb/23/french-far-right-targets-mixed-race-teen-playing-joan-of-arc (accessed January 19, 2019). On Jeanne d'Arc as gay icon, see http://qspirit.net/joan-of-arc-cross-dressing-lgbtq/ (accessed 19 January, 2019).
6 Adrian Shubert, "Women Warriors and National Heroes: Agustina de Aragón and Her Indian Sisters," *Journal of World History* 23, no. 2 (June 2012): 279–313. Agustina de Aragón was also mentioned by some of the tellings of women warriors discussed by Booth.
7 Caitlin C. Gillespie, "Boudica the Warrior Queen," *Aeon*, November 6, 2018, https://aeon.co/essays/boudica-how-a-widowed-queen-became-a-rebellious-woman-warrior (accessed January 19, 2019).
8 On the Mulan story and its place in early-twentieth-century Chinese narratives, see Joan Judge, *The Precious Raft of History: The Past, the West, and the Woman Question in China* (Stanford: Stanford University Press, 2008), 143–186; on Mulan in the 1920s and 1930s and a number of film adaptations of her story, see Kristine Harris, "Modern Mulans: Reimagining the Mulan Legend in Chinese Film, 1920s–1960s," in *The New Woman International: Representations in Photography and Film from the 1870s through the 1960s*, ed. Elizabeth Otto and Vanessa Rocco (Ann Arbor: University of Michigan Press, 2011), 309–330.
9 Irene Castells, Gloria Espigado, and María Cruz Romeo "Heroinas por la patria, madres para la nación: mujers en pie de guerra," in *Heroínas y patriotas. Mujeres de 1808*, ed. I. Castells, G. Espigado, and M.C. Romeo (Madrid: Ediciones Cátedra, 2009).
10 Boyd Cothran, *Remembering the Modoc War: Redemptive Violence and the Making of American Innocence* (Chapel Hill: University of North Carolina Press, 2014), 81–110.
11 Laura Gotkowitz, "'¡No hay hombres!': Género, nación y las Heroínas de la Coronilla de Cochabamba (1885–1926)," in *El Siglo XX: Bolivia y América Latina*, ed. Rossana

Barragán and Seemin Qayum (La Paz: IFEA, 1997), https://books.openedition.org/ifea/7381 (accessed October 13, 2019).

12 Pablo Ortemberg, "Monumentos, memorialización y espacio público: reflexiones a propósito de la escultura de Juana Azurduy," *Tarea* 3, no. 3 (2016): 96–125. http://servicios.infoleg.gob.ar/infolegInternet/anexos/130000-134999/131043/norma.htm (accessed October 13, 2019).

13 Neil Carrier and Celia Nyamweru, "Reinventing Africa's National Heroes: The Case of Mekatilili, a Kenyan Popular Heroine," *African Affairs* 115, no. 461 (October 2016): 600. In South Africa, in contrast, women have not been much recognized. Sabine Marschall, "How to Honour a Woman: Gendered Memorialisation in Post-Apartheid South Africa," *Critical Arts* 24, no. 2 (2010): 260–283.

14 There are many examples of works that have explored this topic for a non-academic audience; see, for instance, David E. Jones, *Women Warriors: A History* (Washington, DC: Potomac Books, 1997); Jeannine Davis-Kimball and Mona Behan, *Warrior Women: An Archaeologist's Search for History's Hidden Heroines* (New York; London: Warner Books, 2003); Robin Cross and Rosalind Miles, *Warrior Women: 3000 Years of Courage and Heroism* (New York: Metro Books, 2011).

15 For a representative sample, including several authors represented in this book, see Harleen Singh, *The Rani of Jhansi: Gender, History, and Fable in India* (Cambridge: Cambridge University Press, 2014); Ruamisai Charumbira, *Imagining a Nation: History and Memory in Making Zimbabwe* (Charlottesville: University of Virginia Press, 2015); Louise Edwards, *Women Warriors and Wartime Spies of China* (Cambridge: Cambridge University Press, 2016); Judge, *The Precious Raft of History*; Gina Martino, *Women at War in the Borderlands of the Early American Northeast* (Chapel Hill: University of North Carolina Press, 2018).

16 Samuel Brunk and Ben Fallaw, *Heroes and Hero Cults in Latin America* (Austin: University of Texas Press, 2006); Ilene V. O'Malley, *The Myth of Revolution: Hero Cults and the Institutionalization of the Mexican State, 1920–1940* (New York: Greenwood, 1986); Miguel Angel Centeno, "War and Memories: Symbols of State Nationalism in Latin America," *Revista Europea de Estudios Latinoamericanos y del Caribe/European Review of Latin American and Caribbean Studies* (1999): 75–105; Maria Theresa Valenzuela, "Constructing National Heroes: Postcolonial Philippine and Cuban Biographies of José Rizal and José Marti," *Biography* (2014): 745–761.

17 Lucy Riall, *Garibaldi: Invention of a Hero* (New Haven, CT: Yale University Press, 2007); Robert Gerwarth, *The Bismarck Myth: Weimar Germany and the Legacy of the Iron Chancellor* (New York: Oxford University Press, 2005); Robert Gerwarth and Lucy Riall, "Fathers of the Nation? Bismarck, Garibaldi and the Cult of Memory in Germany and Italy," *European History Quarterly* 39, no. 3 (2009): 388–413.

18 Sabine Marschall, "Commemorating 'Struggle Heroes': Constructing a Genealogy for the New South Africa," *International Journal of Heritage Studies* 12, no. 2 (2006): 176–193.

19 Graeme Turner, *Making It National: Nationalism and Australian Popular Culture* (Sydney: Allen and Unwin, 1994); Warick Frost, "Braveheart-ed Ned Kelly: Historic Films, Heritage Tourism and Destination Image," *Tourism Management* 27, no. 2 (2006): 247–254; Allison Holland and Claire Williamson, *Kelly Culture: Reconstructing Ned Kelly* (Melbourne: State Library of Victoria, 2003).

20 Gerwarth, "Introduction," EHQ 3; Katherine Aaslestad, Karen Hagemann, and Judith A. Miller, "Introduction: Gender, War and the Nation in the Period of the Revolutionary and Napoleonic Wars—European Perspectives," *European History Quarterly* 37, no. 4 (2007): 501–506.

21 There is much debate over what defines global history as a unique approach or whether it is a distinct approach from older, boundary-crossing approaches such as the history of imperialism and colonialism, migration, and environmentalism. See, for instance, Sebastian Conard, *What Is Global History?* (Princeton: Princeton University Press, 2016); Lynn Hunt, *Writing History in the Global Era* (New York: W. W. Norton, 2014); Diego Olstein, *Thinking History Globally* (New York: Palgrave Macmillan, 2014). In this volume, we pursue a more expansive and comparative look at these seemingly local phenomena in order to understand how stories of women warriors developed at specific moments in global history.
22 On some of these tensions in retelling Joan's story in the Chinese case, see Judge, *The Precious Raft of History*, 162–177.
23 On India, see, for example, the figure of Joan in Shobna Nijhawan, *Women and Girls in the Hindi Public Sphere: Periodical Literature in the Hindu Public Sphere* (Oxford: Oxford University Press, 2011), n.p. (Figure 9).
24 Janet Tuckey, *Joan of Arc: "The Maid"* (New York: Putnam, 1886). Janet Tuckey, Awaya Kan'ichi 粟屋関一 transl. *Futsukoku bidan: kaiten iseki*仏国美談: 回天偉績 (A moving French tale: Changing the world with her glorious achievements) (Tokyo: Hayakawashinzaburō早川新三郎, 1887).
25 Rigiyōru リギヨール (François-Alfred Désiré Ligneul) (1847–1922). *Jyan Daaku* ジャンダーク (Joan of Arc) (Tokyo: Kyōgaku kensan Wa-Futsu kyōkai, 1910). Ligneul arrived in Japan in 1880.
26 Iwasaki Sodō 岩崎徂堂, and Mikami Kifū 三上寄風. *Sekai jūni joketsu* 世界十二女傑 (Twelve world heroines) (Tokyo: Kōbundō shoten, 1902). The collection also includes a number of less martial figures: Catherine the Great (1729–96), Madame de Staël (1766–1817), Lucy Hutchinson (1620–80), Queen Isabella of Spain (1451–1504), Queen Elizabeth of England (1558–1603), Queen Louise of Prussia (1776–1810), and Frances Willard (1839–98).
27 Iwasaki Sodō 岩崎徂堂, and Mikami Kifū 三上寄風, Zhao Bizhen 趙必振, trans. *Shijie shier nüjie* 世界十二女傑 (Twelve world heroines) (Shanghai: Guangzhi shuju, 1903). On the circulation of biographies of Western women from Japan to China, see Xia Xiaohong, "Western Heroines in Late-Qing Women's Journals: Meiji-Era Writings on '"Women's Self-Help' in China," in *Women and the Periodical Press in China's Long Twentieth Century: A Space of Their Own?* ed. Michel Hockx, Joan Judge, and Barbara Mittler (Cambridge: Cambridge University Press, 2018), 236–254. Joan Judge, "Blended Wish Images: Chinese and Western Exemplary Women at the Turn of the Twentieth Century," *Special Issue of Nan Nü: Men, Women, and Gender in Early and Imperial China*, ed. Susan Mann, 6, no. 1 (2004): 102–135.
28 Iwasaki, *Shijie shier nüjie*, 36–37. The translator, Zhao Bizhen, gave two, not fully compatible explanations for Jeanne's faith. The first was because she lived in a time before science had trumped superstition, the second was strategic: Jeanne invoked the gods in order to gain the allegiance of the ignorant masses.
29 Linda Heywood, *Njinga of Angola: Africa's Warrior Queen* (Cambridge: Harvard University Press, 2017), 249.
30 Lisa Yoneyama, "Liberation under Siege: U.S. Military Occupation and Japanese Women's Enfranchisement," *American Quarterly* 57 (2005): 3, "Legal Borderlands: Law and the Construction of American Borders" (September 2005), 891.
31 Heywood, *Njinga of Angola*, 256.
32 Ortemberg, "Monumentos, memorialización y espacio público," 117–120.

33 Alison Booth, *How to Make It as a Woman. Collective Biographical History from Victoria to the Present* (Chicago: University of Chicago Press, 2004), 2.
34 Cicely Hamilton, *Pageant of Great Women* (London: The Suffrage Shop, 1910), 41.
35 Jitka Malecková, "Nationalizing Women and Engendering the Nation: The Czech National Movement," in *Gendered Nations. Nationalisms and Gender Order in the Long 19th Century*, ed. Ida Blom, Karen Hagemann and Catherine Hall (Oxford: Berg, 2000), 306–307.
36 For another case, see Malecková, "Nationalizing Women and Engendering the Nation," in *Gendered Nations. Nationalisms and Gender Order in the Long 19th Century*, ed. Ida Blom, Karen Hagemann, and Catherine Hall (Oxford: Berg, 2000) 293–310.
37 Aziz Tunny, "Martha Christina Tiahahu: The 'kabaressi' heroine of Maluku," https://www.webcitation.org/64EwxOpzW?url=; http://www.thejakartapost.com/news/2008/04/27/martha-christina-tiahahu-the-039kabaressi039-heroine-maluku.html (accessed January 19, 2019).
38 Karla Lewis Gottlieb, *The Mother of Us All, A History of Queen Nanny, Leader of the Windward Jamaican Maroons* (Trenton: Africa World Press, 2000); *Queen Nanny. Legendary Maroon Chieftainess. Study Guide*, www.nannythemovie.com/Queen%20Nanny%20Discussion%20Guide.pdf (accessed January 19, 2019).
39 Heywood, *Njinga of Angola*, 252.
40 Heywood, *Njinga of Angola*, 253–255.
41 Solitude appeared in no historical documents. The only written attestation of her existence came in a single paragraph in August Lacour's *Histoire de la Guadeloupe*, written decades after the events and based on interviews with witnesses. Given the negative portrayal of Solitude, these were undoubtedly whites. Laurent Dubois, "Solitude's Statue: Confronting the Past in the French Caribbean," *Outre-mers*, 93, 2006, 38.
42 http://www.cnmhe.fr/spip.php?article144 (accessed January 20, 2019). Author's translation.
43 https://collectiffemmespantheon.wordpress.com/2013/08/01/au-pantheon-solitude/; https://www.atout-guadeloupe.com/La-mulatresse-Solitude-et-Olympe-de-Gouges-au-Pantheon_a430.html (accessed January 20, 2019).
44 On this trope of reversion in the case of Hua Mulan, see Louise Edwards, *Men and Women in Qing China: Gender in The Red Chamber Dream* (Honolulu: University of Hawai'i Press, 2001), 88; Judge, *The Precious Raft of History*, 153–155.
45 https://www.eldiario.es/norte/euskadi/invisibles-movimiento-clandestino-antifranquista-Vitoria_0_769923744.html; http://www.sabinoarana.eus/Portada/tabid/308/ArtMID/1467/ArticleID/304/language/es-ES/Default.aspx (accessed April 26, 2019).
46 Mike Dash, "Dahomey's Woman Warriors," *Smithsonian.com*, September 23, 2011, https://www.smithsonianmag.com/history/dahomeys-women-warriors-88286072/ (accessed January 22, 2019). The only full-length English-language study of them is Stanley B. Alpern, *Amazons of Black Sparta: The Women Warriors of Dahomey* (New York: New York University Press, 2011).
47 https://www.vogue.com/article/black-panther-dora-milaje-comic-series-preview (accessed January 24, 2019).
48 https://www.syfy.com/syfywire/meet-the-dahomey-amazons-the-inspiration-for-the-dora-milaje (accessed October 13, 2019).

49 http://www.africafundacion.org/spip.php?article23545; http://sabordecafe.com/2018/12/20/%EF%BB%BFcafe-ginga-no-alibaba-da-china/ (accessed January 24, 2019).
50 https://en.unesco.org/womeninafrica/ (accessed April 4, 2019).
51 http://www.unesco.org/new/en/communication-and-information/resources/news-and-in-focus-articles/all-news/news/unesco_launches_women_in_african_history_an_e_learning_t/; https://en.unesco.org/womeninafrica/njinga-mbandi/comic (accessed January 19, 2019).
52 Doris R. Jakobsh, "Marking the Female Sikh Body: Reformulating and Legitimating Sikh Women's Turbaned Identity on the World Wide Web," in *Young Sikhs in a Global World: Negotiating Traditions, Identities and Authorities*, ed. Knut A. Jacobsen and Kristina Myrvold (New York: Routledge, 2016); Anne Murphy, "Objects, Ethics and the Gendering of Sikh Memory," *Early Modern Women: An Interdisciplinary Journal* 4 (2009): 161–168.
53 Fernando Monzón y Eduardo Mendoza, *Agustina*, (Zaragoza: 1001 Ediciones, 2009).
54 John Shannon, "Father of the American Cavalry' may have been female or intersex, documentary claims," *USA Today*, April 6, 2019 (accessed April 6, 2019).
55 See, for instance, her 2018 television ad "Deployed," https://www.youtube.com/watch?v=obHlECvtrEs (accessed April 12, 2019).
56 Derek Wallbank and Greg Sullivan, "Candidates Speak Out about #WhyIDidntReport," *Bloomberg*, September 24, 2018.
57 Helen Cooper, Dave Phillips, and Richard A. Oppel, Jr., "I, Too, Was a Survivor: Senator McSally Ends Years of Silence," *New York Times*, March 26, 2019, https://www.nytimes.com/2019/03/26/us/senator-martha-mcsally-rape-assault.html (accessed April 12, 2019).
58 Karen Tumulty, "Martha McSally's #MeToo Statement Took Guts—Maybe as Much as Her Actions in Uniform," *Washington Post*, March 6, 2019, https://www.washingtonpost.com/opinions/2019/03/07/martha-mcsallys-metoo-statement-took-guts-maybe-much-her-actions-uniform/?noredirect=on&utm_term=.6af295567904 (accessed April 12, 2019).
59 On this contentious issue in the US, see https://www.nbcnews.com/feature/nbc-out/nothing-short-ban-transgender-troops-advocates-react-new-military-policy-n983166 (accessed October 13, 2019).

Part One

Process

1

India's Rebel Queen: Rani Lakshmi Bai and the 1857 Uprising

Harleen Singh

Considered the "most dangerous of all the Indian leaders," the Queen of Jhansi, Rani Lakshmi Bai, challenged the might of the British Empire in the Uprising of 1857 and died on the battlefield with her soldiers.[1] Known as the greatest heroine of Indian history, represented variously in folk song, poetry, novel, and film, and commemorated with postage stamps and statues, the Rani (Hindi for Queen) of Jhansi is a contradictory figure in British history as both worthy foe and rapacious whore. My work looks at the historical and literary representations of a woman warrior to flesh out the myriad and conflicting, and yet often coterminous, narratives surrounding women in the public sphere.[2] For example, Indian mythology has numerous Goddess figures who are invoked in the realm of war, and yet widespread understanding of women's roles in India continues to stem from, and is limited to, the domestic. Similarly, even as Queen Victoria ruled England at the time of the rebellion in 1857, Indian queens were rarely entitled to occupy a parallel position of authority in British policies.

This chapter seeks to question why, despite her bold stance as a rebel against the British, the Rani continues to symbolize, in mythological and historical modes, a politically conservative figure. What about history allows for a safe distance of representation when the same actions in India's present might be demarcated seditious, recalcitrant, monstrously masculine, and even anti-national? Although she was a warrior in the midst of violent battles, the Rani does not disrupt gender and sexual hierarchies and continues also to be represented as a socially conservative figure. How does a woman who upset social, cultural, religious, and gendered norms in both India and Britain end up being represented as a laudable figure of kinship, femininity, and nationalism? It is precisely this domestication, an impulse to bring the Rani under control, that forms the basis of this chapter. My work questions how the woman "warrior" functions as a conservative signifier of a glorious national past, even as representations, ambivalent and ambiguous, of the Rani change over time to accommodate the British Empire and Indian Nationalism. I suggest that the female warrior, whether in the service of the nation (India) or against it (Britain), produces a problem of signification—thus, while a plethora of representations are extant, these figurations essentially distil dominant narratives of femininity, sexuality, and sex to reorder structures disrupted by the female warrior, to reinstate patriarchy's sanctioned categories, and to bring to heel the fighting woman.

The rebellion of 1857, sparked partly by the insubordination of *sepoys* (soldiers) of the East India Company and timed to end one hundred years of British rule, lasted only a year. It resulted, however, in monumental administrative and social policy shifts, as India went from being a mercantile colony to a dominion of the crown, elevating Victoria from Queen to Empress. Known in imperial historiography as the "Mutiny" and in Indian nationalist accounts as the "First War of Independence," the events of 1857 continue to galvanize scholarship and popular representation; the figures of the rebellion, from the sepoy Mangal Pandey to the Queen of Jhansi, Rani Lakshmi Bai, caught both the Indian and British popular imagination. Laws curtailing Indian writing about the rebellion essentially ensured that archives and literature from that period tend to represent the British perspective.[3] It was only during the burgeoning years of Indian nationalism and after independence in 1947 that film, writing, and historiography became available from the Indian side.

> On that discursive battlefield, the Rani embodies an enduring enigma as a character in English romance novels, as a topic of debate in historical narratives, as the mobilizing spirit in the rhetoric of Indian patriotism, and as a celebrated figure in folk ballads and theatre. Doubly articulated as history and metaphor, the Rani is crucial to disciplinary discourses that produce the historical subject within the colonial and postcolonial conceptualizations of gender, political power, and resistance.[4]

History, popularly understood, is the story of the past. In academic, scholarly parlance, one may also read history as a legitimating narrative. Partha Chatterjee calls for an "analytic of the popular" in history—"the domain of the popular had many narrative and performative strategies through which it could tell the story of conquest and tell it to the satisfaction of both victor and the vanquished."[5] Thus, the Rani "doubly articulated as history and metaphor" in the "domain of the popular" embodies one such story told to conservative ends by both the rebel and the oppressor, indicating, of course, that historical representations are "never ideologically or cognitively neutral."[6]

Literature may be read as a collective interpretation that makes history "readable." In colonial narratives, the Rani becomes a sexually excessive, hysterical body, symbolizing the chaos of India in need of masculine British supervision. Similarly, though the Rani is an invaluable symbol of the nationalist project, her disturbance of the male-identified public sphere in India requires her reframing as an acceptable model of symbolic power. Dipesh Chakravarty argued that besides the factual and proven record of history, or the given sensational traction of a debate, the history that takes hold of the popular imagination has at its core a persuasive rhetoric. Rani Lakshmi Bai, in British and Indian literatures, becomes the vehicles for precisely such persuasion, albeit to different ends and yet, with remarkably similar traditional, conformist effects.

> Unlike the usual suspects of colonial and nationalist enterprise—i.e., the subaltern, the prostitute, or the persecuted wife—the Rani is an elite colonial subject whose

refusal to be restrained within the available paradigms necessitates a larger, multilevel project of representation. Neither entirely victim, nor agent, the Rani is objectified by colonial and nationalist discourse to perpetuate sexually, culturally, and politically viable modes of traditional femininity. I rephrase and rework Shahid Amin's concept in typifying the Rani as a "recalcitrant figure"—not because there is a paucity of narrative about her, but because the innumerable ways in which she is textually articulated defy any attempt to uncover a singular historical archive or literary figure.[7]

By moving, as Shahid Amin exhorts, "beyond the territory of the contested fact, the unseen record, from the history of evidence into the realm of narration," I read this symbolic idiom of literary representation to demonstrate how the textual, persuasive sweep of the Rani's story bests the actual military and political history of 1857.[8] A critical evaluation of this figure interrogates women's participation in the public sphere, evaluates the subaltern space of gender, and brings attention to the continued use of sexuality as the primary epistemological lens for women in power.

Her Story

Born Manikarnika, and known affectionately as "Manu," the Rani was the daughter of an ordinary Brahmin in the court of the Peshwa Baji Rao II. She was married to Gangadhar Rao, the aging Raja (King) of Jhansi, in 1842 and given the name Lakshmi Bai. Most Indian sources give her date of birth as 1835, which would make her seven at the time of her marriage and twenty-two in 1857.[9] However, various British records place her birth in 1827 and refer to the Rani as a woman in her thirties.[10] It may have been the advanced years of the King and the lack of natural heirs that recommended, as yet, an ordinary girl to such a prestigious marriage; she had a son who died in infancy. According to the infamous Doctrine of Lapse, the East India Company annexed the kingdom after the Raja's death in 1854 and refused to recognize the adopted son, Damodar Rao.[11] Most sources claim the Rani administered Jhansi at the dispensation of the Company from 1854 to 1857, but threw in her lot with the rebellion once it overtook the area and made her stand against the British in Jhansi and Gwalior in 1858.[12] Details of her life, her friendships, and her daily routine are anecdotal at best as there was no reason before 1857 to record the life of a girl who was not of royal lineage. In both Indian and British accounts, the numerous stories about the Rani are presented as factual history but cast always in the form of legend.

If you visit the famous fort at Jhansi today, most guides will breathlessly pause for effect as they narrate the story of the Rani as she jumped, while astride her horse and with her son tied to her back, from the ramparts of the Jhansi fort, preferring death to surrender. Most historical accounts confirm the Rani was killed in battle against the forces commanded by Sir Hugh Rose, and was presumably cremated by her soldiers before the English could retrieve the body. While Sir Hugh Rose was credited with ending the rebellion, and the death of the "Rebel Queen" fostered Britain's inevitable triumph, the absence of corporeal proof allowed Indian accounts to construct her, and by extension the nation, as a defiant, even victorious, figure—jumping always out of reach of the colonialists. Yet, dying, suitably so, for the nation.

In India, it is no exaggeration to state that every school-going child and literate adult is acquainted with the story of Rani Lakshmi Bai. The refrain from Subhadra Kumari Chauhan's poem, written in 1930, "*Khoob Lari Mardani woh to Jhansi wali Rani thi*" (It was the Rani of Jhansi who fought with the valor of a man), is an aspect of all school curriculum in Hindi. When Subash Chandra Bose created the Indian National Army, bifurcating from the dominant ideology of Gandhian non-violence, he raised the only regiment of female soldiers on any side in the Second World War and named it the Rani of Jhansi regiment. Perhaps it is no surprise that Gandhi, given his particular enthrallment by Indian women's so-called capacity for restraint, passive resistance, and silent suffering, did not refer to this warrior from India's past in his writing and exhortations on Indian nationalism and social reform. For many ardent nationalists, however, the Rani was a harbinger of India's struggle against colonialism, representing as such a dormant feminine energy that lay latent, waiting for the kinesis of twentieth-century masculine nationalism. In the years after independence, stamps were issued; statues were raised; and schools, colleges, and streets were named after her. In England, supplanted perhaps by the villainy of Hitler, she seemed to go out of vogue as a literary and historical character after the Second World War. Nineteenth-century Victorian literature and historiography seemed fascinated by the stories of the "warrior queen," sometimes holding her responsible for the Jhansi massacre and terming her a bloodthirsty Jezebel, and at other times characterizing her as a worthy foe and justified adversary. In Indian and the British narratives, factual details of her biography remain tangential to the ways in which she serves the cause of nineteenth-century colonialism or twentieth-century nationalism. The figure of the warrior queen, in all her myriad representations, poses a threat to the unifying national principles that naturalize colonial and later postcolonial rule. Although Britain and India had obviously competing agendas, both discourses ostensibly produce a singular, coherent version of the Rani, even in negative and derogatory frames, that reproduces her conservative significations. For example, even when she is the "Jezebel" or the "rapacious whore" in British narratives, she is the maternal queen for her kingdom in some stories and thus conforms to the Mother-Whore Judeo-Christian dichotomy. And when the Rani is the Hindu goddess of war or the mother of Indian nationalism, she personifies Hindu mythological representations of "Shakti," the feminine power that is central to all Vedic scriptures. Fight as she might, she is reformulated somehow, in this Anglo-Indian history, from the warrior into the woman.

Victoria(n) Rani

Read slightly creatively, the title above could be interpreted both as "Queen Queen" or the "Queen according to Victorian mores." As the death of the Rani in 1858 brought an end to the rebellion, Queen Victoria issued a proclamation that rescinded the charter to the East India Company and declared India to be a dominion of the British crown. This brought the colony under the direct supervision of Parliament and elevated Indians to the status of Her Majesty's subjects. But, as many statesmen and politicians warned, the rebellion also clearly demarcated the gulf of understanding between Indian cultural mores and the British reformist impulse. For example, the abolition of Sati in 1829 and the Hindu Widows Remarriage Act of 1856, alongside the Doctrine of Lapse, were seen by many Indians as a direct affront to their religious and social values. While colonial administrators seemed to take a step back from cultural reforms, it was also evident to them that structural categories such as caste and religion had to be better defined to serve the imperial machinery.[13]

Thus, mirroring in some ways this new, at odds, colonial policy, the literary representations of the Rani in Victorian fictions ran the gamut from Aryan Royalty to promiscuous eastern whore, and from a worthy foe to a murderous despot.[14] It "provides a site of convergence for the grid of colonial urgency invested in stabilizing the empire after 1857," and "reflects a crisis of authority that had to be resolved both through the continuity of links to the past but also through a recently energized and revamped template of governance for the future."[15] Amidst the taxonomies of loyal races and barbaric despots, a recurring gendered demarcation of masculine and effeminate stereotypes, regional and religious, gained credence. The Rani's story, during Victoria's reign, was as much a parable of disobedient subjects as it was of uppity, unruly queens. Maria Jerinic reads the many colonial stories of the Rani of Jhansi as indicating a "British discomfort with ruling women and consequently with their own queen. This interest in the Rani is tied to an imperialist vision, one that looks with suspicion on all female political involvement, British as well as Indian."[16] Thus, I titled this particular section "Victoria (n) Rani," (Queen Queen) to indicate that not only was the Rani defined by English codes of honor, sexuality, and race but that her story in Britain, a story about the proper place of femininity, was also a disciplining narrative about women who presumed to rule.

Infamous in British history for the Jhansi (Jhokan Bagh) massacre of more than sixty British men, women, and children, the Rani was derisively titled a murderous whore. An eyewitness account from a Mr. Crawford who had escaped the massacre indicated the Rani's culpability:

> It is the general impression that the mutineers after killing some of their own officers and plundering the town, were going off, and it was only at the instigation of the Jhansie Ranee with the object of her obtaining possession of Jhansie state that they attacked the fort the next day together with other armed men furnished by her. The town people are said to have also joined. For this act the mutineers are said to have received Rs. 35,000 in cash, two elephants and five horses from the Rani.[17]

However, in 1943, C.A. Kincaid first raised doubts about the Rani's complicity by providing a contrary eye witness account: an undated letter written to the Rani's adopted son Damodar Rao by a Mr. Martin who had been in Jhansi in 1857, "your poor mother was very unjustly and cruelly dealt with—and no one knows her case as I do. The poor thing took no part in the massacre of the European residents of Jhansi in June 1857."[18] A response followed in the next year, written by R. Burn and Patrick

Cadell, which relied solely on the *Official Mutiny Narratives* for evidence of the Rani's involvement.[19]

Although the alleged perpetrator of the massacre escaped with her army, the British meted out an indiscriminate carnage in which more than three thousand people were killed:

> In Jhansi we burned and buried upwards of a thousand bodies, and if we take into account the constant fighting carried on since the investment, and the battle of the Betwa, I fancy I am not far wrong when I say I believe we must have slain nearly 3,000 *of the enemy*. Such was the retribution meted out to this Jezebel Ranee and her people for the heinous crimes done by them in Jhansi.[20]

Taken from the journal of Dr Thomas Lowe, a member of Rose's contingent, the quote bears testimony to the events that took place in Jhansi. While the loss of sixty British civilians constituted a massacre, "a thousand [Indian] bodies" were simply "retribution." The Governor-General declared the Rani a rebel leader in June 1858 and offered an award of 20,000 rupees for her capture: "His Lordship, however, authorises you to offer such rewards as you think fitting for the capture of these persons, provided that the sum offered for the Nawab does not exceed 10,000 Rs. & that for the Baee[21] is not more than 20,000 Rs."[22] By the time the reward was offered, Sir Hugh Rose had taken Jhansi, and the Rani had fled to Kalpi to join Nana Saheb and Tantya Tope. She regrouped with other rebel leaders and captured Gwalior, but the celebrations were cut short by the British attack and her eventual death in that battle.[23]

Popular novels and newspaper reports sensationally cast the uprising of 1857 as a set of depredations done by Indian men against British women. In this "rape script," popularized by the nineteenth-century colonial romance novel, an English woman under threat by lust-ridden Indian men is rescued by the white colonial male in the eventual, heteronormative triumph of British masculinity. Four particular novels create a rather different rape script with the Rani at their center—Gillean's *The Rane: A Legend of the Indian Mutiny* (1887), Hume Nesbit's *The Queen's Desire* (1893), Philip Cox's play *The Rani of Jhansi* (1933), and George MacDonald Fraser's *Flashman in the Great Game* (1975) depict the Rani of Jhansi as a rapacious whore, effectively "raping" the white British male.

This peculiarly sexually aggressive aspect of the Rani does not just read her rebellion as an extension of her, as yet, untrammeled sexuality but also, in relation, constructs the Indian male as sexually deficient. The colonial logic of gender defined the primary site of British occupation, Bengal, as the home of effeminate, overly educated, artistically prone, heterosexual men subject to the whims of their women, while simultaneously reading the martial ethnicities of the Punjab and the North Western Frontier Province as excessively masculine and homosexual.[24] And thus, the Rani's sexual pursuit of British manhood, when read against these lacking masculinities, is posited as her eventual submission to colonial virility. The eroticization of power, as represented by the Rani's rebellion within these British texts, stands Prometheus-like at the intersection of an inertly feminine Indian masculinity, a sexually aggressive Indian femininity, and the eventual dominion of British manhood over both.

On the other hand, literary representations of the Rani of Jhansi engage with notions of "Aryanism" in Europe and she is defined in Alexander Rogers's novel in verse, *The Rani of Jhansi, or The Widowed Queen* (1895), Michael White's novel *Lachmi Bai of Jhansi: The Jeanne D'Arc of India* (1901), and Norman Partington's novel *Flow Red the Ganges* (1972) as an Aryan queen; a seemingly benevolent readings of Rani Lakshmi Bai as an Aryan model of heroic womanhood comparable to Joan of Arc. The surprising "aryanization" of the Rani illustrated, albeit in a specific example, the unstable context of nineteenth-century racial and colonial theory, which can read the otherwise defined "murderous whore" as a worthy Aryan ruler. Perhaps not so surprisingly, by casting the Rani as an Aryan, these narratives suppress her national and regional context to assimilate her within a larger European hagiography. These two extreme and oppositional characterizations of the Rani of Jhansi in British literature and historiography indicate the instability of such race, sexuality, and gender while highlighting the centrality of such concepts to the business of empire.

Indian Queen

Victorian representations, from seductress to worthy foe, construct a particularly ambivalent framework for the rebel woman. However, Indian representations are not without paradox or ambiguity. As stated before, there are hardly any contemporary accounts of the 1857 rebellion from the Indian side because of colonial censorship, and Indian literary representation begins mostly in the early-twentieth century alongside the rise of Indian nationalism. Subhadra Kumari Chauhan's iconic poem "Jhansi ki Rani" (1930) is arguably the most familiar textual reference for Rani Lakshmi Bai, weaving the Rani's story through the many regional centers of the rebellion, presenting a national geography as in Rabindranath Tagore's song, and now India's national anthem, "Jan Gan Man" (1911). Hindi literature's foremost historical novelist Vrindavanlal Varma published his novel, *Jhansi Ki Rani* (1946), just before India's independence, and the novel's canonization within India's "national" literature clearly links Hindi literature to the historical validation and reconstruction of its national story. In fact, the most famous dialogue attributed to the Rani, "Mein Apni Jhansi Nahi Doongi" (I will not give up my Jhansi), can only be traced to her literary depiction in Varma's novel. Sohrab Modi's *Jhansi ki Rani* (1953), India's first film in technicolor, and based on Varma's novel, reframes the historical "real" in "reel." Notable for its authenticity in depicting historical events, the film remains neglected even in the midst of a great outpouring of scholarship on Indian cinema. In these three examples of poetry, novel, and film, the Rani is invariably compared to and symbolized as a Goddess of freedom and war, and thus indelibly marked as a symbol of India's glorious Hindu past.

However, even as the nationalist celebration of the Rani's story seemed beholden to reinforce her as a "Hindu" queen, postcolonial activist writers such as Mahasweta Devi published a biography, *The Queen of Jhansi*, in 1956. It is the first biography and fictional rendering of the Rani's life written by a woman. Eschewing the academic rendering of a historical narrative that reproduced a chauvinist logic, Mahasweta Devi redrafted the Rani's story as a narrative of the people by engaging memory and folklore as her primary documentation. By telling the story of the Rani through the participation of *Dalit* (untouchable) and indigenous communities in 1857, Mahasweta resists upper-caste, male-identified historiography. By redefining the biographical form and unsettling the historical record, Mahasweta Devi's rendering of the Rani raises the uncomfortable questions about the shared heritage of the postcolonial nation.

These projects of representations engage the power wielded through the body of the rebel woman. Represented as a heroic Aryan, a sexually promiscuous Indian whore, a Hindu goddess of nationalism, or a folk symbol of indigenous resistance, these literary figurations employ a range of tropes that bolster the dominant patriarchal framework where women fight to serve the interests of the family, the community, and the nation. Whereas the Rani embodied a monstrous rendering in the British colonial project, Indian nationalism apotheosized her as a symbol of an enduring India. My reading of these texts, produced from the nineteenth to the twenty-first century, first paid attention to the ways in which the rebel queen unhinges official narratives. However, as I continued to work on this project, even after the book, I began to consider the ways in which this "unhinging" is met with a counter-project, not necessarily cohesive or monolithic, that elevates the Rani as a "proper" national subject: a good wife, a good mother, and thus, a good queen who fights for her nation. The fact that she died provided the perfect *Deus ex machina* for all narratives to follow; neither British nor Indian representations have to deal with the real problem of a woman in power. Of course, each of these stories reduces the Rani in magnitude to colonial and postcolonial preoccupations when such an extraordinary woman ought to serve as the sole motivator of her own story.

These sexual, racial, linguistic, and caste-based formations of the Rani portray the nature and administration of colonial, monarchical, and nationalist authority. My project engages a theory of power articulated by the intersection of the "public woman" with the "public life" of history. "Defiant of a singular theoretical articulation, the woman in the public sphere, elite monarch or nationalist military leader, disrupts the available epistemologies of representation, and power is accessed here through a complex and contradictory matrix of sexuality, race, language, and caste."[25] Spivak has famously argued that feminist thought must return to "measuring silences;" my research seeks a measure of the noise surrounding Rani Lakshmi Bai, the Queen of Jhansi.[26]

The Rani Rides Again: Film, History, and Women's Lives in Contemporary India

"There's always more to tell even when the story has ended."

Mahasweta Devi

I first heard of Rani Lakshmi Bai at the age of six. The national, and only, television channel *Doordarshan* broadcast Sohrab Modi's Hindi film *Jhansi Ki Rani* to mark August 15, India's Independence Day. The next day, I eagerly bought and read the

Amar Chitra Katha comic *Rani of Jhansi*, my first foray into the realm of the popular and the historic. "As a girl growing up in India, I often got into minor brawls with boys of my age. Reprimanding teachers and parents would say, 'Who do you think you are? The Rani of Jhansi?' Fighting boys, or fighting the British, is modulated by social norms. Girls do not fight and legendary warrior queens must not upset the patriarchal, societal order."[27]

The Indian female rebel, monarch and warrior, is a disruptive figure. It is the precise impossibility of reading that motivates and links these assimilative projects for powerful women. If a woman is to be the center of the home, and thus the building block of the society and the nation, she must be a pliable, evident entity—whose functionality can be simply managed. Often depicted as a utopian character of literature, symbolizing freedom, strength, and the poetic worth of outmatched fights, these romantic figurations relegate the Rani to irrelevance; the Rani is worshipped as a goddess or mother figure, but not given the crucial regard she deserves in historical research and political reality. As daughter, wife, mother, and then queen, the Rani holds the home and nation above all else. Thus, she is hardly anomalous in Indian texts where even as a "heroine of unending fascination, her remaking in new forms transforms the marginalized into a bounded, integrated, and meaningful entity."[28]

> It is the precise mobilizing of female icons, such as the Rani of Jhansi, in terms of their "power" that has been undertaken in this project. Thus, gesturing towards the mythically heroic Rani may at first be heard as an inspirational call to the nation's women. Closer reading, however, reveals the extent to which even the historical Rani may be recast as the fictional, poetic, or cinematic configuration of a traditional Indian womanhood that is instead a vehicle for the patriarchal destinies of the nation. Thus, at both colonial and postcolonial junctures, while traditional gender roles are most under scrutiny, the figure of Rani Lakshmi Bai, in equal parts, daughter, wife, mother, and queen, functions as a haven of representation—in which female strength may emerge and yet subside on male-identified and male-dominated whims.[29]

The figure of the fighting Rani has inspired a host of biographies, comic books, historical narratives, films, and even a television serial: *Jhansi Ki Rani,* which aired between 2009 and 2011 and added new twists of familial intrigue, love, and jealousy. The most recent entrant to the field is the film *Manikarnika*, released on January 25, 2019. Almost a year before it was released, the Sarv Brahmin Mahasabha held protests in India against the "indecent portrayal" of Rani Lakshmi Bai. Responding to unfounded reports that the film was based on a novel by Jaishree Mishra, *Rani* (2007), which creates a story of unrequited love between the Rani and the Englishman Major Ellis, the protests were meant to "protect" the chastity and honor of an upper-caste, Brahmin, Hindu woman. The mere hint that this heroic figure of India's glorious past could have had a romance with an Englishman had unsettled the equilibrium for these guardians of caste privilege. "Battling the East India Company, as a queen, mother, and widow, falls squarely within the dictums of *dharma*. Being depicted as a woman with dilemmas or desires, whether fictional or real, calls, however, for the banning of books and films."[30] In an eloquent article on the controversy, the journalist Adrija Roychowdhury writes, "What is certain though is that it is the imaginary depiction of the queen in a banned book that is being feared to disrupt her legendary status."[31] And yet, the Rani's legendary status has everything to do with imaginary depiction. But this impulse to wrest control of women in popular culture has risen in opposition to women's presence as equal participants in India's politics, economy, culture, and society, which has vastly increased since liberalization in 1991. The attacks on *Manikarnika* resonated with the added scrutiny of the female body in the public sphere; the Brahmin Mahasabha was insulted because the film allegedly depicted the Rani in a romantic relationship with a white man. It is not the case, and the filmmakers denied the charge vociferously, but the mere hint of women stepping out of bounds was enough to convulse parts of the country into mob violence.

The author Jaishree Mishra wrote that she created her novel *Rani* (2007) "to find the woman behind the warrior," who was "strangely, a little bit like every woman I knew."[32] Why domesticate the Rani to privilege romance? This motivation to make the Rani a "bit like every woman" is a new iteration of the same domesticating impulse espied in colonial and national novels. Why must the Rani be like anyone else when she is so clearly extraordinary? Why turn the warrior into a woman while dominant narrative continues to make men into warriors? Rani Lakshmi Bai and other female warriors and queens of India—foremost daughters, wives, and mothers—represent a female heroism bound to the family and the nation, a valor that does not detract from masculinity.

This celebration of female power in the service of the nation is complicated, however, by the Hindutva politics in which women have played a significant role militating against other religious and cultural minorities. Amrita Basu's work on women in Hindu nationalism uncovers some of the complex interplay of factors behind women's participation in these chauvinist politics of religion and the nation, which do not challenge "patriarchy in male-dominated societies."[33] Similarly, the high degree of exposure accorded to the Rani of Jhansi is not at odds with the dictates of patriarchal nationalism. Tales of Hindu women in India, whether as military leaders or inspiring mothers, are foundational in the nationalist formations of gender.[34] As recently as 2007, Sonia Gandhi, the leader of the Congress Party, was depicted on a poster as the Rani of Jhansi, with the *adult* face of her son Rahul Gandhi on her back. It was a telling moment in historical recasting to have the Italian-born Sonia Gandhi reconfigured as yet another incarnation of the Rani: "The one who fought with the valor of a man was the queen of 10 Janpath, Delhi." [*khoob lari mardaani woh to 10 Janpath Delhi waali Rani*]

With the talented young actress Kangana Ranaut portraying Rani Lakshmi Bai and directing the film, *Manikarnika* (2019) opened to box office success and a fair deal of press coverage. So much so, that *The Guardian* from England reviewed the film.[35] With the basic framework of Bollywood film, replete with songs and dances, fabulous colors and costumes, the film is a computer-generated image (CGI) bonanza. The landscape, architecture, and military battles of the mid-nineteenth century are portrayed with as much accuracy as one might expect from a commercial release. I was struck however by the use of certain words. Interestingly enough, in Vrindavanlal Varma's seminal

Figure 1.1 Popular Congress Party Election Posters depicting Sonia Gandhi as the Rani of Jhansi.

Hindi novel, *Jhansi Ki Rani*, the Queen repeatedly states that the nation must have *Swarajya* (self-rule). Anachronistic, for sure, as the Rani fought for Jhansi and for the larger Maratha conglomerate of which Jhansi was a vassal state, but there is no evidence that she fought for the nation manifested as India. Replicating Varma's nationalism in the new film, the Rani rips through the Union Jack with her sword, crying out *Azaadi* (freedom). Representing the distance between the concerns and preoccupations of 1946 (Varma's novel) and the language of 2019 (*Manikarnika*, the film), the concepts of *Swarjya* (self-rule) and *Azaadi* (freedom) dovetail with the historical moments of their audience rather than their story. While self-rule is really not a concern for India in 2019, the liberty, dignity, and freedom of women remain under threat.

In a popular trailer for the film, the Rani asserts, "We both want Jhansi, the only difference is that you want to rule it and I want to be of service to those who are mine." In line with the large lettered words that appear on the screen "QUEEN," "MOTHER," and "WARRIOR," the Rani remains, forever, in service to the family, the community, and her kingdom. Even in this new extravaganza, with Hong Kong-style choreographed fight scenes, blood splashing on the camera lens, and with the Rani screaming out her battle cry as she slices the head of a British soldier, it is grief—the loss of her son, her husband, and then her kingdom—that propels the queen into war. The Rani never fights for herself. Aggression in a woman is cleverly showcased as the righteous anger and fire of independence; to allow it simply as an individual trait would legitimate women's anger, which is rarely a spectacle in India, or even the world. Perhaps, in a culture where the idea of sacrifice is so revered, these representations of the Rani may be interpreted as the highest form of praise. And yet, these self-sacrificial modes of representation, clearly in the service of the male-identified nation, are all stories of women whose lives have been erased to make room for their legends. Two major historical films, focused on women, opened in India in the last year or so— *Padmaavat* (2018) and *Manikarnika* (2019). The first is a story based on a poem about a legendary Hindu queen known as Padmini who committed *Jauhar* (ritual self-immolation) along with the other women of her kingdom rather than fall into the hands of the invading Muslim army. The second, of course, as has been discussed, is the story of a warrior queen who dies on the battlefield, in the violent, public arena of war. While the first sacrifice is about protecting the private, domestic realm that is women's bodies, the latter sacrifice is for the larger, public, national dominion of the country, the nation. While diametrically opposed in the actual happenstance of death, both narratives celebrate women dying for the glory of the nation. Thus, in the British imagination of the nineteenth century, the nationalist rendering of the twentieth century, or the popular cultural narrative of the globalized, transnational present, the warrior woman, Rani Lakshmi Bai, Queen of Jhansi, whether celebrated or reviled, presents an abiding conundrum of the warrior woman—whose innate power must be unleashed at strategic moments but then harnessed, locked-up, and brought under control lest she really "cry havoc and let loose the dogs of war," not only on the enemies outside but also on the structures within.

Notes

1 Saul David, *The Indian Mutiny: 1857* (London: Penguin, 2003), 367.
2 Harleen Singh, *The Rani of Jhansi: Gender, History, and Fable in India* (New Delhi: Cambridge University Press, 2014).
3 Lord Canning's "Control of the Press Act" banned publication of political and historical pamphlets in 1858.
4 Singh, *The Rani of Jhansi*, 1.
5 Partha Chatterjee, "Introduction: History and the Present," in *History and the Present*, ed. Partha Chatterjee and Anjan Ghosh (London: Anthem Press, 2002), 1–18. p. 16.
6 Aijaz Ahmad, "Jameson's Rhetoric of Otherness and the 'National Allegory,'" *Social Text*, no. 17 (Autumn 1987): 3–25. p. 6.
7 Singh, *The Rani of Jhansi*, 23. Shahid Amin, "Writing the Recalcitrant Event," a paper presented at the conference "Remembering/Forgetting: Writing Histories in Asia, Australia and the Pacific," University of Technology, Sydney, July 5, 2001.
8 Ibid.
9 Mahasweta Devi, *The Queen of Jhansi*, trans. Mandira and Sagaree Sengupta (Calcutta: Seagull Books, 2000). Originally published in Bengali as *Jhansir Rani* in 1956. Vrindavan Lal Varma, *Lakshmi Bai, The Rani of Jhansi*, trans. Amita Sahaya

(New Delhi: Ocean Books, 2001). Originally published in 1946 in Hindi as *Jhansi ki Rani*. D.B. Prasnis, *Jhansi ki Rani Lakshmi Bai* (in Hindi, trans. From original in Marathi). Allahabad: Sahatya Bhavan Pvt., 1964. Siriniwas Balajee Hardikar, *Rani Laxmibai* (Delhi: National Publishing House, 1968). E. Jaiwant Paul, *Rani of Jhansi, Lakshmi Bai* (New Delhi: Roli Books, 1997).

10 D.V. Tahmankar. *The Ranee of Jhansi* (Bombay: Jaico, 1960). S.N. Sinha, *Rani Lakshmi Bai of Jhansi* (Allahabad: Chugh, 1980).

11 A son was born to the Rani and Gangadhar Rao in 1851 but survived for only a few months, and so the king adopted a five-year-old, Damodar Rao, as his heir. The king's health deteriorated rapidly, and, fearing the worst, he wrote a letter to the East India Company on November 19, 1853, translated by Major Ellis, the political assistant at Jhansi:

> God willing I still hope to recover and regain my health. I am not too old, so I may still father children. In case that happens, I will take the proper measures concerning my adopted son. But if I fail to live, please take my previous loyalty into account and show kindness to my son. Please acknowledge my widow as the mother of this boy during her lifetime. May the government approve of her as the queen and ruler of this kingdom as long as the boy is still under age. Please take care that no injustice is done to her.

The letter expressly stated the king's request that the Company recognize his adopted son as the heir to the throne and the Rani as his regent. The Political Agent at Gwalior, Major D.A. Malcolm, wrote back to say, however, that "the adoption cannot be allowed or recognized without the special authority of the Government of India." The matter was then referred to the Governor-General, Lord Dalhousie, who quoted Sir Charles Metcalfe's words from 1837: "Chiefs who hold grants of land or public revenue by gift from a sovereign or paramount power... the Power which made the grant, or that which by conquest or otherwise has succeeded to its rights, is entitled to limit succession... (and to) resume on failure of direct heirs." With this precedent, Lord Dalhousie argued that since Jhansi had been a vassal state of the Peshwas, whose holdings had fallen to the East India Company, the British held the authority, as the "paramount power" to end the line of succession, and the "Doctrine of Lapse" took effect for Jhansi in 1854. The Doctrine of Lapse, though not formulated by Lord Dalhousie, was primarily implemented under his rule (1848–56). In his words, "Indian Kingdoms lapsed to the sovereign power by total failure of heirs natural." Michael H. Fisher, ed., *The Politics of the British Annexation of India, 1757–1857*.

12 In the ensuing years, the Rani carried out a protracted campaign of diplomacy by writing several letters to the Company, and by engaging a European lawyer named John Lang to argue her case. She invoked Jhansi's long-standing loyalty to the British and sought permission, according to the rights of the Hindu *Shastras* (laws), for her adopted son to be recognized as the rightful petitioner to the throne. In her invaluable *Kharita* (letter) dated February 16, 1854, the Rani initially employed the position of a supplicant in her interactions with the Company:

> I am listing a few precedents of allowing adoption to the widowed queens after the king dies without leaving an heir in various states of Bundelkhand. Because of this permission, their bond of loyalty to the British Government has become

stronger. They are totally happy and at peace. After looking at these examples with your kind consideration, I hope that you would allow the same right also to the widowed daughter-in-law of Shivarao Bhau. Please have sympathy with her helplessness. (Signed and Sealed by Maharani Lakshmi Bai, Translated and signed by R.R. Ellis.)

The letter is evidence that the Rani explored diplomatic channels that were open to her, and neither acquiesced obsequiously to the annexation nor jumped impatiently into rebellion. This is not to say that the Rani was a reluctant participant in 1857, which she may have been, but rather to clarify that in 1854 she was, in fact, carrying on the administrative and diplomatic duties that befell her as a ruler well before she rose to prominence as a rebel leader. The Company, however, ignored the Rani's pleas and stationed a garrison in Jhansi to oversee the administration of the state. In 1857, when the sepoys of Jhansi threw in their lot with the rebellion, the Rani was once again thrust into the limelight.

13 For a larger consideration of these imperatives, see David Washbrook's "After the Mutiny: From Queen to Queen-Empress," *History Today* 47, no. 9 (September 1997): 10–17.
14 Jenny Sharpe, *Allegories of Empire, The Figure of Woman in the Colonial Text* (Minneapolis, MN: University of Minnesota Press, 1993), 58.
15 Singh, *The Rani of Jhansi*, 2.
16 Maria Jerinic, "How We Lost the Empire: Retelling the Stories of the Rani of Jhansi and Queen Victoria," in *Remaking Queen Victoria*, ed. Margaret Homans and Adrienne Munich (Cambridge: Cambridge University Press, 1997), 123–139.
17 "Jhansi rani instigated soldiers to attack the Fort." Letter from S. Thornton, Deputy Collector, to Major W.C. Erskine, *Freedom Struggle in Uttar Pradesh: Source Material. Vol 3. Bundelkhand and Adjoining Territories, 1867–59* (Lucknow: Publication Bureau, 1959), 8–9.
18 C.A. Kincaid, (Charles Augustus), *Lakshmibai, Rani of Jhansi and Other Essays* (London: C. Kincaid, 1941), 12.
19 C.A. Kincaid, "Lakshmibai Rani of Jhansi," *Journal of the Royal Asiatic Society of Great Britain and Ireland*, no. 1 (1943): 100–104. R. Burn and Patrick Cadell, "Rani Lakshmi Bai of Jhansi," *Journal of the Royal Asiatic Society of Great Britain and Ireland* (1944): 76–78.
20 Thomas Lowe, *Central India during the Rebellion of 1857 and 1858* (London: Longman, Green, and Roberts, 1860), 261.
21 "Baee" is the Anglicization of the "Bai" of Lakshmi Bai's name, and is also the usual Maratha honorific accorded to women.
22 National Archives of India, New Delhi. Papers of the Foreign Department of the Government of India, 1858, June 11, 1858.
23 "Because she did not have an army of her own, she requested that Rao Saheb give her one. With Rao's permission, I fought in Kunch under her leadership." Tantya Tope's confession of April 10, 1858. Quoted in G.W. Forrest's *History of the Indian Mutiny 1857-58*.
24 Mrinalini Sinha, *Colonial Masculinity: The "Manly Englishman" and the "Effeminate Bengali" in the Late Nineteenth Century* (Manchester, UK: Manchester University Press, 1995).
25 Singh, *The Rani of Jhansi*, 28.

26 Gayatri Chakravorty Spivak, "Can the Subaltern Speak?" *Economic & Political Weekly* (2013): May 25, 2013.
27 Harleen Singh, "The rani rides again: After Padmavati, Manikarnika's Rani of Jhansi gets caught in the crossfire," https://thewire.in/film/rani-rides-padmavati-manikarnikas-rani-jhansi-gets-caught-crossfire (accessed October 13, 2019).
28 Singh, *The Rani of Jhansi*, 164.
29 Singh, *The Rani of Jhansi*, 167.
30 Singh, "The Rani Rides Again", The Wire, 12 Februray, 2018, https://thewire.in/film/rani-rides-padmavati-manikarnikas-rani-jhansi-gets-caught-crossfire, (Accessed 12 October, 2019).
31 Adrija Roychowdhury, "Manikarnika: A Legendary queen and her fictional depiction in a banned book," https://indianexpress.com/article/research/manikarnika-controversy-rani-laxmibai-jhansi-brahmin-protests-5054740/ (accessed October 13, 2019).
32 Jaishree Mishra, *Rani* (New Delhi: Penguin, 2007), vii.
33 Amrita Basu, "Women's Activism and the Vicissitudes of Hindu Nationalism," *Journal of Women's History* 10, no. 4 (Winter 1999): 104–120.
34 Manisha Sethi, "Avenging Angels and Nurturing Mothers: Women in Hindu Nationalism," *Economic and Political Weekly* 37, no. 16 (April 20–26, 2002): 1545–1552. Kalyani Devaki Menon, "'We Will Become Jijabai': Historical Tales of Hindu Nationalist Women in India," *The Journal of Asian Studies* 64, no. 1 (February 2005): 103–126.
35 Michael Safi, "Bollywood epic Manikarnika tells tale of India's anti-British queen," https://www.theguardian.com/world/2019/jan/25/ill-destroy-you-all-indian-historical-epic-star-critics-manikarnika (accessed October 13, 2019).

2

Historians and Nehanda of Zimbabwe in History and Memory

Ruramisai Charumbira

Beginnings

In continental and diasporic African popular memory, there is a powerful narrative that says the warrior woman Nehanda of Zimbabwe defied the British Empire when it came to her corner of the world. The warrior woman commonly known by the honorific *Ambuya* or *Mbuya* Nehanda, or simply Nehanda, was born Charwe, into the Hwata family of the Eland clan in the Mazowe Valley, *c.* 1862.[1] Oral traditions tell of Charwe as a lifelong resident of Mazowe, initiated for an *Mhondoro* (royal spirit) at a young age. When she came of age, like most young women of her age, she married, had children, and lived the life of a revered *mhondoro* (medium of a royal spirit) of the legendary rain goddess Nyamhita Nyakasikana of Handa (the original Nehanda).[2] This rain goddess is the one who was supposed to rescue the oppressed Africans from colonial rule through its medium Charwe when the British arrived. Instead, the British rendered Nehanda a piddling goddess, and Charwe a commoner who could not escape the British empire's noose at the gallows where Charwe met her death with courageous rage.[3] Charwe, as the Nehanda medium in the Mazowe Valley, initially welcomed the new Europeans thinking, like most people of her time, that they were after gold and other material resources and would return to their native lands.[4] The old Europeans, the Portuguese, were in that part of Africa since the early-to-mid 1500s; although they had not overrun the place with settlers as they did elsewhere in Africa and around the world. Relations between the Africans and the new Europeans soured and deteriorated in the Mazowe, as the British increased immigrants in the African interior, culminating in September 1890 when the Union Jack was officially planted on the territory that became Southern Rhodesia.[5]

The British empire planted that Union Jack by way of Cecil John Rhodes's British South Africa Company (BSAC), a mercantile company granted a royal charter by Queen Victoria in 1889. With that charter, the British Empire effectively outsourced the colonization of southern Africa's interior to business interests. The outsourcing of imperial brutality to ruthless business interests was nothing new in imperial history, British or otherwise. What was particular to the BSAC was that it was one of the last chartered companies in late imperial Britain that gained enough power to rival imperial France whose chartered companies had helped carve a large chunk of northern and western Africa. More importantly, the colonization of what became Northern and Southern Rhodesia (today's Zambia and Zimbabwe) entered British imperial and colonial lore through a popular narrative that claimed the empire was enlarged in Africa without a penny from the British taxpayer. No matter that the taxpayer (and especially the colonized peoples) had paid for the structures and infrastructure of empire that afforded the BSAC its charter and a large pool of immigrants from which to choose future settlers for the Rhodesias.[6]

The BSAC made good on its charter, sending a mercenary army—the British South Africa Police (BSAP)—into the African interior. It arrived in kwaHarare on September 11, 1890, a place christened Salisbury, Southern Rhodesia shortly afterward. The attempted erasure of the African place name kwaHarare with a new name that memorialized a British aristocrat of the British Empire is a critical element in the fuel that burns in the popular narrative of Nehanda of Zimbabwe's defiance of the British Empire. As I will demonstrate in this chapter, the woman popularly known as Nehanda of Zimbabwe was (and is) a symbol of African resistance against European settler colonialism as well as an example of the power of memory over history. Indeed, the story of the warrior woman Nehanda is a compelling narrative popular with anyone seeking a memorable figure in telling the story of African resistance to European hegemony in the mid- to late-nineteenth century. As will become obvious below, the Nehanda story has been bottled like holy water and passed on from hand to hand in books, from mouth to ear in poetry performances, and from mind to body in musical and theater performances.

The importance of the Nehanda spirit and its (mostly female) mediums among the peoples of southern Africa is similar to that of the Dalai Lama among the Tibetan people. Although the name Nehanda is now associated with one woman, Charwe, it is actually a politico-religious title held by many women before and after her. I use the analogy of the Dalai Lama to also highlight the fact that, among other things, there are similarities in the processes of selecting a new Dalai Lama or Nehanda. A new Nehanda, often a child, was chosen through an elaborate collaboration between the living and the dead.[7] Once chosen, that child was then raised with the knowledge that they have to carry on the legacy of their spirits, which, all things being equal, leads to the realm of self-realization. Most people in Zimbabwe and elsewhere do not often recognize the historical name Charwe wokwa Hwata, as they know her as *Ambuya* or *Mbuya Nehanda*, or in English, as Nehanda of Zimbabwe. Professional historians, on the other hand, have filled many gaps in popular memory, but, as will become clear in the rest of the chapter, in the process, they have also participated in making her larger than life as a warrior woman in the histories of Zimbabwe and southern Africa—a symbol of African resistance to colonialism.

Given the contrast between popular memory and professional history, the question that guides my historical analysis is: How do historians (broadly defined) contribute to the restoration, preservation, or making of warrior women in national and global histories? To answer this question, I analyze some key texts from literature and historiography of the 1950s to the 2000s to show the ways historians have consciously and unconsciously viewed Nehanda as nationalist or feminist warrior heroine. I have written at length on Charwe and the legend of Nehanda of Zimbabwe in my book *Imagining a Nation: History and Memory in Making Zimbabwe*, and will refer to that text where necessary to avoid repeating what I have already written on the subject there and elsewhere. My own work on Nehanda-Charwe has left within me a need to have a conversation with other historians about the toll (to the mind and soul) of prolonged exposure to primary sources detailing gruesome and traumatic events. The effects of this prolonged exposure, to my mind, invite a reconsideration of the lives of historians as humans with souls, too.

The texts I have chosen to answer this chapter's key question are: Solomon Mutswairo, *Feso*; Donald P. Abraham, "Early Political History"; Terence O. Ranger, *Revolt in Southern Rhodesia*; Yvonne Vera, *Nehanda*; David N. Beach, "A Woman Unjustly Accused"; and Ruramisai Charumbira, "Nehanda and Gender Victimhood."[8] These texts are not all that has been published about Nehanda-Charwe, but I chose these particular texts because they represent the changing representations of her by prominent historians over time; sometimes agreeing on the subject, sometimes differing markedly in sharp, albeit veiled, terms. The inclusion of my own work is not to say this is the last word on the subject and I hope not self-indulgent. Rather, I see this as an important moment to pause and reflect on a topic of enduring fascination in popular memory and professional history.

A Brief History of the Legend of the Original Nehanda

In chiShona orature, the name Nehanda is attributed to Nyamhita Nyakasikana, daughter of Mutota. Mutota was the man who, with his wife, founded the dynasty that spawned the Mwene Mutapa or Munhumutapa Empire (Mutapa for short) c. 1400–1902.[9] Although historians do not agree on whether the Mutapa Dynasty spawned a state, a kingdom, or an empire, I would argue that the Mutapa spawned both a kingdom and an empire, in part because of the durability of the legend of the Ne-Handa at the heart of that kingdom-empire. The Mutapa was a kingdom whose leadership was bolstered by bloodline, and succession politics insisted on that bloodline, although outsiders sometimes contested and breached that biopolitics.[10] The Mutapa was also an empire in the sense that its tentacles reached out to neighboring kingdoms with varying degrees of imperial success. This is analogous to, but not the same as, the British Empire with the UK at its center and in that center the English as the first among equals. I make this assertion about the Mutapa as kingdom-empire not to valorize empire, African or otherwise, but to highlight the fact that the terms Mwene Mutapa, Munhumutapa, and Mutapa loosely translate to *conqueror*. Understanding the historical context of the Nehanda within the powerful politico-economic entity of Mutapa helps us understand why the legacy of the warrior woman has endured.[11] Reimagined over time and space, the Nehanda legend has been preserved within the narrative of an African kingdom's imperial center and handed down as a powerful female spirit determined to live beyond the fog of history.

Donald Abraham, a professional historian interested in African-centered and African- articulated history in the 1950s and 1960s, collected orature on the Mutapa in northeastern and central Zimbabwe.[12] It is noteworthy that his research emerged at the same time as a mission-educated African elite writing and publishing in African languages, mainly in chiShona and isiNdebele (elaborated in the next section). Some of those (academic and literary) writings found their way into the Native Affairs Department Annual (NADA), first published *c*. 1923. The Native Affairs Department was something similar to the US Bureau of Indian Affairs, a settler colonial arm catering to all indigenous peoples' matters, and administered by the settlers. NADA was initially part chapbook, part magazine, and later became an amateur and professional journal on "native" life in Southern Rhodesia.[13] The orature that Abraham collected became important evidence that many inside and outside the academy relied upon to augment Portuguese records that were and still are the main primary sources for studying that part of Africa from the 1540s to the mid-1800s.[14]

Included amongst the narratives that Abraham collected of the founding of the Mutapa Kingdom was the legend of Nehanda, who appears as the first holder of that title and whose birth names included Nyamhita Nyakasikana, the daughter of Mutota and the founder of the patrilineal Mutapa dynasty. In today's chiShona language (cognizant of regional/dialect differences), the prefix or article Mwene/*Ne/ne* (and *Nya/nya*) means "of," signifying ownership, rulership, or guardianship. *Handa* is also the plural word for lace (from the singular *ruhanda*). The word *mhita* means a bead used as currency, and the word *kasikana* means young maiden or girl.[15] Furthermore, the word *nehanda* could also mean a young girl dedicated to being (or to) a rainmaker, or, paradoxically, the son-in-law of a chief. While historians have not puzzled over the meanings of these names, they are important to our analysis because among vaShona and most African societies of the past (and present) names have deeper meanings and often embody important contemporary and historical details. To that end, I believe it is important to analyze the name Nehanda and in the process gain an appreciation of the fact that the name has endured for at least five hundred years in some form.[16] And though the meaning of those words may have changed over time, it is important that in recent memory they still hold within them ideas of ownership or guardianship of a people's wealth: *mhita* as symbolizing beads as currency; *handa* symbolizing lace, for the veil between the living and the dead; and last but not least, *handa* as representing a geographic place in the eastern corner of the Mutapa kingdom.[17]

The name Nehanda, then, might have been spelled Ne Handa or Ne-Handa, or NeHanda as this rendition signals two things: *Ne* and *Handa*, rather than the contemporary Nehanda now associated with one person, Charwe. As I have argued elsewhere, the histories and memories of the original Ne Handa and all other Nehandas are complicated by the fact that they are histories and memories that often isolate and deprive them of their womanhood and the company of other women. We must reconsider the Nehandas within the culture of women's lives infused into the larger society. More importantly for our purposes here, considering Nehandas as women living within a female culture would clarify the warrior status of the original and subsequent Nehandas. But all told, Abraham's work brought African orature to the fore as an important primary source in the telling of African histories in colonial Zimbabwe read widely by other academics and those within settler society who sought to understand African histories and traditions.

Africans, on the other hand, turned to the few writings by emerging African writers rather than to Abraham's. Among the most popular was the poet, novelist, and cultural nationalist Solomon Mutswairo, the author of the popular novel *Feso*. Born in 1924, he belonged to the bridge generation between the old ways and the new—the generation that came of age in colonial mission schools and had not known any other system of government other than the settler colonial one. In fact, Mutswairo was born after the self-governing referendum of Southern Rhodesia in 1922 that came into effect in 1923, separating Southern Rhodesia from the Union of South Africa.[18] The political climate in Southern Rhodesia, the place where Cecil John Rhodes's last remains are buried, could not be more imperial, as Rhodesians beginning in 1910 sought self-rule in the fashion of Canada, Australia, New Zealand, and the Union of South Africa. It was in this political climate that the retelling of the legend of Nehanda and especially the story of Nehanda-Charwe and her contemporaries' resistance to colonialism took root as an ethnic memory among certain segments of the African population. As will become clear below, the fledgling African nationalist groups found in Nehanda a compelling anchor for their narrative of righteous indignation against European imperialism and colonialism.

Nehanda and Nehanda-Charwe in the 1950s and 1960s

Feso was first published in chiShona in 1957 by the Rhodesia Literature Bureau (RLB). Ironically, the RLB was an arm of the colonial government that often sought to control and limit African access to Western education, which would have engendered competition with new European immigrants settling in the colony at the time. As Donald Herdeck, the novel's English language editor, wrote in his introduction: "Mutswairo's *Feso* became at once the most popular novel and a bestseller, capturing the imagination of many students in [colonial] Zimbabwe from grammar school to the university."[19] More importantly, where Abraham's oral narrative told of a Nehanda trapped in linear time, Mutswairo's Nehanda was a goddess who collapsed time, fusing the past and present. Those who read *Feso* projected the nineteenth-century woman of the Mazowe Valley, Nehanda-Charwe, into the Nehanda of Mutswairo's historical novel, who may or may not have been Nyamhita Nyakasikana. It was the elegiac poetry of *Feso* that

> became particularly popular and was widely anthologized in [Shona] texts. The fact that a recent Nehanda had actually been hanged by the British during the rebellions of 1896–97 ... no doubt added some salt to her namesake's appearance in the novel. Two poems which appear in *Feso* invoke, then, the spirit of the ancestral "*Nehanda who speaks for the oppressed of all times and places.*"[20]

Mutswairo's *Feso* catapulted the legend of Nehanda from ethnic history and memory to national myth, bound together with the figure of Nehanda-Charwe. Mutswairo's writing was so powerful that even though Nehanda was not in the novel's plot, his poetry coiled memory tightly around history, making it easier for people to project Nehanda-Charwe, who lived in the near past, back into the longed for, golden African past where the gods and ancestors could not and did not abide their people's oppression and exploitation.

Ironically, outside Zimbabwe, it was through the English language that the legend of Nehanda gained a wider Pan-African and even global audience. Mutswairo translated his famous novel into English while living in the United States. The book gathered more fans not only because a critical mass of "native Rhodesians" lived in exile in the United States at the time, but also because of the Civil Rights and Black Power movements with their spirit of Pan-African solidarity. Indeed, in the 1960s, the intellectual movement of Afrocentricity gathered steam in the United States, and one of its most visible remnants today is the celebration of Kwanzaa—a celebration of African heritage and pride in the African diaspora. This, of course, is not to suggest that the legend had not previously made its way across the Atlantic and beyond by word of mouth. Rather, what I am suggesting here is that though *Feso* was both legend and protest novel, it was perfectly attuned to the male-centered nationalist narratives of both African and Pan-African nationalisms whose rallying cries were often for the black man's freedom—literally using the word "man" rather than the more inclusive term "human," which would include black women. And to this day, both in Africa and the African diaspora, women have to fight and insist on their rights to the same privileges that nationalism promised and promises to all but tends to deliver mostly to men.

Thus, although Mutswairo's politics in *Feso* was radical for 1950s and 1960s Southern Rhodesia, its radical potential was limited when it came to imagining African women—past and present—as equals. Indeed, Mutswairo could not be blamed as the only sexist male writer; it is obvious in reading *Feso* that his Nehanda and the women of his novel were not much different from those in Western classic literature—until Western women wrote women's histories. After all, in Homer's narratives women often appear in service to men's heroic external journeys to and from war as well as their internal journeys of self-discovery. In literature, women make men's ambitions possible at their own expense—for the sake of the tribe and the nation.

In *Feso*'s case, Nehanda's (male) children needed her in their hour of desperation so liberation could return them to an idyllic past where their masculinity was not challenged by foreign invaders. Indeed, for all its radical ambition as political manifesto, Mutswairo's *Feso*, especially in its English translation, valorized masculinity by invoking a Nehanda-Nyakasikana devoid of her femininity and womanhood, a goddess without a female culture around her. This echoed common practices in nationalist movements at the time—the practice of shunting women and their priorities and needs for freedom to the back burner of nationalist politics.[21] All said, however, it is indisputable that Nehanda, the warrior woman of Zimbabwe (Nehanda-Charwe) owes her popularity in national and global consciousness to Mutswairo's elegiac poetry (to Nehanda) in *Feso*.

Charwe as *the* Nehanda in the 1960s and 1970s

The next decade saw the legend of the warrior woman Nehanda grow. This time it was because more guild historians took it upon themselves to write African histories from the colonial archive to prove that the colonial project was real, and Africans were its undisputed victims, African cultural translators and collaborators notwithstanding.

I am not alone in arguing that the late historian Terence Osborne Ranger, the doyen of Zimbabwean history for many decades, did more for the study of African resistance in Zimbabwe than any other European Africanist scholar in the 1960s and 1970s. Ranger was not, of course, the only historian at the time. What I am arguing here is that his work raised the profile of Charwe wokwa Hwata to heroine status as *the* Nehanda of Zimbabwe. Intentionally or not, Ranger's African-centered historical research into African resistance to European colonialism in Southern Rhodesia was a study that also fell into the center of a growing interest in differentiating between primary nineteenth-century and secondary post-1945 African resistances to European imperialism and settler colonialism.

Ranger's now classic text *Revolt in Southern Rhodesia* was published in 1967. His preface and acknowledgments make clear that in the process of researching and writing that book he made lifelong friends with Africans who held not only Nehanda-Charwe, but also the people of her period, in the highest regard.[22] The Africans, Ranger shows in his study, resisted European settler colonialism because of their own outrage at their oppression. More importantly, the power of the high ancestral spirits fostered the emotional state of the people. Although 99 percent were male spirits or men, the exceptional female spirit was that of Nehanda and her female medium Charwe.

Where Mutswairo's *Feso* had inspired the masses and the nationalists with its Nehanda-Nyakasikana, Ranger's text inspired nationalists and scholars with its deep research in the colonial archive. Most importantly, Ranger's monograph carried a guild historian's argument based on written evidence (colonial written evidence, no less) doing more for the legend of Nehanda and Charwe than had been done by Abraham who relied on orature to tell the history of the original legend and later relied on Portuguese sources as well. To most readers at the time, Ranger's text brought forward the narrative of the Nehanda of popular memory, the Nehanda everyone seemed to remember, because that was the Nehanda who was slowly becoming one nationalist movement's matron saint. She was being pushed forward as the ancestor who would come to rescue, not just the nationalists fighting for liberation, but the ordinary people laboring under an oppressive settler colonial regime. Ranger's Nehanda fit in with the period in which the book was published, showing the ways in which historians also respond to their contemporary moment even as they labor to write objectively about the past and its peoples.[23]

Ranger's *Revolt* has the design of a microhistory, but it tells a macrohistory of African resistance. Moreover, although Ranger's history was African-centered, it tended to boost the history of the majority ethnic group among the oppressed, neglecting everyone else. As historian Clapperton Chakanetsa Mavhunga wrote in his essay memorializing Ranger (who passed away on January 3, 2015), although *Revolt* stood out as African history representing African voices, Ranger's body of work on Zimbabwe was, until later in his life, "Shona history," at the expense of most other ethnic groups and, indeed, Zimbabwean history writ large.[24] While I agree with Mavhunga's assessment, I also see Ranger's work as emanating from his (Ranger's) own history, some of which he elaborated on in his memoirs.[25] To my mind, Ranger had an affinity for the underdog and at the time he wrote *Revolt*, the Shona were the underdogs in much of colonial and settler historiography. Given his background, Ranger consciously and unconsciously took on the cause of the oppressed. And who does not like a delightful story of triumph even if that triumph is death itself?

Now, more than fifty years after *Revolt* was first published, one can see Ranger's passion for the disenfranchised, although seemingly from a position of power as he projects himself as the champion of the voiceless—the white savior complex, some might argue cynically. I say this not to defend or to besmirch the legacy of a towering figure in African (and Zimbabwean) history, I say this because Ranger himself had the capacity for self-reflection which most historians eschew in the name of objectivity. I also write this to show the ways historians are themselves shaped (and sometimes trapped) by the time(s) they live in and how their own personal histories and passions influence their work. To this last point, it matters even more to our understanding of *Revolt* and the Nehanda therein, that Ranger's doctoral supervisor at the University of Oxford was none other than the infamous (or famous, depending on who you ask) Hugh R. Trevor-Roper, the man most quoted for his bigoted remarks on the (non) existence of African history.[26] Trevor-Roper, it would seem, took courage from Hegel's work that describes Africa, and especially Africans, as beneath the serious study of the past, an attitude that still affects even the most astute of scholars who do not ask where terms like sub-Saharan Africa originated and use them with a sense of righteous indignation to champion the underdog "Black Africans."[27]

It also makes sense, to my mind, that Ranger wrote *Revolt* to rebel against the British Empire whose brutal colonial history he investigated when he wrote his dissertation on land issues in Ireland. In Africa, Ranger found in the Shona people's history not only fascinating politics, but also the captivating religious history that introduced the dazzling figure of the high, royal spirit of Nehanda. That was the same Nehanda whose memory was resurgent through Mutswairo and Abraham's work, a memory the colonial government was trying to thwart in an attempt to terrorize the Africans into abandoning their gods and their ways of knowing and of being. The memory of Nehanda that Ranger resuscitated and validated through his archival research added mystical and righteous fire to the nationalists' fight for freedom. As a British citizen and a British-trained historian from the University of Oxford, Ranger gave Nehanda-Charwe a credibility denied her by colonial reports, colonial historiography, and settler propaganda beginning in the 1890s. Ranger's *Revolt* and most of his research and writing on Zimbabwe and southeastern Africa influenced a generation of now illustrious historians, many of whom wrote moving obituaries and tributes when he died in 2015.[28] Many of Ranger's former students have become leading historians themselves, and far fewer have done work centered on African religious and spiritual practices inspired by the kind of religion and spirituality represented by Nehanda-Charwe.

Nehanda-Charwe as Feminist Nationalist in the 1980s and 1990s

The most prominent response to Ranger's *Revolt* in the 1990s came from someone who was not Ranger's student, but who became his dear friend, the late literary scholar and novelist Yvonne Vera. Even before she met Ranger, Vera took *Revolt* and, to turn a phrase, revolted against his historiographical rendition tied to the colonial archive. Instead, in an elegiac and triumphalist novel titled *Nehanda,* she presented a feminist nationalist Nehanda-Charwe. Elegiac because it mourned the cultural and economic plunder of colonialism; triumphalist because the Nehanda in Vera's book was Charwe, the Nehanda of popular memory. She was a Nehanda that resonated with the rise of many women's organizations from the late 1980s to the late 1990s in Zimbabwe and southern Africa. In that time period, Nehanda-Charwe had almost become the exclusive nationalist property of the Zimbabwe African National Union-Patriotic Front (ZANU-PF),[29] and used as a blunt instrument that nationalists, especially men, used to silence women who questioned the exclusion of women from the nation's founding myths—even as Nehanda-Charwe's statue was installed within the courtyard of then Prime Minister Robert Mugabe's offices, a stone's throw away from the nation's Parliament.[30] The masculine nationalist concept of Nehanda had ceased to be the radical champion of freedom and, instead, had become a domesticated saint that all people, especially women, had to emulate postindependence.

Yvonne Vera's *Nehanda* was published in 1993. Once published, the book set her reputation as a rising star in Zimbabwe's literary circles, and the book as an important addition to the canon of Zimbabwe's literature, and postcolonial literature more generally. By my reading, her first novel *Nehanda* is a homage to both Mutswairo's and Ranger's Nehandas. In her rendition, however, she explores more than just the mystical and political figure to tell the story of a female and womanly Nehanda, one surrounded by other women and, yes, in a patriarchal culture no less. Vera's *Nehanda* is a nationalist in the spirit of Mutswairo's nationalist *Feso*, and she is a patriot in the spirit of Ranger's nationalist and patriotic *Revolt*. Vera's *Nehanda* taps into the mythology of a Nehanda born knowing that she has a rendezvous with history and memory.

It is also significant that Vera's *Nehanda* was first published in Toronto where she was studying at York University earning her doctorate in English Literature. She was an active participant in the larger circle of postcolonial writers who gave voice to the histories of their countries of origin. These were places that were once colonies with few spaces for native or indigenous writers with no connections, writers who then found welcoming audiences and platforms abroad. The irony does not escape me, of course, that places like Canada and the United State were where most postcolonial thinkers, academics, writers, and artists (myself included) found acceptance, and were (and are) also places that made (and make) little to no space for the dispossessed First Peoples of North America. Broadly speaking, Native American writers have had to insist that their voices be heard above the din of immigrant voices wanting to be seen and heard in a white, racist Canada or United States. It is, therefore, quite ironic that the telling of global women warrior narratives, like those of Nehanda of Zimbabwe in North America, reached mass circulation before we could read of similar kinds of women and communities from the First Peoples of the Americas as told by a critical mass of indigenous historians, scholars, poets, and writers of all kinds. But, I digress; the point is that the Nehanda that Vera (and even I) wrote about is a figure with equivalents among the First Peoples, women we could see as universal archetypal warrior women, rather than singular nationalist and national figures representing particular ethnic or national histories. What Vera's *Nehanda* did that was different from all that had been previously written about her was to signal to the preceding generations that the new generation was claiming history for itself, with or without archives—and that history would be for all, an inspiration to resist any form of colonization. Women, children, and all those marginalized peoples who had a right to claim Nehanda's legacy and that of their ancestors could do so in this brave new world brought to light by the new generation of historians and writers.

Vera's *Nehanda* firmly established Nehanda-Charwe of Zimbabwe as a warrior woman, a warrior ordinary women could relate to as role model for their own rights now that it was obvious that the Nehanda of (black male) nationalists was a Nehanda who favored her (grand)sons over her (grand)daughters. Vera's *Nehanda* evoked a mystical Nehanda rooted in African ways of knowing and being. She was also performing African feminist-nationalist-speak, some of which was emanating from women's memories of war, especially female ex-combatants. Those women found out the hard way that the hero of war has a male body and face, not a female one, or bodies gendered otherwise.[31] And by gendered otherwise, I mean the ways in which gender operates in African societies, not always gendered by sex, something most have forgotten in the binary gender wars.[32] In *Nehanda*, Vera imagined a woman of uncompromising principle, a woman willing to die for the land of the ancestors, but one who retained her humanity, dignity, and individuality.

The warm international reception of Vera's *Nehanda* happened at about the same time when many women's organizations were forming in the second decade of Zimbabwe's independence. These organizations were often funded by Western national funding agencies, such as the (then) Canadian International Development Agency (CIDA), the Dutch Hivos (Humanist Institute for Cooperation with Developing Countries), and private nonprofit women's organizations such as the US Global Fund for Women, to give just a few examples. Thus, although as singular as Yvonne Vera's *Nehanda* was for its time, it also emerged out of an activist postindependent women's movement in Zimbabwe and southern Africa (except for apartheid Namibia and South Africa until 1990 and 1994, respectively). Vera's *Nehanda* presented to many a truer model of the warrior woman in Zimbabwean history. This personal hero was one each woman could claim and hold unto herself, whereas before, many were afraid to claim Nehanda out loud and in public because that Nehanda belonged to the nationalists, especially men. This was because by the time Vera's *Nehanda* was published, the memory of Charwe as Nehanda had effectively become the ruling party's (ZANU-PF's) Nehanda, a Nehanda no one else had a right to claim. The irony is that by turning an archetypal figure and heroine of resistance into the exclusive property of the dominant nationalist party, the party and the government were simultaneously micromanaging the legacy of a popular figure. In fact, Vera's *Nehanda* came in the wake of the ruling party actively trying to silence any other Nehandas that emerged post-1980 so they could claim the exclusivity of the memory of Nehanda-Charwe as matron saint of the ruling party.

Nehanda as Archetypal Heroine of Resistance Abroad

Internationally, many women, especially black women in the diaspora saw Nehanda-Charwe and other warrior women as role models to be emulated in the fight against the patriarchy in racist societies in the West. Vera's *Nehanda*, much like Ranger's *Revolt*, appeared at a particular moment in history and became a touchstone text that spoke of not just Nehanda-Charwe, but African histories and traditions—and especially the fight for liberation from mental and spiritual colonization in the West. Internationally, before Vera's feminist nationalist *Nehanda* came along, many people had accessed the story of Nehanda through David Sweetman's (now out of print) *Women Leaders in African History*.[33] Sweetman's little book is a collection of mini-biographies of famous African women across the whole continent—most of them, warrior women—against some form of European colonialism. Those with limited or no access to Zimbabwean historiography or patience with academic tomes took to Sweetman's book as the go-to source for a profile of Nehanda and her singular story of defying the British Empire. The effect was that many women, especially black women, took Nehanda as an inspiration, a personal and collective role model of resistance.

Two prominent examples come to mind. The first is that of rap musician Nehanda Abiodun whose birth name was Cheri Laverne Dalton. She fled from the United States to Cuba, where she currently lives, but I do not know enough to comment on her case and its parallels to the historical Nehanda-Charwe.[34] Suffice it to say, she was a black woman, an African American woman who took on the name of another black woman to signal the similarities of their lives being handled by legal systems steeped in capitalist, white supremacist patriarchy, to borrow a phrase from bell hooks. The other prominent example of Nehanda-Charwe in the African diaspora is the Nehanda Black Women's Organization based in London. The organization seems less active now but was quite active in the early 2000s when I met its founder while doing doctoral dissertation research in the UK. Nehanda was the name the organization chose because the founder, a Caribbean woman, had made a personal connection to a Zimbabwean woman in London. The organization's founder, being that she was from the older African diaspora (the older diaspora being those descendent from enslaved African people in the Americas; vis-à-vis the newer diaspora whose migration is by "choice"), relished the idea of a historical continental African woman defying the mighty British Empire. In this, the women's organization was reaching for the archetype of the strong black woman that Nehanda-Charwe represented. Nehanda became the lead name of the organization and her image became their symbol. Nehanda-Charwe had triumphed again in British history, this time through popular memory.

The growing profile of Nehanda on the international stage as well as at home drew a contrarian position from none other than the prominent Zimbabweanist, the late British historian David Norman Beach. I contend that Beach found the Nehanda of popular memory troubling and sought to correct what he considered to be misconceptions about her through an article published in 1998. In that piece, Beach argued that Nehanda-Charwe was a "gender victim." The article sought to reexamine, in his words, "the historical basis of the legend" of Nehanda.[35] The gender victim argument was accompanied by what I call a classic historian's instinct to bicker when memory trumps history. Beach's argument, to be sure, was also a longstanding argument against Ranger's *Revolt*. In the contemporary moment, it was also an argument against Robert Mugabe's ruling party and government's stranglehold on the narrative of Nehanda and its bid to use the National Museums and Monuments of Zimbabwe to memorialize the places where Nehanda and her comrades had fought the British in the 1890s. That exercise was prompted by the realization that many years after political independence most monuments in Zimbabwe were honoring Rhodesian colonial and settler history—even Great Zimbabwe that had been a subject of great archaeological controversy in the history of archeology as a discipline.[36]

No less important to Beach was the popular memory that flooded the streets, including the urban legend of the *hanging tree* in the Avenues neighborhood of Harare.[37] Beach pointed a subtle accusatory finger at Ranger's *Revolt*. Mavhunga's previously mentioned essay better articulated Beach's accusatory finger, a finger, I would add, which was also pointed (by Beach) at most of those who had written about Nehanda as warrior woman of Zimbabwe. Suffice it to say, the stance Beach took as an objective historian in his 1998 article sought to show that others were trying to create a heroine out of thin air. He asserted that the archives did not show Nehanda-Charwe to be a heroine. Instead, he argued the documents showed she was a victim of a patriarchal society (the Shona) and a male-dominated colonial legal system (the BSAC on behalf of the British Crown). Given those gender disparities, Beach argued, Nehanda-Charwe was no heroine; rather, she was a gender victim, and the only reason she rose to the forefront in historical memory as a heroine is because she was the only woman in a sea of men.

Ten years later I published an article refuting that argument in the same journal.[38] There I argued (and still argue) that, though a careful historian who had done much for Shona and Zimbabwean historiography and for southern African history more generally, Beach's women's and gender history, while coopting the language of gender, did not provide a rigorous gender or gendered analysis to show how Nehanda-Charwe was a mere victim and not a real agent of and in history. Instead, I argued that Beach's article and much of his historical research seemed more intent on taking down a heroine than providing a real understanding of not just Nehanda-Charwe, but African women in the 1890s generally and those in the resistance movement to colonial rule specifically. Moreover, Beach's gender victim argument did not consider men and their gendered ways either. Unfortunately, Beach was not alive to continue the historical debate, but his article highlighted the tensions embedded in about five decades of a historiographical interrogation of Nehanda-Charwe—mostly in praise of her but sometimes in contestation.

Endings

The legend of Nehanda turned nationalist and feminist icon as discussed in this chapter brings us to an important conclusion, the uses and abuses of that history and historiography by Robert Mugabe's ruling party (ZANU-PF) and his government

(1980–2018). ZANU-PF, as indicated in this chapter, had a stranglehold on the historical narrative of Nehanda-Charwe, a stranglehold that alienated many people as the 1896–97 African resistance became more politics than history lesson. Nehanda-Charwe, in ZANU-PF hands became a very gendered grandmother, and her title as Ambuya/Mbuya became a literal interpretation of her role as the psychic enforcer of patriarchal values intended to put and keep women in their (subordinate) place. Nehanda-Charwe ceased to be the radical woman who defied the British Empire that offered her and her fellow prisoners (mostly men) Christian conversion in exchange for commuted sentences of capital punishment.[39] Nehanda-Charwe in history was defiant and would not hear of converting to foreign religions, even if for her own freedom. Indeed, as I have written about, she adamantly refused Catholic conversion and that refusal bolstered her credibility as a freedom fighter, the warrior woman defying the British Empire with all its furious gun-powdered cannon power. Thus, when the National Museums and Monuments of Zimbabwe was curating African memory places of resistance to British imperial colonization, which included the Mazowe where Nehanda and her comrades had fought the British in the 1890s, her place in the Mazowe became overt political work than historical monument making and preservation, which often eschews big P politics.

The role of historians in the making of Nehanda-Charwe the warrior woman brought forth a response from Terence Ranger as it became obvious that his scholarship and activism in the 1960s, especially in *Revolt in Southern Rhodesia*, rather than be a resistance History book as intended in the settler colonial era (1960s), it had become elite political weapon for silencing dissent in the postsettler rule period. In his essay "Nationalist Historiography, Patriotic History and the History of the Nation: The Struggle over the Past in Zimbabwe," Ranger sought to re-argue the place of the historian in the making of History and its heroes and heroines considering how ZANU-PF was using his earlier works.[40] Yvonne Vera's historical novel, *Nehanda*, and her body of work more generally, democratized the legend of Nehanda-Charwe and what history means in a former colony deeply marked colonization that still lives in the minds and souls of the former colonized. My own work on Nehanda has brought forth the conversation of not only how the past affects whole societies, but what it does to individual memory, including the memory of the historian whose mind (and soul) is deeply marked by the gruesome details of frontier history. This chapter, then, shows the ways historians can, consciously and unconsciously, participate in the making of real and imagined historical figures.

Notes

1   It is worth pointing out from the outset that the honorific *Ambuya* or *Mbuya* Nehanda that often goes with her name (and its equivalent, *Sekuru* for men) is a title that signals respect for an ancestral spirit as well as for the medium as a respected elder. The honorifics are also used when addressing mere mortals, grandparents, aunts and uncles, and any wise people in the community.
2   There is no definitive scholarly biography of Charwe nor of the legend of Nehanda, the varied works, in English, include: David N. Beach, "An Innocent Woman,

Unjustly Accused? Charwe, Medium of the Nehanda Mhondoro Spirit, and the 1896–97 Central Shona Rising in Zimbabwe," *History in Africa: A Journal of Method* 25 (1998): 27–54; Ruramisai Charumbira, *Imagining a Nation: History and Memory in Making Zimbabwe* (Charlottesville: University of Virginia Press, 2015), 22–27; Stanislaus I.G. Mudenge, *A Political History of Munhumutapa c. 1400–1902* (Harare, 1988); David N. Beach, *The Shona and Zimbabwe 900–1850* (Gwelo: Mambo Press, 1980); David Lan, *Guns and Rain* (Harare: James Currey and Zimbabwe Publishing House, 1985); Solomon Mutswairo, *Mapondera, Soldier of Zimbabwe* (Harare: Longman, 1983).

3 Ruramisai Charumbira, "Nehanda and Gender Victimhood in the Central Mashonaland 1896–97 Rebellions: Revisiting the Evidence," *History in Africa* 35 (2008): 103–131. http://www.jstor.org/stable/25483719 (accessed October 13, 2019).

4 See C.G. Chivanda, "The Mashona Rebellion in Oral Tradition: Mazoe District," in *Revolt in Southern Rhodesia, 1896–7: A Study in African Resistance*, ed. Terence O. Ranger (London: Heinemann, 1967), 390–394.

5 Charumbira, *Imagining a Nation*, 27–40.

6 See among others, Charumbira, *Imagining a Nation*, 77–137; Alois Mlambo, *White Immigration into Rhodesia: From Occupation to Federation* (Harare: University of Zimbabwe Publications, 2002).

7 For the Shona process, see David Lan, *Guns and Rain: Guerrillas and Spirit Mediums in Zimbabwe* (Berkeley, CA: University of California Press, 1985); and for the Tibetan process, including on the current Dalai Lama Tenzin Gyatso, whose birth name is Lhamo Thondup, see Alexander Norman, *The Secret Lives of the Dalai Lama: The Untold Story of the Holy Men Who Shaped Tibet, from Pre-History to the Present* (New York: Doubleday, 2008); and Robert Thurman, *Essential Buddhism* (New York: HarperCollins, 1995).

8 Solomon Mutswairo, *Feso*, ed. Donald E. Herdeck (Harare, Zimbabwe: HarperCollins, [1957, 1974] 1995); Donald P. Abraham, "The Early Political History of the Kingdom of Mwene Mutapa, 850–1589," in *Historians in Tropical Africa* (Salisbury, Rhodesia: University College of Rhodesia and Nyasaland, 1962): 61–90; Ranger, *Revolt in Southern Rhodesia, 1896–7: A Study in African Resistance* (London: Heinemann, 1967); Yvonne Vera, *Nehanda* (Toronto: TSAR, 1993); Beach, "An Innocent Woman," 27–54; Charumbira, "Nehanda and Gender Victimhood," 103–131.

9 Abraham, "The Early Political History"; Mudenge, *A Political History*; Charumbira, *Imagining a Nation*, 22–25.

10 See especially Mudenge, *Munhumutapa*, chapters 1–3.

11 For more, see especially Kingsley Garbett, "From Conquerors to Autochthons: Cultural Logic, Structural Transformation, and Korekore Religion Cults," *Social Analysis: The International Journal of Social and Cultural Practice* 31 (July 1992): 12–43.

12 I borrow from Ngugi wa Thiong'o's definitions of orature, expressed in many writings, especially: "The Language of African Literature," in *Decolonizing the Mind: The Politics of Language in African Literature* (London: James Currey, 1986), 4–33; and "Oral Power and Europhone Glory," in *Penpoints, Gunpoints, and Dreams: Towards a Critical Theory of the Arts and the State in Africa* (Oxford: Clarendon Press, 1998), 103–128.

13 The 1950s and 1960s also saw the rise of African and Area Studies in the Western academy and, in the process, that rise built a critical mass of Africanist and Area Studies experts, those outsider-insiders.

14 For Portuguese primary sources that yield some fragments on the Nehanda and other royal women, see George M. Theal, *Records of South-Eastern Africa Collected in Various Libraries and Archive Departments in Europe*, vols. 1 & 2 (London: William Clowes & Sons Ltd for the Government of the Cape Colony, 1898–1903); and *Documents on the Portuguese in Mozambique and Central Africa, 1497–1840* (Lisbon: Centro de Estudos Histócos Utramarions; Harare: National Archives of Zimbabwe, 1989), vols. VII & VIII.
15 Here I only refer to one Shona dictionary, but there are several others; M. Hannan, *Standard Shona Dictionary* (Harare, Zimbabwe: College Press, 1987). The work of historical linguist Herbert Chimhundu is highly recommended for a critical assessment of Hannan's work on the Shona dictionary.
16 I define the word, not so much to demystify the mystery of the sacred for those who might consider this blasphemy, but rather to affirm its historical meaning so it can be better appreciated. I imagine this to be the same that any decent scholar of the life of, say, Jesus of Nazareth would do when defining the name Jesus and his form of enlightenment as "the Christ." I imagine that scholar is not intent on merely knocking down a puffed-up god, but engaging in honest scholarship that helps us understand the enduring legacy of not just a man, but an idea of transcending the mundane life humans encounter on this planet.
17 An excellent map of this location is in David Lan, *Guns and Rain*, 33; from that map, the interested reader can proceed to less detailed old maps such as the eighteenth-century map of Rigobert Bonne, "*Carte du Canal de Mosambique*" (1780), Stanford University, *Maps of Africa: An Online Exhibit: A Digital Collection of African Maps at the Stanford University Libraries*, https://exhibits.stanford.edu/maps-of-africa (accessed October 13, 2019).
18 See most recently, Abraham Mlombo, "Southern Rhodesia's Relationship with South Africa, 1923–1953." (PhD dissertation, University of the Free State, 2017), 54–55.
19 Donald E. Herdeck, introduction to *Feso*, n.p. Mutswairo translated his own work and Herdeck was the editor, not the translator.
20 Ibid., emphasis added.
21 There is a sizeable historiography on women asserting their rights within nationalist and guerrilla movements in Zimbabwe and southern Africa more generally; here I will only reference ex-combatant women's oral narratives, mainly because some make reference to Nehanda-Charwe as inspiration: Zimbabwe Women Writers, *Women of Resilience: The Voices of Women Ex-Combatants* (Harare: Zimbabwe Women Writers, 2000).
22 Ranger, *Revolt in Southern Rhodesia*, ix–xii.
23 For a critical assessment of Ranger as participant and historian in colonial and postcolonial Zimbabwe, see Luise S. White, "Terence Ranger in Fact and Fiction," *International Journal of African Historical Studies* 44, no. 2 (2011): 325–331.
24 Clapperton Chakanetsa Mavhunga, "Many Terence Rangers," *Africa Is a Country* (January 29, 2015), https://africasacountry.com/2015/01/many-terence-rangers/ (accessed October 13, 2019).
25 Terence O. Ranger, "From Ireland to Africa: A Personal Memoir," *History Ireland* 14, no. 4 (July–August 2006): 46–49; and Terence O. Ranger, *Writing Revolt: An Engagement with African Nationalism, 1957–1967* (Suffolk, UK: James Currey, 2013).
26 Ranger, *Writing Revolt*, 4.
27 Georg W.F. Hegel, *The Philosophy of History*, trans. J. Sibree (Kitchener, ON: Batoche Books, 2001), 109–112.

28 See, for example, an obituary in *The Guardian* newspaper and a collection of obituaries and tributes in the *Association of Concerned Africa Scholars Review* 89 (Spring 2015), http://www.concernedafricascholars.org/wp-content/uploads/2015/04/rangerbulletin1.pdf (accessed October13, 2019); as well as some by Zimbabweans such as Mavhunga above and Percy Zvomuya published online at *Africa Is a Country*, https://africasacountry.com/ (accessed October13, 2019).
29 For a recent take on the founding of ZANU-PF, see among others, Eliakim Sibanda, *The Zimbabwe African People's Union 1961–87: A Political History of Insurgency in Southern Rhodesia* (Trenton, NJ: Africa World Press, 2005).
30 For other ways the ZANU-PF Party, and government utilized the iconography of Nehanda-Charwe, see among others, Ruramisai Charumbira, "Gender, Nehanda, and the Myth of Nationhood in the Making of Zimbabwe," in *National Myths: Constructed Pasts, Contested Presents*, ed. G. Bouchard (New York and London: Routledge), 206–222.
31 For a fuller articulation of a feminist nationalist in Vera's *Nehanda*, see Desiree Lewis, "Biography, Nationalism, and Yvonne Vera's *Nehanda*," *Social Dynamics: A Journal of African Studies* 30, no. 1 (2008): 28–50.
32 See West African examples in Ifi Amadiume, *Male Daughters, Female Husbands: Gender and Sex in African Society* (London: Zed Books, 1987); Oyeronke Oyewumi, *The Invention of Women: Making African Sense of Western Discourses* (St. Paul, MN: University of Minnesota Press, 1997).
33 David Sweetman, *Women Leaders in African History* (African Historical Biographies) (London: Heinemann, 1984).
34 Bret Sokol, "Exiled in Havana," *Miami New Times*, September 7, 2000; she has since died in Havana, Cuba. "Nehanda Obiodun, 68, Black Revolutionary Who Fled to Cuba, Dies," *New York Times*, February 8, 2019.
35 Beach, "An Innocent Woman," 27.
36 Bruce Trigger, *A History of Archaeological Thought* (Cambridge: Cambridge University Press, 1989), 130–135.
37 That tree has since died. See Ruramisai Charumbira, "Zimbabwe's Hanging Tree," *Not Even Past*, https://notevenpast.org/zimbabwes-hanging-tree/ (accessed October 13, 2019).
38 Charumbira, "Nehanda and Gender Victimhood," 103–131.
39 Fr. Richartz, S.J., "The End of Kakubi and Other Condemned Murderers," *Zambesi Mission Record* 1, no. 2 (November 1898): 53–55; "The High Court," *Rhodesia Herald*, March 9, 1898.
40 Terence O. Ranger, "Nationalist Historiography, Patriotic History and the History of the Nation: The Struggle over the Past in Zimbabwe," *Journal of Southern African Studies* 30, no. 2 (2004): 215–234.

3

Women Warriors or Mothers of the Fatherland: Hero Cults and Gender in Basque Nationalism

Nerea Aresti

Basque nationalism emerged as a movement seeking to defend Basque political rights at the end of the nineteenth century. From its inception, it was influenced by contemporary ideas of social hygiene, social Darwinism, degeneration, and race. This new movement was also in part a response to far-reaching local transformations, specifically rapid industrialization and, above all, the mass of migrants from other parts of Spain attracted by employment opportunities in the region's mines, foundries, and shipyards. One response to what some perceived as an invasion of brutish, impious, and racially inferior immigrants was the elaboration of Basque nationalism, a political project in defense of the Basque race and its traditions. This was, in the words of the historian José Luis de la Granja, traditionalist and fundamentalist, anti-liberal and anti-Spanish, anti-socialist and anti-industrialist.[1] Above all, it was Catholic. Basque nationalists defined the Basque race more in moral than in biological terms, and they called on the Basque people to recover the halcyon days when they, as an example of moral and religious rectitude, had enjoyed political independence from Spain and a racial purity free from "Spanish" contamination.

In addition to alleged racial differences, early Basque nationalists built their identity on the basis of the uniqueness of their culture and their language—*euskera* or Basque, a language without Indo-European roots—and long history of resistance against invaders led by heroic patriots. As with other contemporary nationalisms, this narrative resulted from a specific understanding of gender and sexual difference. But despite the fact that Basque nationalists' visions of gender were traditionalist and profoundly misogynistic, some women were given pride of place in the pantheon of national heroes. Far from being a contradiction, this was the consequence of the particular way Basque nationalists understood gender with respect to the nation. Their misogyny prevented real women from participating in the movement, but their privileging of race meant that it could prevail over gender so that Basque women could be regarded as being more the former than the latter. Sharing the virtues of the race,

This research for this chapter was done as part of the Project "La experiencia de la sociedad moderna en España: Emociones, relaciones de género y subjetividades (siglos XIX y XX)," HAR2016-78223-C2-1-P, MINECO y FEDER, and the UPV/EHU research group, GIU17/37.

they were able to transcend their sexual condition and, in some cases, act as women warriors who were then celebrated as virile heroines by a movement of men.

This situation did not last long. In just under two decades, these virile heroines of early Basque nationalism had been feminized and/or expelled from the canon of national heroes. This development was linked to far-reaching changes in gender relations in the context arising from the First World War. Traditionalist misogyny gave way to more essentialist and biological visions of sexual difference that had no room for the exceptional virile heroine. The notion of "motherland" then burst onto the political scene, achieving a huge symbolic relevance. That was when the *sexed* incorporation of women in the Basque nationalist movement came about. The once-celebrated virile warrior women had no place in this new scenario. The changing representations of the fifteenth-century woman warrior Libe between 1902 and 1934 illuminate this process particularly well.

Through the analysis of Basque nationalism, the aim of this chapter is to illustrate two more general ideas. The first is that the national heroine cult—or its absence—had more to do with gender perspectives in each specific context than with the real role that these women had played in past national liberation struggles or in wars against invaders. Accordingly, gender is a category with a huge, but unstable, capacity for generating meaning and creating order in the processes of creating patriotic discourses and symbols.[2] The second idea is that, insofar as the meaning of gender is unstable, it is impossible to establish *a priori* the way in which this category functions to build nations and nationalisms. This mutability thus helps us to understand the transformations over time in the significance of warrior women and in their status as national heroines.

The Virile Heroines of the First Basque Nationalism

The origins of modern Basque nationalism at the end of the nineteenth century are linked to the figure of Sabino Arana, regarded as the political and doctrinal father of the new movement.[3] Even though Arana's thought was framed by Catholic traditionalism,[4] he gave this tradition new meaning in a political project intended to cast modernity in the mold of the past, holding out the promise of an "archaic future."[5] Arana's understanding of sexual difference should be analyzed in the light of his worldview. He firmly believed in women's inferiority to men. Women were flawed versions of the opposite sex, imperfect beings prone to sin.[6] This judgment was grounded in a series of values universally regarded as positive and inherently masculine, according to a sole chain of human perfectibility. The Basque people were, Arana claimed on more than one occasion, a virtuous and virile people, the qualifier "virile" referring here to the qualities of strength, courage, religiosity, and moral rectitude.

For Arana, gender was a less salient factor than lineage and race in determining the identity of a human being. As a result, there could be outstanding women who were not representative of their sex and who shared with men a set of values that placed them above the rest of their gender. They were exceptional cases in which an attribute other than gender defined their character and actions. From this point of view, even though Basque men constituted the fullest expression of the nation, their womenfolk shared the virility inherent in the race[7] without it compromising their femininity, a quality which was less appreciated at the time than it would be later on.

Arana's vision was firmly rooted in an image of strong, vigorous, austere, and even coarse women that had been described by many writers, naturalists, anthropologists, and travelers, from German naturalist Alexander von Humboldt to Spanish novelist Emilia Pardo Bazán. Clear precedents for this vision can be found in many eighteenth- and nineteenth-century works, such as the writings of the Jesuit and philologist Manuel de Larramendi, studied by the historian Bakarne Altonaga. In 1754, in his book *Corografía de Guipúzcoa*, Larramendi claimed that the women of the provinces were "worth more than their sex in general, not as skittish as in other provinces" and that on the battlefield "some and many have performed heroic and very manly exploits." Since antiquity "there had been Guipuzcoan Amazons who knew how to handle weapons, wield the sword and defend the breach, killing and dying while wreaking terrible havoc among their enemies."[8]

Despite conceding that Basque women shared the virtues of the race, Arana was, as already noted, an inveterate misogynist. In a letter written in 1897, on one of the few occasions on which he referred to the fairer sex as a whole, he addressed the defects of what he called the "unhappy half of the human race." Women were vain, selfish, superficial, and inferior to men and had, he wrote, "in the extreme all the weaknesses inherent in human nature." They were also morally weaker, threatened by temptation, and prone to sin: "Therefore, they were the first to fall," he recalled. He did not even accept that the values of sensitivity and sentiment were more highly developed in women. They were simply, "inferior to men in mind and in heart."[9]

Arana's misogyny, somewhat attenuated by the value of Christian respect, led to what José Javier Díaz Freire has dubbed the "absent presence" of women in his discourse,[10] a vacuum that structured the space that it occupied. Significantly, Arana neither dedicated much time to discussing women, nor did he give them a place or specific mission in the Basque nationalist movement. They did, of course, have one in the preservation of racial characteristics through the family and lineage, but not as feminine subjects with an active role in the movement. Women per se were excluded from the manly project he created because of the contempt and distrust that he felt for them and, in these first years, they were prohibited from joining the party. This segregation was also extended, for example, to the theater, one of the most important propaganda vehicles for Basque nationalism since its inception.[11] From a moral perspective, the strictly masculine nature of Basque nationalism was incompatible with the participation in its activities of this potentially corrupting element. The avoidance of mixed spaces and the exclusion of women from the public sphere were deemed more important than the task of disseminating its ideology that, moreover, was vital for a party that aspired to prosper in the new world of mass politics.

Based on the conception of women as morally contaminated inferior beings inclined to sin, Arana considered femininity, above all, a set of debasing effects which he would effectively apply to the Spanish people.[12] In his view, femininity was a fixed quality, an attributable value and, as with masculinity in men, a condition potentially separable from the sexual body. Under this logic, the virile Basque people would serve as a moral counterbalance to what he saw as debased and corrupt nations such as the Spanish, "a people both effeminate and brutish."[13]

There might appear to be a contradiction between the misogyny permeating Arana's writings and his benevolent opinion of Basque women. These ideas would have been contradictory only if his had been a train of thought constructed from a rigid essentialist dimorphism that defined gender as an inalterable kernel attached to a specific biological body. To the contrary, according to Arana's vision, gender did not exhaust the cultural and social meaning of human beings with a woman's body, insofar as other variables, and especially race, could be even more decisive. As a result, Arana's tolerance of exceptional women was much greater than in radically essentialist thought, totally sexualized, in which feminine nature imposed inviolable limits on each and every woman. At the same time, admitting that individual women could be feisty warriors and patriots did not contradict his negative view of the gender as a whole. The same yardstick could be employed when assessing all the Vizcayan—Basque—women, who were exceptional within their gender group.

The role played by exceptional women in Arana's writings allows us to gain insights into this harmonious coexistence between his misogyny and his tolerance of feminine excellence. On the one hand, early Basque nationalism constructed patriotic female figures who spilt their blood or laid down their lives on the battlefield in scenes that occasionally had profound religious connotations; while on the other hand, these brave women could also be guilty of sins of the flesh that gave rise to the decadence of the race, contaminated with Spanish, that is foreign, blood. Those female characters, at once heroines and sinners, reflected the complexity of Arana's gender perspective. The possibility that Arana's outlook provided for Basque women to be recognized as women warriors and heroes of the Basque nation was exemplified in his treatment of Libe, a fifteenth-century figure who was the protagonist of his most famous play.

As in other societies and cultures, some Basque women had stood out for their warlike deeds in oral and written traditions. This was the case of María Pérez de Villanañe, an eleventh-century character from Álava, who was worthy enough to be called *La Varona* [The Manly Woman] by King Alfonso VII of Castile and Leon, in honor of her "manly courage" on the battlefield. During the nineteenth century, the story of *La Varona* was recovered and, in 1885, Ricardo Becerro de Bengoa dedicated a poem to her in his *Romancero Alabés*, in which he evoked the lady "who behaved/like a formidable man."[14] By the same token, the figure of Catalina of Erauso, the *Lieutenant Nun*, whose military career was recompensed by the recognition of Philip IV and Pope Urban VIII at the beginning of the seventeenth century, enjoyed a lengthy popularity.[15]

In Arana's oeuvre, the first of these exceptional women appeared on the scene at the Battle of Arrigorriaga in the ninth century, her intervention being decisive for the favorable outcome of the first of the "four patriotic glories." Arana recounts how, having lost his helmet "fighting against a vigorous Biscayan woman, the Leonese commander Ordoño received such a hefty blow on the head from her axe that he fell, lifeless, to the ground." This rendered his soldiers so downcast that, losing their courage and strength, they ended up retreating "like bucks" along the same route they had taken "to launch the criminal conquest of a pacific nation."[16] The heroine of Arrigorriaga was thus construed as an autochthonous Joan of Arc or Agustina of Aragon.[17] Incidentally, it is interesting to note how Arana shifted the limelight from the nobleman and military chief Juan Zuria ("The White Lord"), a mythical Lord of Biscay who, throughout the nineteenth century, was regarded as *the* hero of this decisive Battle of Arrigorriaga.[18]

The figure that best expressed Arana's particular gender perspective was the main character of his play *Libe*. This historical melodrama, which recreated the Battle of Munguía in 1471, was written in 1902 and published a year later.[19] Three different dimensions can be glimpsed in the character of the young female protagonist: the human being endowed with free will who has to choose between virtue and sin; the Eve, a weak woman incapable of resisting the temptation that would condemn her people; and the exceptionally vigorous woman, a splendid example of patriotism and the architect of the victory against the Spanish enemy.

In Arana's telling, the feeble Libe falls in love with a Castilian count, thus betraying her people and hastening them down the path of racial degeneration. "Our blood, always pure: we do not give our daughters away to foreigners," one of his characters cautioned.[20] The danger of sin, of the loss of racial purity, came yet again from women who, using the expression of Mercedes Ugalde, were apparently the "weak link in Basque society."[21] By becoming the heroine responsible for the victory on the battlefield of Munguía against the troops of the Count of Haro, Libe would redeem her original fault. Thanks to her, the fatherland was freed from the invading enemy. Over the dying Libe, who has already raised herself above her status as a woman on the battlefield, her father exclaims, "Oh, Libe! God is taking you for Himself to heaven, because there is no man worthy of you on earth!" The Biscayan troops then kneel before the venerable body of Libe, who will always leave her people with "the memory of a woman who has laid down her life for the fatherland."[22]

Even though Arana did not devote much space in his writings to discussing women, female characters did feature in the most significant and influential literary passages of an epic narrative in defense of the imaginary free and independent Biscayan nation. As he was creating Basque nationalism, Arana gave courageous and virile women warriors pride of place in its pantheon of heroes. This type of discourse and these icons overshadowed the dignifying effect of maternity and the very figure of the "mother country." For Arana, of course, Basque women played a central role in the reproduction of the lineage and the Basque nation-family. Biscay was built on family ties and women there acquired a greater political significance than they would have done in a nation considered to be an aggregate of individuals. But in Arana's thought, this positive significance of maternity was incapable of prevailing over a misogyny in which virtue was, by definition, masculine.

The Mothers of the Fatherland

Between 1914 and 1930, gender perspectives evolved apace in Spanish society as a whole. This evolution also affected Basque nationalism in a particularly powerful way because the baseline was the traditionalism inherent in Sabino Arana's time. The changes had a direct impact on the Basque heroine cult. Broadly speaking, this evolution was linked to the context emerging from the First World War—notwithstanding Spain's neutrality in the international conflict—a new scenario in which the presumption of male superiority was challenged. In that context, women defied, in practice, longstanding preconceptions with their daily activities and, occasionally, with feminist political action. The conviction that women were incapable of undertaking many physical and intellectual activities clashed with reality. The nationalist press warned against the new dangers, claiming that "the offices, the schools, the chemist shops, the medical profession, the courts are already being invaded by the enemy. It is a terrible avalanche before which we are yielding ground day after day."[23] New fears and anxieties about the future of gender relations, often associated with the figure of the "modern woman," led to changes in the discourses of Basque nationalism, changes that had significant implications for the figure of the woman warrior.

The response to these fears and new uncertainties was the reaffirmation of sexual difference, based on the idea of complementarity between men and women. In reality, this idea had always been present in nationalist discourse, coexisting with more radically hierarchical perceptions. What occurred during the first decades of the twentieth century was that the notion of women as inferior beings, defective versions of men, lost ground to one in which women were essentially different from them. The reaffirmation of the existence of a feminine nature defined in biological terms went hand in glove with a radicalization of the maternal mandate, denying other possibilities of dignified femininity—the pious spinster or the case of exceptional women. As a result of this evolution, women were incorporated into the nationalist political project, albeit always as women and mothers. New exclusively feminine organizations emerged within the movement. And when they were not simply erased, national heroines were feminized, modifying their symbolic and political meaning.

A nationalist organization strictly for women, the so-called *Emakume Abertzale Batza* (EAB) [Association of Women Patriots],[24] was created in 1922, an initiative strongly influenced by the experiences of the Irish women's organization *Cumann na mBan*.[25] Its advent should be understood not only as a product of the movement's change of tack, but also as a personal achievement of women who challenged the prohibition on their participation in the movement and thus created a new space of social and political participation.[26] From the birth of the EAB, different women nationalist leaders became involved in the movement's activities, with a growing number of female speakers occupying the squares of the Basque Country.[27] A chain of events during the Second Republic (1931–36) favored a level of acceptance of female militancy which would have been impossible before, even though no women occupied leadership roles in the Basque Nationalist Party nor did any appear in its lists of candidates for elected office.[28]

It is true that the activity of these women was often of a conservative nature and limited to social welfare, cultural dissemination, Christian charity—the distribution of food, clothing and toys—and even the defense of the family following Catholic principles. More often than not, however, their activity encroached on the field of politics in such a way as to challenge the gender order. Women were plainly visible at decisive moments, such as at the demonstration held on the occasion of the first

Aberri Eguna (Day of the Basque Nation) in Bilbao on March 27, 1932. At these multitudinous events, they demonstrated their capacity to mobilize and occupy public spaces, some of the young protagonists and speakers becoming well-known figures of the movement. The imprisonment of Haydée Aguirre, Polixene Trabudua, and Miren Nekane Legorburu for their propaganda activities had a huge impact, and the political work of young nationalist women earned them the respect and recognition of their peers. On another occasion, in 1933 the premises of the EAB were closed by the government authorities.[29] And the women's demonstration that they themselves organized against the president of the Republic's visit to Biscay on April 30 of that same year is yet another compelling example of the prominence achieved by nationalist women.

Their political drive, their clashes with the police, and the arrest of some of their most prominent propagandists allowed nationalist women to take center stage in party activities, becoming, according to the nationalist press, "the spearhead of the movement."[30] Finally, in 1933, thirty-eight years after its founding, the leaders of the Basque Nationalist Party had to accept women as full members of the party.

As women were allowed to participate in the Basque political sphere, however, the heroic actions of exceptional female patriots became less important. In the words of EAB leader Sorne Untzueta, while all the races had their great women warriors, such as Joan of Arc and Agustina of Aragon, Euzkadi had no need for such occasional figures. The "total volume of work undertaken" by Basque women was more "interesting."[31]

These changes were reflected in re-evaluations of Arana's Libe. In 1922, Carmen de Errazti, the first chairwoman of the EAB, rejected the idea that his Libe was a woman warrior like others.

> Sabino did not conceive [the character] as a woman warrior, like that woman from Álava called '*Barona*', about whom medieval stories speak, or that other woman Catalina of Erauso from San Sebastián, famous for her manly deeds and, therefore, known by her sobriquet 'the Lieutenant Nun', about whom Spanish critics and poets have spoken so much [...].
>
> No; Libe, the main character of Sabino's melodrama, was a young Basque woman, full of beauty and candour, possessing in her noble and simple heart all the virtues of the race; the most sweet and gentle of Basque maidens, but who, when the time came, would know how to answer the call of the blood of a hundred Basque generations flowing through her veins.[32]

The reassessment of Libe culminated in the 1934 adaptation of Arana's play by nationalist writer Manuel de la Sota.[33] His *Libe* embodied the new understanding of gender difference in Basque nationalist circles. Arana's youthful main character: weak, virgin, martyr, and virile warrior all in one, was turned into a tender eighteen-year-old maiden "at the peak of her femininity," from whose breast sprung strong passions and emotions.[34] In this new version, Libe was feminized, but without that femininity implying an ignominious weakness, and sexualized, but without that sexuality being sinful per se. Libe, described by de la Sota as being "all love," was nonetheless faced with the dilemma of two conflicting emotions: "love for the enemy or love for her

Figure 3.1 Polixene Trabudua and Haydée Aguirre, both EAB leaders, leaving Larrinaga Prison, Bilbao, 1933.

Figure 3.2 Cover of Manuel de la Sota's *Libe: Melodrama histórico* (1903, Bilbao: E. Verdes Achirica, 1934).

fatherland." Where Arana had made Libe's love for her fatherland irreconcilable with her love as a spouse and required that she rise above her position as a woman and renounce earthly love, the new Libe could have both and offers her heart to the young Basque Andima: "Yours is my love, Andima, yours are my soul and my body, for you are risking your life to defend the sons that we, the Biscayan women, are expecting in our wombs."[35] The 1930s heroine was converted into a "fearless soldier of the Basque hearth,"[36] a loving wife and fertile mother who fought for her children and cared for and dressed the wounds of the future representatives of the race. The image on the cover of the published version certainly had very little of the warrior about it. The play's final scene, in which De la Sota deviated from Arana's original work by introducing a new character, evokes, as Miren Llona has rightly pointed out, the image of the Mother of Sorrows.[37] A child named Kuliska, who has lost his mother during the war, rushes around on stage disconsolately calling out for her until, attracted by the dying Libe, he approaches her. Then, drawing her last breath, Libe addresses all the children of the Basque fatherland, warning them about the dangers of being orphaned and abandoned, like the boy remaining at her side, who is comforted by her words despite his grief.[38]

The Spanish Civil War and After

Following Francisco Franco's coup d'état against the government of the Second Republic, the Spanish Civil War (1936–1937 in the Basque Country) did not contribute any new female figures to the pantheon of heroes of the Basque fatherland even though many women did take up arms in defense of the republican cause. The most outstanding examples did not hail from the ranks of the Basque nationalists, but from the political groups most committed to bringing about far-reaching social change. Cases such as that of the brave Guipuzcoan anarchist Maximina Santa María[39] made the front pages during the first months of the conflict, in a climate of anti-fascist fervor. But the figure of the *miliciana* (militiawoman), as the women who fought against Franco's army were called, was rapidly discredited on the republican front itself. As part of its effort to rebuild the army that had been shattered by the military uprising, on October 29, 1936, the Republican government restricted the recruitment of soldiers to men aged between twenty and forty-five. The authorities thus regained control of the army, heading off the threat of social revolution and sexual disorder represented by the militiawomen. As historian Mary Nash has noted, within a few months the image of the *miliciana* disappeared from republican propaganda, subjected to a smear campaign that even went so far as to identify militiawomen with prostitutes.[40] Forty years after having fought on the front defending the Peñas de Aya, Kasilda Hernáez fended off the humiliating accusations that had been leveled at her at the time: "It was a tough lesson of the war. And the people belonging to the republican camp who have said that we mountain women were little less than sluts are liars who I shall never forgive."[41]

In the Basque case, the attempt to re-establish order in the context of the war was, moreover, related to a nation-building project that required different things from men and women.[42] Basque nationalists saw the strict division between the masculine front and the feminine rear guard as both important and necessary. This did not mean, however, that as with the sexes themselves, they were theoretically defined by any pecking order: "In any war there are two places from which it is possible to serve the cause being defended. These places are the front and the rear guard. Neither is it necessary, nor is there any reason, to say which is the most important, for even though one of them is more important, it cannot endure without the other."[43] The metaphor of the motherland that, "torn by pain," selflessly offered up her sons in sacrifice became a leitmotif in the nationalist press during the war.[44] The idea that patriotism should not be confused with politics, and that nationalist women had "enlarged the household until transforming it into a national one," were platitudes in a discourse that explicitly appealed to the post-Arana domestic version of Libe, symbol of female patriotism in support of the husband, the son or the fiancé.[45]

Despite the foregoing, it is interesting to note how the Basque mother continued to be seen as a sober, strong, and restrained woman. For the sake of preserving the difference, the value of tenderness and sentimentality commonly associated with maternity were sacrificed, a sacrifice that became both an existential demand and a source of pride for Basque women. As Polixene Trabudua explained in January 1937, "perhaps because they do not know how to kiss, embrace or cuddle," when they saw off the *gudaris* (Basque soldiers) leaving for the front, the women confined themselves to turning their "humble gaze, bearing the full force of their affection as of Basque mothers" on them.[46] Manly self-restraint and emotional austerity marked the character of the Basque woman and distanced her from the gestural voluptuousness and expressive eloquence of her Spanish counterpart. The renewed political importance of maternity was thus endowed with its own distinguishing features.

Notwithstanding this division of wartime roles, some nationalist women also performed highly courageous acts. In recent years, under the impetus of the movement for the recovery of historical memory in Spain, some of these names that had been invisible during the Franco dictatorship and neglected by historians have begun to receive the attention that they deserve. This is the case, for example, of the so-called "Álava Network," an anti-Franco resistance group essentially composed of women that functioned from 1937 to 1940 thanks to the clandestine action of four women: Bittori Etxeberria, Itziar Mujika, Delia Lauroba y Tere Verdes, whose names are only just beginning to be known and recognized. They went to great lengths to assist those persecuted by the dictatorship, helping them to cross the frontier with France, saving lives and carrying out espionage even in the context of the Second World War, when they gathered information on the movement in and out of Bilbao of ships supplying the German army and of U-boats on the coasts of the Bay of Biscay.[47]

The legitimacy that the actions of the heroines of the "Álava Network," framed in the struggle against Francoism and Nazism, currently enjoy is a clear example of the continual rewriting of the past. A different context turns episodes of treachery into glorious exploits and villains into heroes, and we historians play a fundamental role in these processes of symbolic transformation. This same phenomenon, but in an opposite sense, has affected the perception of the role played by women in the activities of the armed group ETA (*Euskadi Ta Askatasuna*, Euskadi and Liberty), a military organization created in 1958 within the Basque nationalist movement, in a break with the mainstream.[48] Social support for ETA fell drastically over the following fifty years, as a result of changes in the political context—with the transition from a military dictatorship to a parliamentary democracy—and in the strategy of the armed organization—targeting objectives in an increasingly more indiscriminate fashion. At the height of ETA's armed activity in the 1970s, women accounted for between 10 and 15 percent of its members. Nonetheless, in spite of the existence of a number of prominent women activists, the image of the organization has always been overwhelmingly male. The figure of the suffering mother of the male militant has been central to the symbolic construction of ETA's world. But it is worth mentioning that, as the historian Carrie Hamilton has noted, the organization's militants tended to be totally against identifying femininity with specific values, such as peace, a greater sensitivity, or a greater rejection of violence. As one of her respondents said, "ETA violence is absolutely political. It's not about testicles. It's not about ovaries. [...] It's a way of fighting because unfortunately there are moments when the enemy doesn't understand any other... Violence has nothing to do, absolutely nothing to do, with the feminine sex, or with the masculine."[49] In all likelihood, the legendary image of the strong and emotionally restrained Basque woman helped to build an identity compatible with the exercise of political violence.[50]

✶✶✶

In this chapter, I have attempted to shed some light on the symbolic construction of national heroes from a gender perspective. The history of Basque nationalism shows that the incorporation of women in the pantheon of national heroes has depended more on the different visions of sexual difference than on the real role played by these characters in the nation's past. Thus, when the conditions were right for constructing these symbols, fictional heroines demonstrated their full capacity to create a feeling of belonging and to trigger patriotic emotions. These characters, real or apocryphal, evoke concrete ways of understanding gender differences, femininity, maternity, and the inter-relationships among them. The Basque case thus confirms that the metaphor of the mother nation has not been a constant image in modern nationalisms and, above all, cannot be taken for granted.

The first Basque nationalism, from its traditionalist perspectives and its strict Catholic anthropology, participated in a regime of truth averse to biological determinism as the basis of sexual dualism. Seen from this perspective, Arana's doctrine would dovetail with an explicitly hierarchical model in which the gender condition was chiefly moral and performative, and virility constituted a universally positive value. This vision fostered the construction of virile heroines such as Libe, who rose above their status as women in such a way that the misogyny permeating that thought combined harmoniously with the glorification of their military exploits. Nor was maternity presented as a necessary or more worthy condition for women. For women—and for men—chastity was the state closest to human perfection.

The changes that occurred in the first decades of the twentieth century responded, in part, to the needs of a nationalist movement that aspired to create a political community in the context of a society of masses. But the symbolic elements employed to integrate women into the movement point to profound transformations in perceptions of gender. The complementarity between the sexes, the strictly biological definition of sexual difference, and the complete identification of women with maternity made the narratives of virile heroines unfeasible. That was when the notion of the mother nation began to play a more prominent role. Political exclusion was replaced by a controlled presence, even though a new generation of young activists defied the political limits imposed on them by the party leadership. The experience of the Civil War consolidated sexual difference and restored the gender order at a particularly volatile moment.

Notwithstanding such significant changes, the legendary image of the strong and austere, coarse and severe Basque woman has survived until this day. This image served to give consistency to political identities to which violence was central, as was the case with the women of ETA. But beyond the now fortunately defunct armed struggle, the power of some narratives, such as that of a supposed ancient Basque matriarchy created by the philosopher Andrés Ortiz-Osés in the 1970s,[51] was felt in other political milieus. Specifically, some currents of feminism in the 1970s looked to the past in search of egalitarian origins, strong women, powerful goddesses, and witches possessing ancestral knowledge of the Basque land, mythical references they could revive and endow with meaning relevant for a new political scenario. Similarly, some elements of Basque nationalism looked to these historically dubious theories as a way of claiming that the Basques were fundamentally different from "Spaniards." The political vitality of those images and their capacity to shape identities contrasted with their brittle historical foundations. A contrast that invites us to contemplate gender as one of most important and versatile materials with which nations have been built, traditions have been invented, communities have been imagined, and nationalisms banalized.

Notes

1   José Luis de la Granja, *El Nacionalismo Vasco: Un Siglo de Historia* (Madrid: Tecnos, 1995), 14. For a similar definition, see Santiago de Pablo and Ludger Mees, *El Péndulo Patriótico. Historia del Partido Nacionalista Vasco (1895–2005)* (Barcelona: Crítica, 2005), 16.
2   Without denying this capacity, Anne McClintock has warned against the danger of reducing the gender analysis of these phenomena to the inversion of the prevailing notions of power. In Anne McClintock, *Imperial Leather: Race, Gender and Sexuality in the Colonial Contest* (London: Routledge, 1995), 15.
3   Sabino Arana (1865–1903) was the author of the body of doctrine of Basque nationalism, the leader of the movement founded in the 1890s, and the inventor of its main symbols and emblems. He also created a historical narrative that would—along with the Basque language and a moral definition of the Basque race—provide the basis for the forging of a new national identity.
4   José Luis de la Granja, *El Nacionalismo Vasco*, 14; Santiago de Pablo and Ludger Mees, *El Péndulo Patriótico*, 16.

5   José Javier Díaz Freire, *La República y el Porvenir: Culturas Políticas en Vizcaya durante la Segunda República* (San Sebastián: Kriselu, 1993), 224–225.
6   Sabino Arana, "Los pseudo-ángeles y el pseudo-arte," *Euzkadi*, no. 4 (1901), in Sabino Arana, *Obras Completas de Arana Goiri'tar Sabin* (San Sebastián: Sendoa, 1980), vol. III, 1984.
7   José Javier Díaz Freire, "Cuerpos en Conflicto: La Construcción de la Identidad y la Diferencia en el País Vasco a Finales del Siglo XIX," in *El Desafío de la Diferencia: Representaciones Culturales e Identidades de Género, Raza y Clase*, ed. Mary Nash and Diana Marre (Bilbao: Universidad del País Vasco, 2003), 78. See the same author, "Nacionalismo Vasco y Redención del Pasado," in *Historia e Identidades Nacionales: Hacia un Pacto entre la Ciudadanía Vasca*, ed. Mercedes and Pilar Pérez-Fuentes (Bilbao: SRB, 2007), 127. Galician women were also often portrayed as strong, but in this case the feminizing images of the Galician nation and its menfolk represented the downside. Helena Miguélez-Carballeira has analyzed the endeavors, after 1916, to create a narrative of masculinization aimed at countering these feminizing metaphors. Helena Miguélez-Carballeira, *Galicia, a Sentimental Nation: Gender, Culture and Politics* (Cardiff: University of Wales Press, 2013), 101–134. See also Paula Pérez Lucas, "Género, Literatura e Identidad Nacional: Discursos Políticos y Culturales del Proto-Nacionalismo Gallego," *Arenal* 15, no. 1 (2008): 342.
8   Bakarne Altonaga, "Mujeres Viriles en el Siglo XVIII: La Construcción de la Feminidad por el Discurso Foralista de Manuel de Larramendi," *Historia Contemporánea* 52 (2016): 11 and 28.
9   Sabino Arana, *De Fuera Vendrá… Comedia en Tres Actos* [edition and historical study by José Luis de la Granja] (1898, San Sebastián: Haranburu Editor, 1982), 145. For this quote and the previous ones appearing in the paragraph.
10 José Javier Díaz Freire, "Cuerpos en Conflicto," 74.
11 See a specific study on the issue performed by José Luis de la Granja, "El Teatro Nacionalista Vasco de Sabino Arana," in *Eskual Antzertia. Le Théâtre Basque*, ed. Pierre Bidart and Txomin Peillen (Bayonne: Université de Pau et des Pays de l'Adour, 1987), 19–37. See also Santiago de Pablo and Ludger Mees, *El Péndulo Patriótico*, 10. José Javier Díaz Freire has emphasized the role of theater as a generator of emotions. In José Javier Díaz Freire, *La República y el Porvenir*, 213. Arana himself noted the need to place "the Biscayan before his own eyes, clearer than in the most life-like portrait, and make him feel, moving his most delicate fibre." Sabino Arana, "El Teatro como Medio de Propaganda," *Bizkaitarra* 17 (1895), in Sabino Arana, *Obras Completas*, vol. I, 471.
12 José Javier Díaz Freire, "Cuerpos en Conflicto," 78.
13 Sabino Arana, "La Ceguera de los Bizkaínos," *Bizkaitarra* 30 (1894), in Sabino Arana, *Obras Completas*, vol. I, 365–366. This characterization had a dual purpose: to debase the Spanish by feminizing them, on the one hand, and by reducing them to mere nature, to a brutal and uncivilized masculinity, on the other.
14 Ricardo Becerro de Bengoa, *Romancero Alabés* (1885, Bilbao: Amigos del Libro Vasco, 1985), 104.
15 On the historical significance of this figure from the point of view of constructing sexual difference, see Nerea Aresti, "The Gendered Identities of the 'Lieutenant Nun'. Rethinking the Story of a Female Warrior in Early Modern Spain," *Gender & History* 19, no. 3 (2007): 401–418.
16 Sabino Arana, *Obras Completas*, vol. I, 113.

17 The nationalist poet Esteban Urkiaga, *Lauaxeta*, would see in Libe "our Joan of Arc." Cited in José Luis de la Granja, *El Siglo de Euskadi. El Nacionalismo Vasco en la España del Siglo XX* (Madrid: Tecnos, 2003), 181.
18 Coro Rubio and Santiago de Pablo, "Before and after the Nation: Basque Patriotic Heroes, 1834–1939," *Studies on National Movements* 2 (2015): 5–6.
19 Sabino Arana, *Libe* (Bilbao: Tipografía Universal, 1903).
20 Sabino Arana, *Libe* (1902), in Sabino Arana, *Obras Completas,* vol. III, 2016.
21 Mercedes Ugalde, who in her pioneering work already identified Libe with Eve, has offered an interpretation of this play in which she has placed the accent on the instrumental role given to women in Basque nationalism. In Mercedes Ugalde, *Mujeres y Nacionalismo Vasco. Génesis y Desarrollo de Emakume Abertzale Batza, 1906–1936* (Bilbao, Universidad del País Vasco, 1993), 44.
22 Sabino Arana, *Libe* (1902), in Sabino Arana, *Obras Completas,* vol. III, 2035 and 2037. Miren Llona has drawn interesting parallels between the character of Libe created by Arana and the symbolic reference to the Immaculate Conception, underscoring the emphasis placed on female purity. In Miren Llona, *Entre Señorita y Garçonne. Historia Oral de las Mujeres Bilbaínas de Clase Media (1919–1939)* (Málaga: Atenea-Universidad de Málaga, 2002), 192. José Javier Díaz Freire, for his part, has stressed the tension between the feminine nature of the character and her participation in the Basque identity, dispelled by the former's rectification and its confirmation in the tragic final scene. In José Javier Díaz Freire, "Cuerpos en Conflicto," 79.
23 R. de I., "Notas al Margen. Feminismo," *Euzkadi,* January 14, 1930: 1.
24 The organization was established in Bilbao in April 1922 by fifty founding members. Beforehand, in 1907 some women had already used charity or welfare organizations as a political springboard, and in 1908 they created the "Ropero Vasco" [Basque Wardrobe]. After its activities were resumed in 1931, the EAB had over 25,000 members at one juncture. Mercedes Ugalde, *Mujeres y Nacionalismo Vasco,* 471–492; Mercedes Ugalde, "Dinámica de Género y Nacionalismo: La Movilización de Vascas y Catalanas en el Primer Tercio de Siglo," *Ayer* 17 (1995): 143–146. González i Vilalta attributes a greater political initiative to female Basque nationalists than to their Catalan counterparts. Arnau González i Vilalta (2005): "Mujer y Nacionalismo Conservador (1931–1936). Análisis Comparado de Dos Casos: Las Emakumes del PNV y la Secció Femenina de Lliga Regionalista," *Historia Contemporánea* 31 (2005): 625 and 632.
25 The Irish influence on Basque nationalism, which has been considerable since the end of the twentieth century to date, has been analyzed by Xosé M. Núñez Seixas in the "Irlanda" entry in Santiago de Pablo, José Luis de la Granja, Ludger Mees and Jesús Casquete, *Diccionario Ilustrado de Símbolos del Nacionalismo Vasco* (Madrid: Tecnos, 2012), 547–562. See also Alexander Ugalde Zubiri, *La Acción Exterior del Nacionalismo Vasco (1890–1939): Historia, Pensamiento y Relaciones Internacionales* (Bilbao: IVAP, 1996).
26 Miren Llona, *Entre Señorita y Garçonne,* 161–207.
27 Isaac de Etxebarria, "Nuestras Oradoras," *Euzkadi,* July 9, 1932: 1.
28 Leire Arrieta Alberdi, "Emakume," in Santiago del Pablo et al., *Diccionario Ilustrado,* 208.
29 Begoña Bilbao, Gurutze Ezkurdia, Karmele Pérez, and Josu Chueca (coords.), *Emakumeak, Hitza eta Bizitza* (Bilbao: Universidad del País Vasco. Servicio Editorial, 2012).

30 Egizale (Alberto de Onaindia), "Homenaje de Admiración. A las Emakumes del 3 de Mayo," *Euzkadi*, May 6, 1933: 1.
31 (Unsigned), "Los Actos Organizados por EAB. Conferencia de la Señora Untzueta de Errazti," *Euzkadi*, December 4, 1931: 3.
32 Carmen de Errasti, "Juventud Vasca de Bilbao al Maestro. En el XIX Aniversario de su Muerte" (speech delivered by the chairwoman of EAB), *Abeŕi*, November 30, 1922: 5–6.
33 Manuel de la Sota, "Libe. Unas Consideraciones sobre su Realización Dramática," *Euzkerea* 6, no. 1 (1934): 39–53.
34 Ibid., 46.
35 In Manuel de la Sota's own words, as the person adapting the work, he had allowed himself "to be generous with Andima and give him the love of Libe." Ibid., 53.
36 Sabino Arana (adapt. by Manuel de la Sota), *Libe: Melodrama Histórico* (1903, Bilbao: E. Verdes Achirica, 1934), 100.
37 Miren Llona, *Entre Señorita y Garçonne*, 193.
38 Sabino Arana (adapt. by Manuel de la Sota), *Libe: Melodrama Histórico*, 108.
39 El reportero de guerra, "Una Mujer Lucha por la Libertad y la Justiciar en la Avanzadilla de las Peñas de Aya," *Frente Popular. Diario de la República*, August 10, 1936: 1.
40 Mary Nash, *Rojas: Las Mujeres Republicanas en la Guerra Civil* (Madrid: Taurus, 1999), 96–97 and 160–170. For an analysis of the attitude of the nationalist forces in the Civil War, see José Luis de la Granja, *República y Guerra Civil en Euskadi. Del Pacto de San Sebastián al de Santoña* (Oñati: IVAP, 1990); Santiago de Pablo and Ludger Mees, *El Péndulo Patriótico*; Xosé M. Núñez Seixas, "Los Nacionalistas Vascos durante la Guerra Civil (1936–1939): una Cultura de Guerra Diferente," *Historia Contemporánea* 35 (2007): 559–599.
41 Luis M. Jiménez de Aberasturi, *Kasilda, Miliciana. Historia de un Sentimiento* (San Sebastián: Txertoa, 1985), 44.
42 Miren Llona, "From Militia Woman to Emakume: Myths Regarding Femininity during the Civil War in the Basque Country," in *Memory and Cultural History of the Spanish Civil War. Realms of Oblivion*, ed. Aurora G. Morcillo (Leiden-Boston: Brill, 2014), 194.
43 Eduardo de Nafarrete, "Labor de Retaguardia: Lo que Corresponde a Nuestra Emakume," *Euzkadi*, September 27, 1936: 5.
44 Germán M. de Iñurrategi, "Ciudadanía y Confianza," *Euzkadi*, November 8, 1936: 1. The approaches of the women of the non-denominational and liberal nationalist organization Basque Nationalist Action, a splinter group of the PNV, were more advanced. They encouraged women to follow the example of the physical and spiritual strength showed by the *Lotta Svärd* (a Finnish voluntary auxiliary paramilitary organization for women) during the First World War. In Delfi Bonet, "La mujer en la guerra," *Tierra Vasca*, December 30, 1936: 8. Also in the number of January 1, 1937.
45 Garbiñe, "Labor de Emakume," *Patria Libre*, February 18, 1937: 3.
46 Polixene, "Amatxu," *Patria Libre*, January 14, 1937: 1.
47 Josu Txueka Intxusta, "Emakumes Presas, las Primeras en la Resistencia," *Hermes: Pentsamendu eta Historia Aldizkaria* 44 (2013): 20–27; Ane Azkue Zamalloa, *Red Álava 1936-1947. La red de mujeres invisibles* (Vitoria: Diputación Foral de Álava, 2018).

48 The organization's armed actions since 1961 until the announcement of the permanent ceasefire in 2011 claimed the lives of 854 people.
49 Carrie Hamilton, (2017): "Political Violence and Body Language in Life Stories of Women ETA Activists," *Signs: Journal of Women in Cultures and Society* 32, no. 4, 2007, 923. See also Carrie Hamilton, *Women and ETA: The Gender Politics of Radical Basque Nationalism* (Manchester, UK: Manchester University Press, 2007).
50 Paradoxically, Dolores González Catarain, *Yoyes*, the first female leader of the organization at the end of the 1970s, is conceivably the militant who has enjoyed the greatest popularity and symbolic significance. *Yoyes* was murdered by the organization itself in 1986, under the accusation of betrayal, after she had decided to abandon the armed struggle and return to civilian life.
51 Andrés Ortiz-Osés and Franz Karl Mayr, *El Matriarcalismo Vasco: Reinterpretación de la Cultura Vasca* (Bilbao: Universidad de Deusto, 1980).

Part Two

Violence

4

Murderous Daughters as "Exemplary Women": Filial Piety, Revenge, and Heroism in Early Modern and Modern Japan

Marcia Yonemoto

Whether served cold and calculatingly or run hot through the mad blood of the moment, revenge and the violent actions it motivates—most notably the vendetta, the blood feud, and the duel—have captivated observers, scholars, and popular audiences in Europe, the Americas, and Asia from medieval to modern times.[1] Although details differ according to historical and cultural context, vendettas in the past generally involved the planned killing of an individual (or individuals) who had wronged one's family or one's lord or master, and they tended to be undertaken by relatives, vassals, servants, or spouses in the cause of rectifying family or household honor. Vendettas were socially and (for the most part) politically sanctioned, and little if any punishment accrued to those who committed them. Most often the perpetrator, if not the mastermind of the vendetta, was male. Over time vendettas became highly ritualized, subject to their own protocols, and took on a certain procedural rationality. And yet, they remained a viscerally human and therefore irrational act of vengeance that could provide a singularly satisfying conclusion to the otherwise messy and imprecise business of filling the void of loss.

While it has deep roots in southern European honor cultures and popular resonances in American-underworld mafia violence, the vendetta has come to be closely associated with Japan; indeed it is a central component of a death-centered, samurai-obsessed exoticism that continues to exert a peculiar appeal to Japanese and non-Japanese alike. The tale of the forty-seven loyal retainers and their undying devotion to samurai honor ending in self-sacrifice (immortalized in the eighteenth-century play *Chūshingura*, or *The Treasury of Loyal Retainers*, and its countless later adaptations on stage, film, and in print) is among the most durable of the masculinist, militarist tropes signifying "Japanese culture."[2] Although historians tend to take it for granted that the vendetta was an elite (read: warrior) male practice, during the Tokugawa period (1603–1868), commoners and women also undertook vendettas. In fact, by the late-eighteenth and early-nineteenth century commoner-led vendettas began to exceed in number those led by samurai, and about half of the one dozen recorded cases of vendettas led by women were accomplished by commoners.[3]

This chapter focuses on the story of one such vendetta. It was said to have been undertaken by two young peasant girls on behalf of their father, who was unfairly murdered by a samurai. This particular story became hugely popular in the early modern period, and the filial daughters, far from being criticized or punished for their violent act, were celebrated as virtuous, loyal, and even "exemplary" women.[4] The daughters' vendetta remained in the public consciousness well into the twentieth century and continues to resonate today. The present chapter explores the vendetta's history, and its shifting meanings and implications as a way of understanding how a particular instance of women's violent autonomous action could be lauded as heroic by official state regimes that were, by turns, conservatively Confucian and pacifist (the early modern state, c. 1600–1868); stridently militarist, nationalist, and imperialist (the modern prewar and wartime state, c. 1868–1945); centrist-democratic and capitalist (the postwar state, c. 1945–present); and, in all cases, decidedly patriarchal and patrilineal. As several of the chapters in this volume show, before the twentieth century public acts of violence by women had to be justified and legitimized by (male) authorities in order to be deemed acceptable. As Gina Martino's chapter shows, eighteenth- and nineteenth-century American observers validated women's martial activities by placing them in the context of maternal defense of home and hearth. Sakis Gekas shows how nineteenth-century Greek warrior women had to be portrayed as victims in order to make their heroism publicly acceptable. While in these two examples the tropes of maternal fortitude and female vulnerability render violence by women acceptable, in the case of early modern Japan, I argue that filial devotion was the legitimizing force. Expressions of extreme loyalty by daughters to parents, especially fathers, transformed violence into devotion by subsuming it within the encompassing hierarchy of the family and, ultimately, of nation (the latter in Japanese is *kokka*, literally "state-family").

The Vendetta Itself

Any avid reader, theater-goer, or purveyor of gossip in the nineteenth century would likely have heard some version of the story of the two filial daughters of the peasant Yotarō from Sendai. A version recorded in 1801 goes as follows: in the early decades of the seventeenth century, there was a peasant in Sendai named Yotarō who had two young daughters named Taka and Haru.[5] One day, when the older sister was sixteen years old and the younger thirteen they were out working in the fields with their father when a samurai named Shiga Danshichi passed by. A bunch of grass the younger girl tossed in the air accidentally fell on Danshichi, who flew into a rage and, even though Yotarō and the girls apologized profusely, Danshichi cut down Yotarō with his sword. The girls ran home and told their mother, who was already ill, of their father's murder; the mother then fell into despair and died. Subsequently, the family's land was sold and the proceeds turned over to the orphaned daughters, who were put in the custody of their aunt in another village. From there, the girls set out to avenge their parents' deaths by waging a vendetta against Shiga Danshichi. They informed their aunt that they wanted to go into service in Fukushima, but from there they surreptitiously went on to Edo (present-day Tokyo), where they searched for a teacher of swordsmanship, eventually finding their way to the house of renowned sword master Yui Shōsetsu (1605–51). Although he was initially resistant, they implored Shōsetsu to take them in as servants and to teach them martial arts so that they might avenge their father's death by killing Shiga Danshichi. Shōsetsu was moved by the sisters' resolve and he put them in the care of the women of his household. At this time, Shōsetsu also bestowed upon the girls' new names: the elder became Miyagino and the younger Shinobu.

After five years of intense training, Shōsetsu pronounced the girls ready to undertake their vendetta, and with his blessing they journeyed back to Sendai, where they went straight to Shiroishi Castle, then the headquarters of Sendai domain, and appealed to authorities to sanction their vendetta. The sisters received permission and Danshichi was called to duel with them. The duel itself took place in an arena marked off by bamboo fences, with both bakufu and domainal officials in attendance. Miyagino and Shinobu entered the battleground dressed in funereal white kimono, assisted by the three samurai who accompanied them from Yui Shōsetsu's residence in Edo. At the signal of the drum roll, the battle began. The girls took turns fighting with Danshichi, Shinobu wielding her halberd (*naginata*) and Miyagino attacking with the fighting sickle (*jingama*). Miyagino succeeded in immobilizing Danshichi's arms with her weapon and then called for Shinobu, who cut off his arms with the halberd. Miyagino delivered the final blow, severing Danshichi's head with her sickle. In the end, the sisters offered the head to their deceased father's spirit.[6] Their long sought-after goal attained, Miyagino and Shinobu then attempted to commit ritual suicide, but the officials prevented them from doing so—the officials wanted to praise the girls' filiality, and their suicide would imply that their vengeance was shameful. Instead, the girls decided to cut their hair and declared that they would become Buddhist nuns and "offer prayers for their deceased parents and Danshichi for the rest of their lives." The daimyo rewarded the sisters, posthumously granting their father stipend lands in excess of 100 *koku*, so that the sisters could support themselves in their religious life.

But the sisters' filial acts did not end with the vendetta, for in 1651 their teacher Yui Shōsetsu committed ritual suicide by disembowelment (*seppuku*) in the wake of a botched coup attempt against the bakufu, and his severed head was put on public exhibition. The tale recalls how the sisters—now nuns—surreptitiously stole their master's head and re-interred it properly at a temple. As in many other filial revenge tales, the story of Miyagino and Shinobu ends with a just resolution, embodied in the death of the antagonist, the avenging of parents *and* teacher, and the establishment of the female protagonists in suitable new lives.

There are many aspects of this story worth commenting on: the role of women and girls as vendetta leaders, the sanctioning of violence by a regime that otherwise rigorously enforced peace, the refusal by officials to sanction the vendetta leaders' self-sacrifice, and the motivating cause of filial piety, which nonetheless resulted in the extinction of the family line, since the daughters never married or had children. I will return to these issues later in the chapter, after addressing the spread and popularization of this particular tale.

Dramatizations and Representations of the Vendetta

Playwrights and novelists were quick to recognize the dramatic potential of the peasant sisters' vendetta. The story of Miyagino and Shinobu inspired two early-eighteenth-century plays *Amidai Yui no hama de* (The nuns at Yui Beach) in 1729 and *Taiheiki kikusui no maki* (The Kikusui volume of the *Taihei Chronicles*) in 1759.

In 1780 the sisters attained starring roles in "Go Taiheiki Shiroishi banashi" (The Tale of Shiroishi and the *Taihei Chronicles*). Shiroishi was the town in Sendai from which the sisters ostensibly hailed, and the play was performed in Edo in bunraku (puppet theater) and kabuki versions.[7] The play became among the most famous of the many that were made on the vendetta theme. Like many fictional works seeking to avoid the shogunate's ban on representations of "current events," the plot was significantly altered. At the play's beginning, Miyagino is a high-ranking courtesan in a large brothel in the Edo licensed pleasure quarters of Yoshiwara, and she is said to be a descendant of the fourteenth-century imperial-loyalist samurai Kusunoki Masashige, her father a loyal retainer of the Kusunoki clan who, after the noble death of his lord, fell to the status of masterless samurai (*rōnin*), and then to peasant, which was his lowly standing at the time of his murder. Giving Miyagino—a peasant's daughter in the original tale—a samurai heritage allows the play to exploit her sense of warrior honor, which undergirds and makes logical her later, extreme filial behavior. When Shinobu appears, it is as a newly hired servant at Miyagino's brothel, a country girl with a comically heavy northeastern accent. Only through the typically hyper-dramatic discovery of an amulet she wears, the twin of which is possessed by Miyagino, do the two women realize they are sisters. The two then begin to plot revenge, inspired by the ur-revenge tale, the *Soga monogatari* (The Soga tales, *c.* 1266). Adaptations of the Miyagino-Shinobu story were not limited to the theater. In 1780, the same year that *The Tale of Shiroishi and the Taihei Chronicles* was first produced, Santō Kyōden wrote a comic novel (*kibyōshi*) entitled *Musume katakiuchi kokyō no nishiki* (Hometown brocade of a daughter's vendetta).[8]

The exploits of Miyagino and Shinobu became the subject of visual representations as well, most of them depicting not the tale itself, but later dramatizations of it. An 1844 print by Hiroshige from the series "Chūkō adauchi zue" (Illustrations of Vendettas Filial and Loyal) retells the "tale of Shiroishi," and shows the sisters gazing lovingly at each other, an image of filial devotion—the younger sister gazing upward in admiration at the elder—that would be reproduced in variant versions in subsequent years.[9]

In 1873 the new Meiji government proclaimed that a vendetta was a form of murder, and though it was "a natural expression of the deepest human feelings, it is ultimately a serious breach of the law on account of private enmity, a usurpation for private purposes of public authority, and cannot be treated as other than the crime of willful slaughter."[10] In spite of this official condemnation, the sisters and their filial vendetta continued to be depicted on stage and in prints well into the Meiji period and beyond. In 1881 the artist Tsukioka Yoshitoshi included a print showing the courtesan Miyagino's reunion with her sister Shinobu in his series *Kōkoku nijūshikō* (Twenty-four paragons of filial piety in imperial Japan).

Figure 4.1 Utagawa Hiroshige, "Shiroishi banashi (The Tale of Shiroishi)," from Chūkō adauchi zue (Illustrations of Loyalty and Vengeance), 1844–45. Photograph © 2019 Museum of Fine Arts, Boston, William Sturgis Bigelow Collection.

Figure 4.2 Tsukioka Yoshitoshi, "The Courtesan Miyagino and Her Younger Sister Shinobu," from *Kōkoku nijūshikō*, 1881. Photograph © 2019 Museum of Fine Arts, Boston, William Sturgis Bigelow Collection.

This image echoes Hiroshige's print from almost thirty years earlier, placing the sisters in similar poses, which evoke a highly emotional tenor. In this series of prints, Miyagino and Shinobu are depicted along with other filial men and women from the recent and distant past, such as Wake no Kiyomaro (733–99), the upstanding and loyal imperial official; Katō Kiyomasa (1562–1611), a loyal vassal of the great late-sixteenth-century warlord Hideyoshi; Hideyoshi himself; Tokiwa Gozen (1138–80), mother of the great warrior Minamoto Yoshitsune, who sacrificed her virtue for her family's safety. But notably, unlike previous versions of the twenty-four paragons, this Meiji-era print is part of a series lauding filial virtue in the Japanese empire, and in so doing it places Miyagino and Shinobu firmly among a panoply of historic, national, imperial heroes and heroines. The sisters were thus historicized and glorified at the same time.

Miyagino and Shinobu's story even spread to the foreign community in Japan as an apt illustration of key aspects of Japan's history and "civilization." In 1877, the Yokohama-based *Japan Weekly Mail* included a detailed account of the vendetta as part of a profile of the sword master Yui Shōsetsu. The article ran under the heading "Some Other Chapters of Japanese History."[11] The French writer Antoine Rous (the Marquis de la Mazelière) included a short account of the event in his multi-volume study *Le Japon, Histoire et Civilization*, published in Paris in 1907.[12] Although both accounts contain some errors of fact and discrepancies in detail, they share an emphasis on the valiant deed accomplished by the sisters, in order to properly avenge, as the *Japan Weekly Mail* put it, the "dastardly murder of their father."[13]

In sum, by the late Meiji period, Miyagino and Shinobu seem to have shifted from the realm of folklore to the status of folk heroines, their filial deed of vengeance commemorating not only a personal cause, but a national, even imperial one. In order to become true heroines, however, Miyagino and Shinobu had to be just that—true and real. And here their story encounters an unexpected, perhaps unwelcome plot twist, because Miyagino and Shinobu were neither true nor real—*they never existed*, at least not in any way even vaguely resembling the story that was told and retold over the course of more than one hundred years. How did this happen, and why, and in what ways did their story continue to resonate with readers and viewers in the twentieth century, and even today?

Veracity and the Sisters' Vendetta Tale

With regard to the veracity of the Miyagino and Shinobu tale, to date, no archival historical sources have been found that prove unequivocally that the sisters lived in Sendai, certainly not in the early-seventeenth century. Several different written accounts of the Miyagino and Shinobu incident are held by the temple Sennenji in Shiroishi, but they are compilations of local hearsay and embellished stories performed orally for popular audiences by chanters (*kōshakushi*) in the eighteenth century.[14]

The closest we get to a written source verifying a vendetta by two peasant daughters is an account from 1723, in a volume of reportage entitled *Getsudō kenmonshū* (Getsudō's collection of things seen and heard), by the author Motojima Chishin. This genre of text, popular in the later Tokugawa period, was, as its title suggests, a compilation of newsy current events, much of it gleaned from hearsay passed on in writing or orally. In an entry for the fourth month of 1723, Motojima writes that: "from Sendai we hear a report about a vendetta (*adauchi no koto*). In 1718, there was a peasant named Shirōzaemon who lived in Adachi village, which was in the territory of Katakura Kōjūrō, a *karō* in the service of the lord of Rikuoku, Date Yoshimura. The sword teacher to Kōjūrō was a samurai with a 1000-koku stipend named Tanabe Shima. One day when Tanabe was going around inspecting the territory, Shirōzaemon got into an argument with him, and Tanabe cut him down. At the time Shirōzaemon had two daughters, ages eleven and eight, who upon his death were left homeless, so the lord of Rikuoku (Date Yoshimura) ordered his sword master Takimoto Denpachirō to accept the girls as his servants. For six years the girls served Denpachirō, and during this time they secretly observed how swordsmanship was taught, and they learned the techniques themselves. One day Denpachirō was passing by the servant girls' room and he heard the sounds of wooden swords being wielded and so he opened the door and found the two sisters practicing. When he asked what they were doing, the girls said they were planning a vendetta against their father's murderer. Denpachirō was so moved by their resolve that he began to teach them swordsmanship properly, passing on to them his secret techniques."[15]

Eventually Denpachirō told his lord Date Yoshimura of the vendetta plan, and it was arranged that the sisters would engage in a duel with Tanabe Shima in a fenced-off area in front of the Hakuchō Daimyōjin shrine in Sendai in the third month of 1723 (about a month before Motojima recorded his account). With a group of samurai from Sendai observing and officiating, the sisters took turns fighting their opponent, and succeeded in slashing Shima crosswise from shoulder to armpit. Finally, the older sister came in and finished him off.

Date Yoshimura then appealed to the assembled vassals, asking if any wished to take the two sisters into their households. But the girls adamantly refused, saying that "the fact that we killed our father's enemy Shima does not negate our crime; for this we should be punished." Upon hearing this, "everyone was moved." Then Denpachirō intervened, telling the girls that he had been their master and taught them swordsmanship, and out of respect to him they should accept the Lord's decision. After hearing his impassioned plea, the sisters accepted the plan. The older sister was at that time sixteen years old and she was taken in by the lord of Aki, Date Muranari. The younger sister, who was thirteen, was taken in by a vassal named Ōshōji Gonkurō.

There are some key differences between Motojima's account of the sisters' vendetta and later versions, including the one enshrined as fact in Shiroishi. One is of course the timing. The event described by Motojima took place in 1723, and so of course the sword master in question was not and could not have been Yui Shōsetsu, who died in 1651. Also, in Motojima's account, the cause of the vendetta is somewhat less noble—the father got into an argument with the samurai Tanabe, and was cut down for his rudeness (which was, of course, the prerogative of any samurai with regard to any commoner). And in the end, the sisters—who in Motojima's account are never named—do not retire to the religious life; they are taken in, de-facto adopted, by vassals of Date Yoshimura who, like the other individuals named, was an actual Sendai samurai official at the time.

There are also similarities between the earlier and later versions of the story: both tell of the sisters' crafty initiative in learning swordsmanship, the patronage of the sword master, the official sanctioning of the vendetta, and its ritualized public performance that results in the killing of the samurai. And of course, there is the sisters' moral rectitude and attention to justice when, upon completing their vendetta, they seek to be punished themselves. Although Motojima Chishin's version seems on the face of it more believable than later elaborations of the tale, at the end of his account, Chishin writes: "we don't know if this story is true or not, but it has spread out of Sendai and has become a popular tale among the people."[16]

If Motojima was equivocal in assessing the truthfulness of the Miyagino-Shinobu story, other notable local observers at the time declared flatly that the tale was false. The travel writer Furukawa Koshōken (1726–1807), whose accounts of his journeys throughout Japan in the late-eighteenth century gained him considerable notoriety and official recognition, wrote in his report on Sendai that "there is no evidence and the local people say it is a false story... I don't know if in other provinces there was another Yotarō who was avenged by his daughters, but here in Shiroishi it is safe to say there was no such vendetta."[17] The prominent Edo-born woman writer Tadano Makuzu (1763–1825), who married a Sendai samurai official, lived for years in the northeast and wrote widely about the culture and history of her adopted home, dismissed the tale of the filial sisters as inflated hearsay: "The women's vendetta and the talk about Miyagino and Shinobu, the two daughters from Shiroishi, absolutely did not happen. Both stories are made-up and have been passed around by people."[18] Mitamura Engyō, a prominent twentieth-century scholar of Edo culture, was of the same opinion, calling the tale "a complete lie."[19]

And yet, despite the absence of evidence and refutation of the story by contemporary observers, not only the tale has persisted through time, it has been re-made as true over the past century, most notably in the girls' "hometown" of Shiroishi, but elsewhere as well. The sisters' story has become, to borrow the satirist Stephen Colbert's phrase, not so much true as "truthy," that is, true enough to be rhetorically, ideologically, or practically useful to someone or some agenda. Like many apocryphal accounts, the Miyagino and Shinobu story clearly served and continues to serve particular purposes at particular times, purposes that made the story's meaning transcend fact. What were these purposes and meanings? To answer this question, we need to turn to the broader context framing the sisters' purported actions.

Vendettas in Historical Context: The Changing Perspectives on Filial Vengeance and Female Heroism

What did early modern readers, writers, and observers find compelling about the sisters' vendetta? As noted above, by the late-eighteenth century the sisters' story appeared in collections of biographies of "exemplary women," individuals culled from Japan's distant and recent past, whose lives and actions were meant to inspire virtuous behavior in women of all classes.[20] That this honor was accorded Miyagino and Shinobu at first glance seems strikingly odd, given that they were commoners who planned and executed a violent act of private vengeance, both of which were grave offenses to the status-bound and pacifist Tokugawa regime. But, however counterintuitive it may seem, it was precisely the exceptional nature of the sisters' (invented and embellished) act that made it simultaneously newsworthy to the public at large and laudable by the authorities. The non-elite public audience in the later Tokugawa period could focus on the underdog nature of the young orphaned sisters' campaign against their callous samurai enemy.[21] Elites and officials could point to their attention to proper procedure in pursuing their vendetta, and their willingness to feel remorse, accept punishment, and atone for the killing.[22] And elites and commoners alike, steeped as both populations were in the ideology of filial piety, could celebrate the girls' devotion to their father.[23] In a time of crisis and challenge for both families and state, the sisters' story provided a satisfying narrative arc of completion and resolution for all concerned.

The focus on Miyagino and Shinobu's vendetta as an act of filial piety is at one level unsurprising, given the importance accorded that value in Confucian thought, in canonical texts as well as in popular literature. But of course, it was only one of many moral and ethical values in a crowded discursive field. A comparison to the contemporary Chinese case reveals some important differences in degree of emphasis on filial piety as compared to other Confucian values for women, such as chastity and conjugal fidelity. In Qing China (1644–1911), there was a significant upsurge in the number of suicides and incidents of self-mutilation by women in order to display their devotion to deceased husbands.[24] In theory, widows, and even young women who were engaged to be married when their fiancés died, were expected to remain chaste, living with and serving their late husbands' families. Widows who remarried were subject to punishment and criticism.[25] Although there was considerable debate among Qing literati over the ethics of widow suicide, and the Kangxi Emperor (r. 1661–1722) and the Yongzheng Emperor (r. 1723–35) attempted to ban the practice, for many women, overt acts of conjugal loyalty could honor both the dead and the living. Offering a sacrifice of one's life or one's body consoled the spirit of the deceased and burnished the worldly reputation of the family who remained.[26]

In early modern Japan, however, commoner women were expected to have sexual experiences before marriage, and among all classes—even samurai—divorce and remarriage were frequent.[27] Furthermore, women divorced and remarried, sometimes repeatedly, with the full knowledge and endorsement of their families. The reason was simple: the Japanese stem family was relatively small, and its existence was tenuous. Every family needed an heir—and only one heir—so strategic, even multiple marriages were one way to better the odds of producing a viable candidate.[28] A corollary of this was that, unlike Qing China or Chosŏn Korea, continuity of patrilineal blood relations was not as important in early modern Japan as continuity of the nominal familial line of descent, through whatever means necessary. Women were key players in this process of lineage management, as mothers, wives, and daughters, but also as conduits between natal and married families and kin networks.[29] In this context, it makes sense that in Japan filial piety tended to be more often and more forcefully emphasized than conjugal fidelity.

By being extraordinarily filial daughters, Miyagino and Shinobu were thus cast, far and wide, as heroines. They were also represented as deeply loyal, not only to their murdered father but also to an ideal of family that transcended the honor theoretically restored by their successful vendetta. For after avenging their father's death, the sisters were completely bereft of actual family. Their mother was gone, they had no other siblings, and, having taken Buddhist vows, they foreswore marriage and childbearing. Ironically, by performing the ultimate act of filial devotion the fictional sisters also committed the greatest offense against their parents and ancestors: they extinguished the family line. Their devotion was thus to a vanished reality, and only their subsequent fame and (invented) legacy served to perpetuate the ideals they were said to have upheld. Early modern readers would have easily related to these two commoner women who were bound not to masculine warrior-centered ideas of loyal service, whose object was a flesh-and-blood overlord, but instead to a transcendent, fragile, and widely understood ideal of the family.

If the importance of family was the framework through which early modern Japanese readers and observers were encouraged to interpret the Miyagino and Shinobu story, modern readers learned to see them through different eyes. We can see most clearly the modern manipulations of the sisters' story in Shiroishi itself. By the early-twentieth century, Miyagino and Shinobu's act of loyalty and filial piety—developed and spread through print and performance—had become a cornerstone of local history and identity, and any doubts about the veracity of their story were simply suppressed. In Hachimaida, the area where Shiga Danshichi supposedly cut down the peasant father Yotarō, a commemorative temple called the Kōshidō, the "Hall of Filial Children" was built in 1926, dedicated to honoring the sisters and their valorous act. Today, a neat row of eight miniature rice paddies is maintained at the site.[30] Informational signage explains the history of the event, noting the killing of Yotarō by Shiga Danshichi in 1636 ("in this spot"), and the daughters' subsequent revenge. The only hint that this explanation is not a recitation of historical fact comes at the end of the long second paragraph, which ends with the phrase "*to tsutaerarete iru*," a passive construction meaning, roughly, "it has been said." At about the same time that the Kōshidō was built in the 1920s, local officials put on an exhibit of the weapons used by Miyagino and Shinobu in their battle with Shiga Danshichi, and local historians wrote numerous articles tracing the etymology of local place names back to people and events related to the vendetta as a way of showing how the event was embedded in the very physicality of the town itself.[31]

In the postwar period, the aura of factuality around the incident persisted. A stone monument standing next to the rice paddies at Hachimaida, erected in 1966, commemorates the 330th anniversary of Yotarō's "death." Another monument in Shiroishi, put up the following year to mark the purported location of the girls' duel with Danshichi, stands several miles away at the riverside site of Ropponmatsu. And despite the staunchly pacifistic attitudes of the generation who survived the horrors of the Asia-Pacific conflict, Miyagino and Shinobu's legacy of legitimate violence in the name of honor and redemption seems to have been harnessed to the cause of honoring the war dead, as two more stone memorials were erected at the Hachimaida site in the postwar decades, one honoring the town's fallen soldiers and the other praising the loyalty of the imperial army.

Miyagino and Shinobu appear elsewhere in present-day Shiroishi as well—one can go on historical walks to sites related to the vendetta, including a stroll on "Danshichi dōri," and one can visit the Shiroishi Castle Exploratory History Museum (Shiroishi-jō Rekishi Tanbō Myujiamu), which on a daily basis screens, in 3-D, a brand-new historical drama called "Shiroishi-banashi Miyagino, Shinobu Monogatari" (The tale of Shiroishi: The story of Miyagino and Shinobu). This film, shot in Shiroishi and completed in 2017, is a retelling of the local version of the sisters' vendetta. In March 2018 the film premiered with much fanfare at the new and sleekly modern Shiroishi White Cube Community Center, which also put on exhibit numerous Miyagino-Shinobu artifacts, and held a panel discussion with the actors and director. In the supporting and promotional materials for the film, the vendetta is described as having occurred in the seventeenth century. In these materials, descriptions of the vendetta as an example of "daughterly revenge" are clearly highlighted—there is no suggestion that it is a legend, not an historical event. Only in the museum's second-floor documents room can one find, at the end of a description of the "tale of Shiroishi," an acknowledgment that "there is no record [of the event] in the *Katakura dai dai ki* (Great records of the Katakura family) the documents of the Katakura family, lords of Shiroishi Castle... one cannot confirm [its occurrence] in the historical record." The description goes on to state:

> however, the story of the daughters of a peasant who were able to publicly take revenge for a samurai's unjust act and plan a vendetta spread throughout the country during the Edo period, was taken up as a theme in kabuki and *jōruri* dramas, and became very popular. Today, from Aomori Prefecture [in the far northeast] to Okinawa Prefecture [in the far southwest], folk songs, *obon* dances, Shinto ritual music and dance, stage performances and the like, [the story] has spread in various forms.[32]

This statement in many ways brings us full circle with regard to the issues of veracity and violence. Edo-period writers and artists, not to mention their audiences, did not seem to be much bothered by the question of whether vendetta stories were actually true or not—indeed it hardly mattered, given that the drama (and the dramatization) was what interested them, and the topic of vendettas became hugely popular in fiction as well as in stage drama by the turn of the century.[33] The exercise of violence, in reality forbidden to commoners and highly regulated even among samurai, exerted a powerful appeal to audiences of all classes when represented in fiction or performance. And as for the compilers of early modern "exemplary women" and filial piety tales, their mission was to teach and preach about the importance of honoring one's parents, so embellishment and invention—making the story really compelling—were part of their job. Violent revenge and sacrifice of self and others, especially when committed by women, upped the dramatic ante, and if facts fell by the wayside, so be it. In short, the sisters' story clearly resonated with a wide range of people in the eighteenth and nineteenth centuries, and it seems to continue to resonate with a perhaps smaller sector of the population today, regardless of whether it really happened or not. And in any case, the "truthiness" of the Miyagino and Shinobu story seems to have filtered upward, for even reliable standard scholarly reference works refer to the vendetta as if it actually occurred.[34]

The sisters' story and its modern fate remind us that people have always cared about the reputation of their native places—in the early modern period localities proudly promoted local people who were deemed virtuous, especially if the designation came from on high. In terms of acts of filial piety, the epitome of such official recognition was inclusion in the *Kankoku kōgiroku* (The records of filial piety) compiled by the shogunate in the early-nineteenth century.³⁵ Likewise today, the people of Shiroishi and its environs are trying their best to create or recreate Miyagino and Shinobu as local notables, but not necessarily for official recognition. They are primarily trying to market the sisters as a tourist attraction on a very crowded domestic market. And for that, they need to suggest that they were real people who did real things in real historical time.

Perhaps they also want to rally the locals to focus on a valiant act from the distant past, especially now, and especially in the Tōhoku, where the memory of the 2011 tsunami disaster is much more present and real than it is anywhere else in Japan, and individuals and families feel the precarity of modern life especially acutely. In addition, it doesn't hurt that Miyagino and Shinobu are shown to be powerful women, in a time when the Abe government is exhorting women to "shine" in the workplace and as mothers and wives at home, so as to heroically help lead Japan on to renewed glory, even if it means exhausting themselves by working around the clock doing paid and unpaid labor. Miyagino and Shinobu remind us how women's life stories from the past, however "tall," fabricated, or violent, have been strategically deployed at different historical moments, at different spatial scales, to serve ideological needs. At key historical junctures, including perhaps the very recent past, official encouragement has transformed acts of rage and revenge by women into laudable and legitimate acts of female heroism, albeit in the service of a "higher" calling defined by people other than women themselves.

Notes

1   The literature is vast, but notable studies include: Kevin McAleer, *Dueling: The Cult of Honor in Fin-de-Siècle Germany* (Princeton, NJ: Princeton University Press, 1994); Edward Muir, *Mad Blood Stirring: Vendetta and Faction in Friuli during the Renaissance* (Baltimore: Johns Hopkins University Press, 1993); Donald Weinstein, *The Captain's Concubine: Love, Honor, and Violence in Renaissance Tuscany* (Baltimore: Johns Hopkins University Press, 2000); Barbara Holland, *Gentlemen's Blood* (Harrisonburg, VA: Bloomsbury USA, 2004); William Ian Miller, *Bloodmaking and Peacemaking: Feud, Law, and Society in Saga Iceland* (Chicago: University of Chicago Press, 1997); Eiko Ikegami, *The Taming of the Samurai: Honorific Individualism and the Making of Modern Japan* (Cambridge, MA: Harvard University Press, 1997); D.E. Mills, "Kataki-Uchi: The Practice of Blood-Revenge in Pre-Modern Japan," *Modern Asian Studies* 10, no. 4 (1976): 525–542. For a comparative study of dramatic representations of vendettas, see K.J. Wetmore, Jr., ed., *Revenge Drama in European Renaissance and Japanese Theatre: From Hamlet to Madame Butterfly* (New York: Palgrave Macmillan, 2008).

2   During the Tokugawa period, samurai were not only obliged, they were legally required by the state to avenge an attack on a family member. On the centrality of the vendetta to the "way of the warrior" (*bushidō*), see Taniguchi Shinko, *Bushidō kō: kenka, katakiuchi, bureiuchi* (Tokyo: Kakokawa Gakugei Shuppan, 2007). For a case study of two late-Tokugawa vendettas, see Mills, "Kataki-Uchi."

3   Precise numbers of vendettas are difficult to come by, and some went unregistered and unreported, but the shift away from samurai-led vendettas and toward those led by commoners seems clear. According to Hiraide Kōjirō there were around 103 registered and recorded vendettas committed over the course of the Tokugawa period. Of these, thirty-three took place between 1609 and 1703, and of those twenty-nine were led by samurai and four by commoners; between 1703 and 1804, thirty-five vendettas took place, of which nineteen were led by samurai and sixteen by commoners; between 1804 and 1865 there were thirty-five vendettas, sixteen led by samurai and nineteen led by commoners, for a total of sixty-four samurai-led vendettas to thirty-nine commoner-led vendettas over the course of the entire Tokugawa period. See Hiraide Kōjirō, *Katakiuchi* (Tokyo: Saigetsusha, 1975), 129–136; see also Mills, "Kataki-uchi," 531. Makita Isao counts twelve vendettas led by women between the late-seventeenth and mid-nineteenth century, of which four were led by peasant women, one by the daughter and wife of a mountain ascetic/priest (*yamabushi*), and two were of unknown status; see Table 1 in Makita Isao, "Kinsei josei to katakiuchi: 'Ōshū Shiroishi onna katakiuchi wo chūshin ni," in *Rekishi to minzoku ni okeru kekkon to kazoku*, ed. Miyachi Takahiro and Mori Kenji (Tokyo: Daiichi Shobō, 2000), 59. Eiko Ikegami would attribute the shift toward commoner-led vendettas to the "taming," or pacification and bureaucratization of the samurai status group over the course of the Tokugawa period; see Ikegami, *The Taming of the Samurai*. On legal regulations and attitudes toward vendettas, see Taniguchi Shinko, *Kinsei shakai to hō kihan* (Tokyo: Yoshikawa Kōbunkan, 2005), esp. 194–235.

4   Biographies extolling the virtues of "exemplary women" (Ch: *lienü*, J: *retsujo*) served as prescriptive texts and circulated widely in pre- and early modern East Asia; see Dorothy Ko, JaHyun Kim Haboush, and Joan Piggott, eds., *Women and Confucian Cultures in Premodern China, Korea, and Japan* (Berkeley: University of California Press, 2003); Joan Judge and Hu Ying, eds., *Beyond Exemplar Tales: Women's Biography in Chinese History* (Berkeley: University of California Press, 2011); Marcia Yonemoto, *The Problem of Women in Early Modern Japan* (Berkeley: University of California Press, 2016), 22–25.

5   The version recapitulated here is based on Matsudaira Yorinori, *Daitō fujo teiretsu ki* (The Record of Exemplary Women in the Great East [1801]), in *Kinsei joshi kyōiku shisō*, vol. 2 (Tokyo: Nihon Tosho Sentaa, 1980), 1–56. For a more detailed analysis, see Yonemoto, *The Problem of Women*, 21–37.

6   Matsudaira, *Daitō fujo teiretsu ki*, 40.

7   The kabuki version can be found in: Ki no Jōtarō, Utei Enba, and Yō Yōtai, *Go taiheiki shiroishi banashi*, in *Shinpen Nihon koten bungaku zenshū*, vol. 77: *Jōruri shū*. (Tokyo: Shōgakukan, 2002); for an English translation of the scene that continues to be performed in kabuki today, see "The Tale of Shiroishi and the Taihei Chronicles," in *Kabuki Plays On-Stage*, vol. 2, ed. James Brandon and Samuel L. Leiter (Honolulu: University of Hawaii Press, 2002), 83–106.

8   Santō Kyōden, "Musume katakiuchi kokyō no nishiki," in *Santō Kyōden zenshū*, vol. 1, ed. Santō Kyōden Zenshū Henshū Iinkai (Tokyo: Perikansha, 1992), 37–54. Their exploits also merited their inclusion in a major collection of biographies of

exemplary women, the aforementioned *Daitō fujo teiretsu ki* (Record of Exemplary Women in the Great East, 1801).

9   In addition to their depiction in a variety of woodblock prints, beginning in the late Edo period Miyagino and Shinobu also appear on satirical *banzuke*—the rankings or ratings of all manner of things and people, displayed in the format of a multi-tiered *sumo* wrestling-meet schedule, and which served as popular commentary on trends of the time. In these texts, different vendettas (*adauchi*) are ranked, with the more important or significant ones—like the forty-seven loyal retainers made famous by *The Treasury of Loyal Retainers*—represented as if they were the highest-ranked "wrestlers" (*ōzeki*), placed visually at the top of the list, and lesser vendettas represented as mid-ranking wrestlers (*maegashira*), and placed in the middle and lower tiers of the list. In one *banzuke* from the early Meiji period (1871), "Miyagino, Shinobu" are positioned in the first tier, although they are relegated to middling *maegashira* standing. On the representation of vendettas in *banzuke*, see Setoguchi Ryūichi, "Mitate banzuke ni miru adauchi to o-ie sōdō," in *Banzuke de yomu Edo jidai*, ed. Hayashi Hideo and Aoki Michio (Tokyo: Kashiwa Shobō, 2003), 258–275.

10 Quoted in Mills, "Kataki-Uchi," 525.

11 *Japan Weekly Mail*, Yokohama, November 3, 1877, 982–983.

12 Marquis de la Mazelière (Antoine Rous), *Le Japon: Histoire et Civilization*, vol. 3, *Le Japon des Tokugawa* (Paris: Librairie Plon, 1907). The Marquis writes of "deux paysannes, Miyagino et Shinobu, de Shiraishi [sic] en Ôshû, qui obtiennent de leur suzerain la permission de venger leur père assassiné par un kerai."

13 *Japan Weekly Mail*, 982.

14 Although such accounts are referred to generically as *jitsuroku bon* (lit. books of true accounts), this is not because they are factual—the term is an abbreviation of the longer term *chūjitsu kiroku*, or "records of loyalty and devotion," and thus do not necessarily connote accuracy or truthfulness per se. See Makita, "Kinsei josei to katakiuchi," 61.

15 The entire account can be found in Motojima Chishin, *Getsudō kenmonshū*, in *Kinsei fūzoku kenmonshū*, vol. 2 (Tokyo: Kokusho Kankōkai, 1925), 618–619.

16 *Getsudō kenmonshū*, 619. Other textual sources telling of a vendetta by two peasant daughters are of similarly dubious veracity. Ōta Nanpō (1749–1823), a shogunal official turned popular writer and poet, recorded in his miscellaneous writings a letter that relates an unverifiable story resembling that of Miyagino and Shinobu, and which was likely a compilation of several different vendetta tales; see Makita, "Kinsei josei to katakiuchi," 65–66.

17 Furukawa Koshōken, *Toyū zakki* (Miscellaneous records of a journey to the East), quoted in Makita, "Kinsei josei to katakiuchi," 74. On the travel writing of Furukawa, see Marcia Yonemoto, *Mapping Early Modern Japan: Space, Place, and Culture in the Tokugawa Period (1603–1868)* (Berkeley: University of California Press, 2003), 69–100.

18 Makita, "Kinsei josei to katakiuchi," 74; on Makuzu's life and writings, see Bettina Gramlich-Oka, *Thinking Like a Man: Tadano Makuzu (1763–1825)* (Leiden: Brill, 2006).

19 Mitamura Engyō. "Katakiuchi no hanashi," in *Mitamura Engyō zenshū* (Tokyo: Chūō Kōronsha, 1976), 305–51.

20 Earlier texts in this genre featured mostly elite women (imperial noblewomen and women of the samurai class) but later texts focused on commoner women as well. See Yonemoto, *The Problem of Women*, 21–50.

21 It is difficult to know to what degree the sisters' vendetta appealed to commoner audiences who wanted to see a samurai receive his comeuppance. Edward Muir raises the question, in *Mad Blood Stirring*, of whether private vendettas in early modern Friuli served to resolve disputes that could not be adjudicated through official channels, and thus served as a means for people to demonstrate the ineffectiveness of the state. However, the comparison with Japan is imperfect, and in any case, dramas centered on samurai heroics (like *The Treasury of Loyal Retainers*) remained popular throughout the Tokugawa period.

22 All vendettas, whether perpetrated by samurai or commoner, had to be registered and approved by authorities in order to be considered lawful. See Ikegami, *The Taming of the Samurai*, esp. pp. 241–253.

23 The importance of filial piety spread widely through educational texts used in commoners' schools (*terakoya*). A total of 126 separate texts on filial piety and loyalty were published between the mid-seventeenth and late-eighteenth century and official edicts enforced proper observance; see Taniguchi, *Kinsei shakai to hō kihan*, 131–149.

24 See Tian Rukang, *Male Anxiety and Female Chastity: A Comparative Study of Ethical Values in Ming and Ch'ing Times* (Leiden: Brill, 1988). See also Qian Nanxiu, "*Lienü* versus *Xianyuan*: The Two Biographical Traditions in Chinese Women's History," in *Beyond Exemplar Tales: Women's Biography in Chinese History*, ed. Joan Judge and Hu Ying (Berkeley: University of California Press, 2011), 55–87.

25 On the unmarried yet chaste "faithful maiden" phenomenon, see Weijing Lu, *True to Her Word: The Faithful Maiden Cult in Late Imperial China* (Stanford: Stanford University Press, 2008).

26 On the debates over suicide and extreme displays of female chastity, see Joan Judge, *The Precious Raft of History: The Past, the West, and the Woman Question in China* (Stanford: Stanford University Press, 2008), esp. pp. 29–59, and Janet M. Theiss, *Disgraceful Matters: The Politics of Chastity in Eighteenth-Century China* (Berkeley: University of California Press, 2004).

27 On marriage practices and sexuality among rural Japanese women, see Anne Walthall, "The Life Cycle of Farm Women in Tokugawa Japan," in *Recreating Japanese Women, 1600-1945*, ed. Gail L. Bernstein (Berkeley: University of California Press, 1991), 42–70, and "Masturbation and Discourse on Female Sexual Practices in Early Modern Japan," *Gender & History* 21, no. 1 (2009): 1–18. On the frequency of divorce in the early modern period, Harald Fuess writes: "Statistical records confirm that divorce was a common feature of life in the Edo period. From about 6 percent to 40 percent of marriages ended in divorce. Divorce was least likely among samurai in western Japan and most likely among village outcastes. Although the evidence is not conclusive, there is a hint that the frequency of divorce rose during the Edo period." See Fuess, *Divorce in Japan: Family, Gender, and the State, 1600-2000* (Stanford, CA: Stanford University Press, 2004), 24. See also Kurosu, Satomi, "Remarriage in a Stem Family System in Early Modern Japan," *Continuity and Change* 22, no. 3 (2007): 429–458.

28 Adoption—which in the Japanese case meant adoption of kin-related or unrelated women, men, adults, children and even couples—was another strategy for ensuring family continuity. On the importance of adoption for heirship within the samurai class, see Yonemoto, *The Problem of Women*, 164–192, and Yonemoto, "Adoption and the Maintenance of the Early Modern Elite: Japan in the East Asian Context," in *What Is a Family? Answers from Early Modern Japan*, ed. Mary Elizabeth Berry and Marcia Yonemoto (Berkeley: University of California Press, 2019), 47–67; Satomi

Kurosu and Emiko Ochiai, "Adoption as an Heirship Strategy under Demographic Constrains: A Case from Nineteenth-Century Japan," *Journal of Family History* 20, no. 3 (1995): 261–288; Ann Waltner, *Getting an Heir: Adoption and the Construction of Kinship in Late Imperial China* (Honolulu: University of Hawai'i Press, 1990); Kuentae Kim and Hyunjoon Park, "Family Succession through Adoption in the Chosun Dynasty," *Reproduction in East Asian Historical Demography* 15, no. 4 (2010): 443–452.

29 See Yonemoto, *The Problem of Women*, 164–192.
30 Hachimaida means "eight rice paddies," and their presence can also be seen to connote an essentialist connection to rice as Japan's "national" crop and rice cultivation as Japan's sustaining civilizational practice.
31 Makita, "Kinsei josei to katakiuchi," 67–68.
32 From "About Ōshū Shiroishi banashi," unpublished, collection of Shiroishi Castle Exploratory History Museum (Shiroishi-jō Rekishi Tanbō Myujiamu).
33 On literary representations of vendettas, see David Atherton, "Valences of Vengeance: The Moral Imagination of Early Modern Japanese Vendetta Fiction," PhD dissertation, Columbia University, 2013.
34 A vendetta against the samurai Tanabe Shima by two peasant sisters in 1723 is included along with actual vendettas in a comprehensive table listing Tokugawa-period vendettas in the *Kokushi daijiten*, an authoritative multivolume historical encyclopedia; see entry for "*katakiuchi*," in *Kokushi daijiten*, vol. 3 (Tokyo: Yoshikawa Kōbunkan, 1979), 348–352. It is also mentioned as a representative vendetta in another major historical dictionary, Nihon shi daijiten; see "*katakiuchi*," in *Nihon shi daijiten*, vol. 2 (Tokyo: Heibonsha, 1992), 229–231; also cited in Makita, "Kinsei josei to katakiuchi," 67. Brandon and Leiter write that the play is based on "an incident that occurred in Ōshū in 1723." See Brandon and Leiter, *Kabuki Plays On-Stage*, vol. 2, 84.
35 On the *Kankoku kōgiroku*, see Sugano Noriko, *Edo jidai no kōkōmono: Kōgiroku no sekai* (Tokyo: Yoshikawa Kōbunkan, 1999), and "State Indoctrination of Filial Piety in Tokugawa Japan: Sons and Daughters in the Official Records of Filial Piety," in *Women and Confucian Cultures in Premodern China, Korea, and Japan*, ed. Dorothy Ko, JaHyun Kim Haboush, Joan R. Piggott and Joan Piggott, 170–189. For the original text, see Sugano Noriko, ed., *Kankoku kōgiroku* (Tokyo: Tōkyōdō Shuppan, 1999).

5

Women Warriors and the Mobilization of Colonial Memory in the Nineteenth-Century United States

Gina M. Martino

A curious domestic tableau appears in Joel Dorman Steele's 1885 textbook, *A Brief History of the United States*. The image, entitled "New England Kitchen Scene," portrays a seventeenth-century woman sitting at her spinning wheel while minding a young child.[1] Mother and child are the focus of the image, safely ensconced in a Victorian domestic sphere and separated from a pile of cut firewood and an axe on the floor by a wide fireplace and a bubbling cauldron. A translucent valance and a potted plant on the windowsill establish the space as civilized, cultivated, and feminine, signifying the triumph of colonialism to a nineteenth-century audience. But the artist of the engraving understood that the image required another element to effectively portray the colonial woman of the remembered past. And indeed, next to the potted plant— and partially obscured by the dainty window hanging—is the unmistakable shape of a blocky, early-modern musket within arm's reach of the woman at the spinning wheel. This striking combination of military readiness and luminous domesticity suggests something unexpected: the remembered foremothers of nineteenth-century Americans were warriors in a larger colonial project.

The celebrated martial woman of the "New England Kitchen Scene" had an Indigenous mirror image in New England's historical memory. Readers paging through another work of American history, William Cullen Bryant and Sydney Howard Gay's 1883 edition of *A Popular History of the United States*, would have undoubtedly noticed a terrified, barely clothed Indigenous woman who appears to be falling out of a striking, full-page engraving as she plunges face first into a river.[2] The woman in the river is Weetamoo, Sachem of the Pocasset, one of the most powerful female leaders in seventeenth-century North America.[3] Although Weetamoo helped organize and lead the anti-colonial Indigenous coalition in King Philip's War (1675–76), the engraving, "Weetamoo Swimming the Matapoisett," depicts her drowning in a desperate attempt to escape enemy forces. Silhouetted in the background, a small battle rages between a disciplined English force and disorderly Indigenous troops. Weetamoo, who memorably appears as Mary Rowlandson's proud and lavishly dressed mistress in Rowlandson's 1682 captivity narrative, *The Soveraignty* [sic] *and Goodness of God*, has been stripped of the fine English clothing and wampum that Rowlandson described and that had identified Weetamoo as a woman of rank and power.[4] Her long hair, decorated with a single feather, flows past her waist but offers no cover. And although the artist allows Weetamoo a skirt to protect the audience's sensibilities, she is naked from the waist up, her bare breasts framed by her hair and subject to the colonial gaze as her body succumbs to the forces of nature.

The moody atmosphere of "Weetamoo Swimming the Matapoisett" and the sunlit, wholesome kitchen of the colonial mother reflect two distinct nineteenth-century visions of remembered martial womanhood. They also reveal the significant work performed by authors and artists who shaped those memories. Such portrayals served to legitimize the seventeenth-century conquest of New England and encourage the ongoing colonization of the American West in the nineteenth century. In histories, textbooks, poems, plays, sculpture, and images, authors and artists depicted Anglo-American women embracing their role as "pioneer mothers of the republic," feminine civilizers and fierce maternal protectors who conquered the New England wilderness and Indigenous peoples.[5] Indigenous women—particularly female sachems of the seventeenth century—appeared in these sources as foils for celebrated martial white

Figure 5.1 "New England Kitchen Scene" in Joel Dorman Steele, *A Brief History of the United States* (New York: A. S. Barnes & Company, 1885), 94.

Figure 5.2 "Weetamoo Swimming the Matapoisett" by C.S. Reinhart and S.A. Schoff in William Cullen Bryant and Sydney Howard Gay, *A Popular History of the United States*, vol. 2 (New York: Charles Scribner's Sons, 1883), 405.

women. Portrayed as doomed queens, some of whom had failed to maintain "proper" domestic arrangements, female sachems' relationships to civilization and nature stood in opposition to those of English women. Theirs was a more "primitive" manifestation of womanhood, and it rendered them vulnerable to a wilderness they supposedly embraced. In these colonialist narratives, female sachems would become ghostly figures, separated from the land they ruled and vanishing before the power of the civilizing martial mothers of the United States.[6]

In exploring the relationship between nineteenth-century memories of colonial women's war-making and ideologies of gender and expansionism, I first identify the origins of those narratives before turning to characterizations of colonial martial women in nineteenth-century histories and literature. Historiographic interpretations of the meanings of these women's military participation changed as the United States adopted increasingly aggressive colonial policies in the early-to-mid nineteenth century. Such narratives reflected and reinforced newer gender ideologies that stressed the centrality of motherhood and domesticity to women's performance of citizenship. Although chroniclers of the border wars writing during the seventeenth and eighteenth centuries had acknowledged when women warriors were also wives and mothers, those roles neither defined nor explained women's martial activities. By the late-eighteenth and early-nineteenth centuries, historians had found new value in those stories, recasting martial women as delicate inhabitants of a domestic, private sphere who fought to defend their children, their homes, and by extension, larger colonialist projects. At the same time, writers of fiction began to infuse plays, short works of prose, and poems about historic figures with tropes that highlighted Anglo-American women's roles as domestic martial mothers and female sachems' troubled relationships to nature and household order. I conclude with an analysis of how nineteenth-century writers reworked stories of three well-known martial women of the seventeenth and eighteenth centuries: Hannah Dustan and Hannah Bradley of Haverhill, Massachusetts, and Weetamoo, Sachem of the Pocasset. All three women's wartime activities are unusually well documented and, as a group, their stories represent the most violent and dramatic examples of colonial martial womanhood. Over time, they became larger-than-life figures whose names and exploits remained part of New Englanders' shared historical memory and through whom authors and artists could explore changing ideas about violence, colonialism, and ideologies of gender throughout the nineteenth century.

Martial Womanhood in Colonial New England

The two Hannahs of Haverhill were the best-remembered martial women of the colonial era during the nineteenth century. Hannah Dustan became famous for killing ten Indians during her successful escape from captivity, while Hannah Bradley was known for her two wartime captivities and for her attempt to prevent her second captivity by killing one man and maiming another with boiling soap.[7] Both women gave birth immediately before or during their captivities, and both watched their infants die at the hands of their captors. One nineteenth-century historian even enthused that "the story of Mrs. Dustin's exploit and escape reads like a romance."[8] Accounts of their grim experiences of maternal anguish, acts of war, and captivity in the wilderness would serve as "flagship" stories of a violent, resolute motherhood packaged for eager nineteenth-century colonists and armchair expansionists.

No female sachem featured as prominently in New England's historical memory as Weetamoo of the Pocasset. Indeed, Weetamoo was such a rich, familiar, and complex historical figure that authors used her as a character when their work called for an Indigenous woman of high rank, regardless of time or place. Although most modern readers recognize her only as Mary Rowlandson's antagonistic mistress in Rowlandson's famous captivity narrative of 1682, Weetamoo remained known for her role as an influential sachem into the early decades of the twentieth century.[9] Even before King Philip's War began in 1675, English and Native leaders competed to sway Weetamoo to their side. As a sachem who led three hundred men and controlled some of the most desirable land in southern New England, Weetamoo's early alliance with Philip (Metacom) demoralized colonial officials. Moreover, her bold rescue of Philip's forces in a nighttime amphibious operation prevented the destruction of the anti-colonial coalition in its infancy. But as the engraving "Weetamoo Swimming the Matapoisett" reveals, successive generations also remembered Weetamoo's dramatic drowning at the end of the war, an event that seemed to highlight how the Indians' "backward" connection to the natural world would contribute to their inevitable disappearance as distinct peoples.[10]

The print "New England Kitchen Scene" presents its seventeenth-century subjects and setting using visual cues derived from nineteenth-century separate spheres ideologies that venerated women's roles as wives and mothers in a private, domestic sphere while reserving the realm of the public sphere for men. But despite this anachronistic veneer, the image of a colonial woman performing household duties yet able to quickly assume a martial role reflects historical reality for a surprising number of seventeenth- and eighteenth-century women. As I have previously argued, English, French, and Native women in the borderlands of northeastern North America played significant military roles in the wars that engulfed the region. English women in remote vanguard towns defended homes that served double duty as military installations. Women also petitioned colonial governments for troops, kept watch, and administered garrisons. Other women, particularly those in relatively safer towns closer to Boston, attempted to alter military strategy through protests and mob action. Because New England's expansionist policies depended on families of settlers founding fortified towns progressively outward from larger towns such as Boston, women who made war or otherwise demonstrated support for the larger colonial project and its patriarchal foundations were accepted and even acclaimed by religious and political officials who used accounts of women's martial activities as propaganda to influence policy and bolster morale.[11]

In contrast to the private, domestic sphere portrayed in the "New England Kitchen Scene," early modern gender ideologies did not separate public and private spheres. As historian Mary Beth Norton has argued, in the seventeenth and early-eighteenth centuries, the concept of "private" applied only to personal affairs, not the home as a domestic or gendered space. Instead, early modern Europeans and Anglo-Americans understood an individual's activities as falling within formal and informal spheres of public action. Men and women were public actors, and although men in official governmental or religious positions dominated the more formal public sphere, women and men shared the community-oriented informal public sphere.[12] In most cases, women made war as public actors supporting the goals of the larger community, and their participation in New England's wars was rarely considered transgressive.

Although it would surely be an overstatement to say that there was a woman warrior cult in colonial New England, in the seventeenth and eighteenth centuries, chroniclers, politicians, and ministers routinely held up martial women as examples to their fellow colonists. Some women, including Hannah Dustan, became celebrities and received financial rewards.[13] Groups of women who defended fortified communities also received commendations, as colonial officials reinforced the idea that other women in similar situations should emulate that behavior. Norton suggests that following England's Glorious Revolution of 1688–89, new gender ideologies emerged that contributed to the development of gendered public and private spheres.[14] This ideological transformation happened slowly, however, over several decades in the first half of the eighteenth century. Indeed, women who fought in New England's wars through the end of the Seven Years' War in 1763 would continue to receive broad social approval even as their fellow subjects lamented the necessity of women's martial contributions.

Native women, too, participated in their polities' wars during this period, serving as sachems, spies, combatants, and leaders in ritual torture. Colonists sought to understand Indigenous women's war-making through the lenses of European formal and informal public spheres, ultimately leading to a split in colonists' perception of Native martial womanhood. Familiar with traditions of female leadership in their own societies, English colonists largely accepted female sachems' claims to authority, viewing these leaders' actions as falling within the formal public sphere. This was particularly true in the seventeenth century when powerful female sachems such as Weetamoo of the Pocasset and Awashunkes of the Saconet ruled coveted land in southern New England and colonists lacked the desire or perhaps the strength to attempt to reorder Native leadership along gendered lines. Colonists' acceptance of Indigenous women's martial roles rarely extended beyond female sachems, as nonelite Native women who attacked colonists and who participated in ritual torture and adoption would always appear disorderly to English observers.[15]

Making (and Remaking) the Remembered Martial Woman

In the decades following the American Revolution (1775–83), local and regional historians labored to craft narratives that would form a foundation for emerging national, state, and local identities. Many of these authors included accounts of seventeenth- and eighteenth-century Anglo-American heroines in their histories. That women in colonial New England had been effective combatants was indisputable. Indeed, local historian Benjamin Mirick noted that in his hometown of Haverhill, "some of the most heroic deeds accomplished by the inhabitants of this town, were performed by women."[16] But in order to safely celebrate these heroines, authors first had to reconcile colonial martial womanhood with the early republic's embrace of separate spheres ideologies.

Placing stories of colonial women warriors within the nascent ideology of "republican motherhood" offered an ideal solution to the problem of the martial woman.[17] The ideology of republican motherhood rejected the idea that women and men were inherently equal, embracing instead the notion that a woman's primary contributions to the republic stemmed from her roles as wife and mother. The model female citizen, then, would be a civilizing, nurturing figure who raised virtuous republican sons. When grappling with colonial women's violent wartime roles, Americans opted to embrace the republican mother's vigilant martial counterpart: the "resolute mother." The figure of the resolute mother first appeared in historical accounts of women's war-making during the mid-to-late eighteenth century, coinciding with the ascendance of separate spheres ideology in Anglo-American culture. And much as the republican mother became the dominant model for women's political participation in the late-eighteenth and early-nineteenth century United States, Americans recast colonial women who defended fortified communities or attacked their captors not as necessary combatants but as resolute mothers who overcame their feminine weaknesses to protect their children and the private, domestic sphere with a steely but temporary resolve.[18]

In accounts that employed the discourse of resolute motherhood, authors emphasized women's identities as mothers and attributed their wartime violence to an almost preternatural calmness and resolve that a woman might adopt in defense of the domestic sphere. Even if women's "limbs were not made to wield the weapons of war" and if their "sphere of usefulness, of honor and of glory, was in the precincts of the domestic circle," as Benjamin Mirick wrote, they could shed their "giddy, thoughtless, and fearful" qualities to don "a steadiness, a thoughtfulness, and a courage, which equals, and oftentimes surpasses, the same qualities in man."[19] The ideological construction of resolute motherhood was ingenious, allowing New Englanders to imagine their colonial grandmothers temporarily assuming positive masculine qualities and rising to the occasion only in times of great need. That era had passed, memory-makers argued. Women who made homes rather than war in the new republic would help create a more genteel future, their martial natures dormant.

Yet even as New Englanders proclaimed that the colonization of the Northeast had concluded, their work colonizing the rest of the continent had only begun. The active domesticity that characterized representations of historical martial women would be needed to provide cultural space for nineteenth-century women who made war while cultivating and civilizing the wilderness. As historian Amy Kaplan has shown, antebellum expansionism relied on the development of the ideology of "manifest domesticity." The domestication of the "wilderness" and the "foreign" provided a "feminine counterbalance" to male violence and incorporated the private sphere as "a mobile and mobilizing outpost that transformed conquered and foreign lands into the domestic sphere of the family and nation."[20] Combining tropes of the wilderness and American exceptionalism with the language of resolute motherhood and manifest domesticity, nineteenth-century historian William Fowler described the colonial woman as "a soldier and a laborer, a heroine and a comforter" who performed those roles "in a peculiar set of dangers and difficulties such as are met with in our American wilderness." In response to those dangers, "women with muskets at their sides lulled their babes to sleep" in defense of their families, communities, and the colonial project. Women's multiple roles as mothers, fighters, Christians, educators, and workers were, in the words of Fowler, essential in all times to "the building of a new empire upon our dark and distant borderland."[21]

Complementing the gender ideologies and imagery that supported women's roles as agents of manifest domesticity were discourses of innocence and redemption. Indeed, the development of redemptive narratives of violence performed by an innocent nation and an innocent people was essential to the justification and perpetuation of colonialism. According to historian Boyd Cothran, throughout early American history, violent acts—particularly in colonial contexts—were permissible when performed "for reasons of moral outrage, self-preservation, or naïveté or paradoxically to maintain one's innocence." And in the era of Manifest Destiny, when white Americans celebrated westward expansion as a righteous and inevitable development of the republic's growth, how could violence in the defense of this project be anything other than innocent? If colonizers faced violent anti-colonial military campaigns along the way, surely the perpetrators of violence against a youthful nation must be at fault. Indigenous people and polities, then, were cast as "irrational aggressors and violators of a civilized nation's just laws," the natural enemies of the resolute and innocent colonizing mother.[22] By portraying heroines of the past as civilizing practitioners of this redemptive colonialism, Americans could construct precedents for ideologies of expansionism and acts of violence.

Authors in the late-eighteenth and early-nineteenth centuries had largely shied away from depicting the resolute mother as especially violent or vengeful, omitting or sanitizing the most gruesome aspects of martial womanhood. But as anti-Indian sentiment grew along with the violence that accompanied a more aggressive program of expansionism beginning in the 1830s, some American writers began recovering and even embracing the more violent aspects of resolute mothers' martial activities in support of the nation's expansionist project. Historian Barbara Cutter has observed that Americans began to read accounts of Hannah Dustan's violence less selectively. Authors writing before the 1830s had struggled to reconcile her role as a mother with the more gruesome aspects of the killings, especially the murder of children. For some, the solution was to shift their focus to the grief Dustan felt when she saw her own newborn killed, incorporating maternal loss and an instinctive desire for revenge as part of a "defense of the 'natural' ideals of motherhood." Dustan's extreme violence allowed mid-century Americans to view their own violence against Indians as "justified, innocent, defensive violence."[23] And while Hannah Dustan was surely the most popular English colonial heroine in the mid-to-late nineteenth century, her story was only one part of a larger literary project highlighting Anglo-American and Indigenous women's active roles in colonization.

In contrast to the sustained popularity of accounts of colonial women's war-making, Native women who made war in the northeastern borderlands largely disappeared from local and regional histories during the later-eighteenth and early-nineteenth centuries. But as national enthusiasm for westward expansion and the maturation of the cult of domesticity contributed to the creation of more active resolute mothers in historical memory, the same authors and artists who adapted memories of the resolute mother to support a new era of aggressive colonialism and manifest domesticity turned to a second group of martial women from the seventeenth century: female sachems. Rather than tending a garden of civilization in the wilderness, the female sachem in fiction and histories had a dangerous, overly close relationship to nature. And portrayals of female sachems' "primitive" and reportedly mystical connection to the natural world complemented notions that Indigenous women lived in disorderly, uncivilized households that would give way to the cultivating force of resolute mothers' orderly domesticity.[24]

New Englanders' recovery of the memory of the female sachem developed alongside a larger cultural movement that scholar Gordon Sayre has described as a "craze for Indian stage tragedy" and a fascination with the image of the Indian chief as a tragic hero. Audiences in the 1830s began clamoring for works that presented the fall of Indigenous "nobility" and their "noble houses." These tragedies focused on leaders such as Tecumseh, Pontiac, and King Philip, who fought bravely and with honor but whose "capture and sacrifice" would ultimately serve to "replace frontier violence and the bitter racial and political factions behind it with a new civil order."[25] The tragic Indian chiefs who captured the imaginations of Americans in the nineteenth century were almost exclusively male, and Indigenous women who appeared in these works more often played the role of a romantic interest or an intercultural intercessor.[26] One notable exception was Weetamoo of the Pocasset.

Occupying an unusual cultural space as a wife and female sachem who led troops in battle before succumbing to the forces of nature, Weetamoo's memory offered New Englanders the opportunity to explore tropes about Indigenous leadership. Distinguishing the trope of the "tragic Indian chief" from that of the "vanishing Indian," Sayre argues that the tragic chief elicited a relieved yet exhilarated catharsis from readers and audiences, while the vanishing Indian encouraged emotions such as melancholy and nostalgia. As we will see, portrayals of Weetamoo borrowed from both tropes while constructing a model of vulnerable Indigenous femininity that, although not truly innocent, was by turns ingenuously naïve and grimly prophetic. And despite Weetamoo's ominous soliloquies' superficial resemblance to the dying curses uttered by tragic chiefs against European enemies, they more often predicted suffering for all parties and the eventual ruin of her people.[27] In the colonial imagination, where images of innocent, civilizing, resolute mothers mingled with fantasies of doomed female sachems, Weetamoo would serve as an ideal figure for New Englanders engaged in the production of historical memory.

Memories of Innocence and Violence at the Dawn of a New Colonial Era

One of the first nineteenth-century authors to depict the opposing qualities of English and Indigenous martial women from the colonial era was the writer and abolitionist John Greenleaf Whittier. The author of dozens of historically inspired stories and poems, Whittier penned works featuring Hannah Dustan and Weetamoo as protagonists, both of which reflect emerging ideas about the relationship between colonialism, innocence, domesticity, and nature in the early-nineteenth century. Hannah Dustan's appearance in Whittier's 1831 short history, "The Mother's Revenge," explores the limits of acceptable female violence in war while introducing stronger themes of motherhood and nature to her story. It also serves as the basis for future, more overtly violent portrayals of Dustan's actions.[28] Like other nineteenth-century New Englanders, Whittier was aware that Dustan's war-making differed only in degree from those of other colonial women, noting that "it was not infrequently [a woman's] task to garrison the dwelling of her absent husband, and hold at bay the fierce savages in their hunt for blood." As if agreeing with Benjamin Mirick's assessment of women's martial roles in the contemporaneous *History of Haverhill*, Whittier observes that "in the dangers and the hardihood of that perilous period, woman herself shared largely" before acknowledging that in less perilous times, a woman's "sphere of action is generally limited to the endearments of home."[29]

Having established both historical context and the gender expectations of his own time, Whittier goes on to describe how a woman of "milder and purer character" might transform into a resolute wife and mother. According to Whittier, women had displayed "astonishing manifestations of female fortitude and power; manifestations of a courage rising almost to sublimity," a description that at first glance appears to resemble the calmer resolute mother of earlier decades.[30] Whittier's use of the word "sublimity" when describing other remembered martial women is telling and helps introduce his audience to a more brutal Hannah Dustan. The aesthetic concept of the sublime was gendered masculine, and although it shared similarities with resoluteness in its associations with acts of courage in the face of extreme danger, it more famously explored the incomprehensible and overwhelming wildness of the natural world.[31] By suggesting that other women in New England's history had nearly achieved sublimity in their war-making, Whittier leaves space for Dustan's transformation from a good wife to a vengeful, blood-thirsty mother. Whittier does not address Dustan's experience in childbirth in his tale; his interest lies in her identity and transformation, and Dustan had already been a mother for many years. While the veteran mother watches her newborn dashed against a tree by her captor, she metaphorically dies as "her very heart seemed to cease beating." As Dustan is reanimated in the wilderness, her innocent, maternal heart is consumed by "the thirst of revenge ... an insatiate longing for blood." Whittier explains that the extremity of the situation had prompted an "instantaneous change [that] had been wrought in her very nature; the angel had become a demon."[32]

In keeping with the sublime quality of the experience, Dustan's transformation into a vengeful mother occurs in nature, far from the domestic, private sphere. Following the bloodletting and the cathartic release of Dustan's maternal redemption, she begins her transformation back from the wild state of sublime violence. As if passing through a stage characterized by resolve rather than sublimity, she abruptly stops, "reflecting that the people of her settlement would not credit her story" without proof, and "deliberately scalp[s] her ten victims." Although Whittier describes the scalps as "fearful evidence of her prowess," Dustan has already begun her journey back to her domestic life, stealing a canoe and "travel[ing] through the wilderness to the residence of her husband." Her act of redemptive violence complete, she returns peacefully to the private, feminine sphere of historical memory.[33]

In the poem "The Bridal of Pennacook," which describes the marriage and death of Weetamoo, Whittier takes far more liberties with historical accuracy than he does in his account of Hannah Dustan's captivity. Indeed, the Weetamoo of his poem shares few similarities with the historical Weetamoo. It appears as though Whittier, like other nineteenth-century writers, saw Weetamoo as a tragic Indigenous noblewoman whose name was familiar to readers. Whittier relocates the fictional Weetamoo closer to the rugged, wilder mountains of New Hampshire, far from her more "civilized" historical home along the border of Massachusetts and Rhode Island. But the intriguing change of setting aside, Whittier's portrayal of "Weetamoo" is perhaps most historically valuable for how it considers—and ultimately rejects—the possibility of Indigenous domesticity, implicitly removing Native peoples from contention in colonial contests. He describes Weetamoo as a "dusky," naïve young bride, a "savage" but "a woman still," who pursued "maiden schemes" and "young dreams" before being drawn to "the light of a new home—the lover and the wife!" Although the historical Weetamoo had deftly negotiated several political marriages, Whittier's ingenuous Weetamoo becomes a homebody following her marriage to the sachem Winnepurkit. To use the author's sentimental words, "her heart had found a home."[34]

Of course, as an Indigenous woman in a nineteenth-century New Englander's sentimental poem, Weetamoo's marital bliss would not last. In his portrayal of the breakdown of her marriage, Whittier channels white Americans' longstanding disapproval of Indigenous gender roles and family structures.[35] Although Weetamoo is content with her life, Whittier describes her as "a slave" to her husband, in keeping with the "maxims of her race." The following spring, Weetamoo visits her father at his request, but as winter nears, she becomes worried when her husband fails to send for her. Whittier reveals to the reader that Winnepurkit had taken a lover before leaving to "trod the path of war." Weetamoo, "Heedless of peril, the still faithful wife," sets out to find her husband in a small canoe on the roaring Merrimack River. In one of the poem's few nods to history, Weetamoo drowns in the river. Unsuccessful in her attempt to form a domestic life in the wilderness, she inevitably vanishes into nature. A mournful chorus of Indian women whom Whittier dubs the "Children of the Leaves" reinforces this theme, singing, "The path she is treading Shall soon be our own; Each gliding in shadow Unseen and alone!" Their voices resemble "Nature's wild music, — sounds of wind-swept trees," as they lament Weetamoo's death and their own eventual demise.[36]

Although the ideologies and tropes present in Whittier's work and in the histories of New England discussed earlier remained remarkably consistent during the nineteenth century, larger social and political developments did shape authors' preoccupations and agendas.[37] If Whittier's earlier, wistful works of the first half of the century contemplated a romanticized loss of tradition in a modern world, authors writing in later decades celebrated westward expansion even as they worried about the loss of regional identity, wondering if emigrants would spread the region's ideals or leave without a backward glance.[38] What role would New England, its history, and culture play in shaping the future of an industrialized, ethnically diverse nation that spanned a continent? Later histories and plays describing the exploits of women warriors would reflect these concerns, insisting that martial women's influence reached beyond New England and served as a national inheritance for a colonialist republic.

"Like an Heirloom": Martial Motherhood and the Inheritance of Colonialism

A collection of plays and short histories about the wars of the seventeenth century, published in 1886 and curiously titled *Battle of the Bush*, offers a unique opportunity to examine how late-nineteenth-century writers refashioned the memory of early American women's war-making. In these colorful works by Robert Caverly, historical accuracy was not required—and was perhaps not even desired. Among the dramatic personae is a predictable group of historical figures, including Weetamoo, Mary Rowlandson, Hannah Dustan, Hannah Bradley, and King Philip, along with the somewhat unexpected additions of Oliver Cromwell, Princess Elizabeth, and King Charles I. Rounding out the collective ensembles are two Quaker insurers (in two separate plays), a troupe of Irish actors, and two Indian Necromancers (also in separate plays). The dialogue is provincial, florid, and peppered with late-nineteenth-century representations of Indigenous speech. Nevertheless, the *Battle of the Bush* is a collection of fantasies that, through its excesses and overreliance on stock tropes and characters, offers one of the most explicit presentations of the ideologies and imagery of remembered martial womanhood in the final decades of the nineteenth century.[39]

Considering Weetamoo's usefulness to earlier colonialist authors as an all-purpose female sachem, it is unsurprising that Caverly deploys her as a figure divorced from time and place throughout the collection of plays.[40] She first appears in *The Last Night of a Nation*, a drama about the Pequot War (1637), which occurred when Weetamoo was likely a child or perhaps not yet born. Billed as "The Squaw Sachem" and "'Potent Princess'" in the cast of characters, Weetamoo wanders onstage at the end of the play as the ally of the Pequot leader Sassacus. Curiously, the play makes no mention of her until this point, and she seems to simply materialize for the sake of drama. The fictional alliance between Weetamoo and Sassacus, two doomed, Indigenous leaders of different genders references Weetamoo's partnership with Philip in 1675. Indeed, the phrase "Potent Princess" that appeared in the description of Weetamoo's character at the beginning of the play cites a passage from a wartime narrative that introduced Weetamoo to readers in London as "Potent a Prince as any round about her."[41] According to the stage directions, the audience first sees Weetamoo traveling with Sassacus "in agony... skulking, several Indians following them." Shortly after this undignified entrance, Sassacus escapes, while Weetamoo moves to join him but chooses to remain instead.[42]

It appears, then, that Caverly included Weetamoo in his play about the Pequot War primarily because he needed someone to play the role of the tragic Indian chief. Sassacus could not serve that function as he had fled to the Mohawk only to be killed, his hands and head sent to the English. As his character dies offstage and is therefore unavailable to perform the obligatory dramatic, noble death scene near the end of the play, Weetamoo will do in a pinch. Having decided to stay and die, Weetamoo soliloquizes: "Fly? Fly from where? From this me own heritage? From the green hunting-ground of me father; from the wild wigwam of my dear, dear old mother, slain? No! no! Sassacus may fly. I'll *never, never* fly. No! No! *No!*"[43] Weetamoo's speech displays an unsophisticated purity that reinforces her closeness to the land but sidesteps the question of whether she is eligible to perform redemptive violence in the name of her people. The answer to that unspoken question is, of course, "no." Weetamoo instinctively understands that having chosen to remain in her homeland, she must ultimately vanish.

Intriguingly, Weetamoo does not curse the colonists in the mode of the tragic male chief; rather, she predicts the demise of Native peoples. After expressing her attachment to her fictional ancestral home, Weetamoo sings of her intention to return to her tent and "lie down in despair." As is fitting in a play entitled *The Last Night of a Nation*, she performs a feminized mourning ritual, declaring, "I'll paint me in black, and will sever my hair!" Finally, Caverly uses Weetamoo's grief to address popular ideas about the relationship between Indians and nature as she announces her plans to "sit on the shore, where the hurricane blows" and pour out her "woes" to "the god of the tempest." Her mourning and communion with the natural world is tinged with "bitterness," as her "kindred have gone to the mounds of the dead." Yet the cause of their death was not natural, a fact that Weetamoo takes pains to point out. Rather, she cries, "They died not of hunger, nor wasting decay, The steel of the white man hath swept them away." The honorable Weetamoo who "[met] her Fate valiantly" would play a more prominent role as a tragic female sachem in another Caverly play, *King Philip*, alongside English martial women Hannah Dustan and Hannah Bradley.[44]

The play *King Philip* offers the most robust—and unintentionally farcical—juxtaposition of the relationships between Englishwomen and female sachems with nature, motherhood, and colonialism. Caverly even titles the play's accompanying "Legend" (historical introduction): "Weetamoo, Queen of Potasset [sic] ... and Mother Duston."[45] Although Weetamoo had died twenty-one years before the two Hannahs' exploits, the dramatic possibilities of their literary union with the Pocasset sachem must have been too tempting to resist. Praising Dustan's actions in the play's Legend, Caverly likens the memory of her slaughter of ten Indians to "an heirloom" that would "impart inspiration, an inspiration which, diffusing itself, shall tend to elevate the aspirations of the descendants of our old New England mothers throughout many ages yet to come." His "Mother Duston" stands in as the spiritual resolute grandmother of all women involved in New England's colonial project, her violence illuminating and redeeming the wilderness. To her spiritual granddaughters, she bequeaths a memory of martial motherhood as an heirloom, an object of high civilization representing the orderly transfer of treasured property through a family line dedicated to colonialism.[46]

Although Caverly opted to provide readers with a more refined Hannah Dustan in his Legend, he portrays the martial women in *King Philip* as a bit rougher around the edges. The play references both Hannahs, although only Hannah Bradley has a speaking role. Never one to let historical accuracy stand in the way of drama, Caverly ignores the fact that Dustan's and Bradley's martial moments occurred twenty years after King Philip's War, and the women's plotline is still unresolved when Philip dies. Caverly calls upon the character Dumkins, "an aged hunter," to provide exposition in a scene meant to tie up loose ends after Philip's death. Dumkins and a Native ally, Lightfoot, walk through a forest as Dumkins reminds the audience of Dustan's maternal grief and potential for violence, predicting that "old Mother Duston is a hard brick for um; a hard nut, they'll find her to be, I'll promise ye. She'll remember her dead baby; and if a chance happens, she'll make game of um."[47] In his characteristic plain speech during a later scene, Dumkins observes that "there's heroism, there's true metal in those two old Hannahs," casting both as seasoned frontierswomen. Turning to their roles as mothers, Dumkins praises New England's martial women as more effective warriors than its men, opining that "if the fathers had spunk, half so much as those old English mothers, why they'd drive these infernal red-skin murderers into the sea."[48]

Perhaps hesitant to actively depict Dustan massacring Indians onstage, Caverly arranges for Dumkins and Lightfoot to observe Dustan from a distance. Stage directions call for the actors to represent the massacre in a tableau, "ten Indians all lying at length asleep, and Duston, Neff, and Leonardson [her conspirators], with tomahawks uplifted, in the act of killing them."[49] As the audience gazes at the motionless players, Dumkins recites a poem describing the sleeping Indians' "guilty heads." By choosing to present Hannah Dustan's famous moment as a tableau, Caverly preserves the distance between the innocence of the constructed "Mother Duston" and the historical figure's brutality.

The script even includes an illustration of the scene, suggesting how the players might arrange the tableau. In this image, Dustan and her fellow captives have not yet killed. The figures appear calm and Dustan, hatchet in hand, strongly resembles her famous statue in Haverhill. This layered presentation of Dustan thus doubly separates the viewer from her historical violence, as the frozen actor poses as a nineteenth-century memorialization of Dustan, a more innocent beacon of light "diffusing" throughout the nation.

Hannah Bradley, who was pregnant in 1704 when she killed a man using boiling soap before giving birth in captivity, is a more provincial resolute mother than Dustan in *King Philip*. As Dumkins and the governor discuss the whereabouts of both Hannahs while under attack, two Indian men enter with news. One of the men, Peter, announces that Hannah Bradley had not been taken to Canada and that she "Got away from Injuns. Got home. She be in garrison here. Injuns be trying to get her agin [*sic*]." When a group of Native men demand Bradley's return, Peter replies, "Old Hannah is safe. Get you gone. Can't have her. She'll fight ye." In a show of defiance, Bradley briefly appears onstage, "extending her head from the window" and shouting, "Begone, ye murderous hell-hounds, begone! Never, never shall ye capture me again. Get out! begone!" According to additional stage directions, having fought off another stint as a captive, Bradley performs an act of redemptive violence as she "lets blaze a blunderbuss from the window; and Bampico, the Indian leader, falling dead, the Indians take flight." As the men carry their leader's body away, "old Hannah at the same time giv[es] them another blast from her blunderbuss."[50] As Haverhill's lesser-known Hannah, Bradley's noisy defense of her garrison allows Caverly to preserve the aura of gentility he created for Hannah Dustan while still including an earthy yet virtuous martial woman in the play.

If the memory of Hannah Dustan's massacre serves as a civilizing light to brighten the American wilderness for her descendants in *King Philip*, Weetamoo instead acts as a prophet, issuing warnings of doom under gathering storm clouds. Throughout the show, she assumes various aspects of the role of a feminized tragic Indian chief. The play opens with the trial of three of Philip's advisers for the murder of the Christian Indian John Sassamon. The trial was one of the last in a series of events that led to the outbreak of war in 1675. Following Philip's denouncement of the proceedings and his dramatic exit, the curtain falls to the sounds of a mournful dirge. When the curtain again rises, Weetamoo enters in a sumptuous costume, accompanied by "her attendants." Although she addresses the court, her words serve more to introduce the presence of the supernatural to the audience than to influence the ongoing legal proceedings. As she speaks, the remaining Indians in the courtroom kneel while lightning and thunder create a mood of foreboding and a connection between Weetamoo and the natural world.[51] The condemnation of Philip's advisers, Weetamoo warns, "brings darkness gathering o'er this western sky." Her powers of prognostication grow in strength as the storm approaches, and she warns of "the bloody sunset, and that forked light which breaks the curtain of this fearful night." It is here that Weetamoo speaks to the civilized maternal power of English women. Fully employing her supernatural abilities, she describes mothers who "hug sweet infants with tenacious arms" as the storm "foretells gross carnage of successive years, and devastation in a land of tears." In a play that also features the dramatic martial activities of Hannah Dustan and Hannah Bradley, Weetamoo—who will ultimately succumb to the natural forces she channels in the courtroom—reminds the audience that a resolute physicality accompanies colonizing mothers' gentle nurturing.[52]

Conclusion

In 1876, a New England gallery exhibited a painting depicting an attack on a colonial garrison house. William Fowler, a nineteenth-century historian who viewed the painting, described how "women, old and young, some of whom are firing over the parapet," defended the fortified home from the roof. The artist of this unknown work also included a scene from the garrison house's interior, revealing "women engaged in casting bullets and loading fire-arms which they are handing to the men." Risking her safety by leaving the fortifications, "a brave girl is returning swiftly to the garrison, with buckets of water." Of this scene of martial manifest domesticity, Fowler concludes, "this is indeed a faithful picture of one of the frequent episodes of colonial life in New England!"[53] Yet for all the importance the painting places on women's martial activities in the seventeenth and eighteenth centuries, the wartime roles the painting portrays would largely disappear from history and historical memory over the course of a few decades.

Why have Americans forgotten these women? It is doubtful that (white) Americans simply lost interest in gripping tales of women's roles in colonization. It is possible that the disappearance of this hero cult is in part related to academic historians' rejection of local history and studies considered "antiquarian." Although local historians produced surprisingly well-sourced histories proclaiming the importance of women in war into the first decades of the twentieth century, academics would ultimately claim the history of the New England town for their own. By the mid-twentieth century, social historians were producing town histories that focused on the experiences of ordinary townsfolk (mostly men) as reconstructed through statistics and other social science methodologies. As historian Alan Taylor has observed, local historians emphasized the special contributions of an individual town while incorporating that town's uniqueness into a common heritage. In contrast, twentieth-century social historians understood the New England town as a "laboratory" for extracting data to help explain larger social patterns. In other words, antiquarians wove their town histories into a larger tapestry depicting a shared regional (albeit proudly colonialist) past, while later data-driven town studies removed those communities from their larger networks.[54] Historians of women and gender have likewise tended to underestimate women's martial contributions in colonial New England. For scholars establishing the field during the 1970s and 1980s, struggling against the perception that their work lacked intellectual rigor, local history with its antiquarian reputation was perhaps something to be approached warily. Complicating matters, seventeenth- and eighteenth-century sources describing women's wartime activities are highly dispersed throughout the historical record, making it difficult to corroborate nineteenth-century local historians' claims of a tradition of martial womanhood. As a result, women's historians as well as social historians focusing on town studies would likely dismiss a single town's lone record of a female-led garrison defense as anomalous rather than part of a larger phenomenon.

Only a hazy memory of early American martial women survives today, sustained by nineteenth-century monuments and borrowed place names. Unsurprisingly, these commemorations also reflect distinct notions of New English and Indigenous martial womanhood. Hannah Dustan's statues in Haverhill, Massachusetts, and Boscawen, New Hampshire, memorialize her actions while attempting to ensure the permanence of her conquest by shaping stone in her image. New Englanders more commonly seized female sachems' identities for use as the names of natural features and landmarks in both northern and southern New England. As one scholar has noted, the retention of Indian names for place names served to "situate Indians securely in the past, separating them neatly as part of nature instead of culture."[55] Indeed, the names of female sachems such as Awashunkes and Weetamoo have been appropriated to identify natural features along the border of Rhode Island and Massachusetts. Awashunkes's name was given to at least one pond, two swamps, and two marshes. Weetamoo is not as well represented in southern New England, although there is a Weetamoe Street and a Weetamoo Woods in Fall River and Tiverton, respectively. Instead, it is in the wilder White Mountains of New Hampshire that Weetamoo's name most frequently appears. Weetamoo Rock, Weetamoo Trail, Weetamoo Cliff, and Mount Weetamoo all testify to the cultural staying power of Whittier's poem and its depiction of Weetamoo's association with the forces of nature.

As time passes, the memory of New England's martial women will likely continue to fade. Visitors to the mountains, forests, and bodies of water that bear Indigenous women's names will simply see another "quaint" place name of obscure Algonquian origin. Of the celebrated Anglo-American women who made war in these conflicts, only the memory of Hannah Dustan remains. And despite the efforts of one historical society to preserve her memory in the form of a bobblehead doll, her figure is fading from New England's landscape. Statues meant to inscribe her memory and tell her story through carvings on their bases have aged, and in the case of the New Hampshire statue, weathering and vandalism have taken their toll. Perhaps there is an argument to be made that there is some justice in forgetting these women, especially the violent Hannah Dustan. Yet this would be a shame, for remembering these women and the generations that memorialized their war-making is neither an antiquarian exercise nor the fetishization of warrior women. Rather, these women's martial activities and the role their memories played in promoting westward expansion offer Americans opportunities to engage with an uncomfortable history of colonialism and learn how societies—past and present—produce and reproduce colonialist ideologies.

Notes

1 Joel Dorman Steele, *A Brief History of the United States* (New York: A. S. Barnes & Company, 1885), 94.
2 William Cullen Bryant and Sydney Howard Gay, *A Popular History of the United States*, vol. 2 (New York: Charles Scribner's Sons, 1883), 405.

3   Weetamoo ruled territory in and around the present-day cities of Tiverton, Rhode Island, and Fall River, Massachusetts.
4   Mary Rowlandson, *The Sovereignty & Goodness of God* (Cambridge, MA: Samuel Green, 1682), 57.
5   William Worthington Fowler, *Woman on the American Frontier: A Valuable and Authentic History of the Heroism, Adventures, Privations, Captivities, Trials, and Noble Lives and Deaths of the "Pioneer Mothers of the Republic"* (Hartford, Conn.: S. S. Scranton, 1886).
6   Scholars have often overlooked female sachems, though studies that do consider these leaders include John A. Strong, "Algonquian Women as Sunksquaws and Caretakers of the Soil: The Documentary Evidence in the Seventeenth Century Records," in *Native American Women in Literature and Culture*, ed. Susan Castillo and Victor M.P. Da Rosa (Porto, Portugal: Fernando Pessoa University Press, 1997), 191–214; Ann Marie Plane, "Putting a Face on Colonization: Factionalism and Gender Politics in the Life History of Awashunkes, the 'Squaw Sachem' of Saconet," in *Northeastern Indian Lives, 1632–1816*, ed. Robert S. Grumet (Amherst: University of Massachusetts Press, 1996), 140–165; Gina M. Martino, *Women at War in the Borderlands of the Early American Northeast* (Chapel Hill: University of North Carolina Press, 2018); Lisa Brooks, *Our Beloved Kin: A New History of King Philip's War* (New Haven: Yale University Press, 2018); and Ethan A. Schmidt, "Cockacoeske, Weroansqua of the Pamunkeys, and Indian Resistance in Seventeenth-Century Virginia," *American Indian Quarterly* 36, no. 3 (Summer 2012): 288–317.
7   Samuel Penhallow, *The History of the Wars of New-England, with the Eastern Indians* (Boston, 1726), 10–11; Cotton Mather, *Humiliations Follow'd with Deliverances* (Boston, 1697). Studies that consider Dustan in her own time include Teresa A. Toulouse, *The Captive's Position: Female Authority, Male Identity, and Royal Authority in Colonial New England* (Philadelphia: University of Pennsylvania Press, 2007); Pauline Turner Strong, *Captive Selves, Captivating Others: The Politics and Poetics of Colonial American Captivity Narratives* (Boulder, CO: Westview Press, 1999); Martino, *Women at War*; and Laurel Thatcher Ulrich, *Good Wives: Image and Reality in the Lives of Women in Northern New England, 1650–1750* (New York: Vintage, 1991).
8   Fowler, *Woman on the American Frontier*, 81.
9   Rowlandson, *The Sovereignty & Goodness of God*.
10 Jean O'Brien has argued that nineteenth-century New Englanders used the genre of local history to justify colonialism while "seizing indigeneity" for themselves. In doing so, they also positioned themselves as the first modern New Englanders, while "Indians remained rooted in nature." O'Brien, *Firsting and Lasting: Writing Indians out of Existence in New England* (Minneapolis, MN: University of Minnesota Press, 2010), 50.
11 Martino, *Women at War*. Historians have been largely unaware of the martial roles women played in these wars. One work that briefly considers the topic is Ulrich, *Good Wives*, 167–183.
12 Mary Beth Norton, *Founding Mothers and Fathers: Gendered Power and the Forming of American Society* (New York: Vintage, 1997); Mary Beth Norton, *Separated by Their Sex: Women in Public and Private in the Colonial Atlantic World* (Ithaca, NY: Cornell University Press, 2011).
13 Barbara Cutter, "The Female Indian Killer Memorialized: Hannah Duston and the Nineteenth-Century Feminization of American Violence," *Journal of Women's History* 20, no. 2 (Summer 2008): 10–33; Sara Humphreys, "The

Mass Marketing of the Colonial Captive Hannah Duston," *Canadian Review of American Studies* 41, no. 2 (August 2011): 149–178; Kathryn Zabelle Derounian-Stodola, "The Captive as Celebrity," in *Lives Out of Letters: Essays on American Literary Biography and Documentation, in Honor of Robert N. Hudspet*, ed. Robert D. Habich (Madison, NJ: Fairleigh Dickinson University Press, 2004), 65–92; and Martino, *Women at War*, 92–102.

14 Norton writes that this development was tied to the elevation of gender over rank as a basis for social organization as "consenting men" became the foundation of the polity. Norton, *Separated by Their Sex*.
15 Ann Little, *Abraham in Arms: War and Gender in Colonial New England* (Philadelphia: University of Pennsylvania Press, 2007), 56–90; Theda Perdue, *Cherokee Women* (Lincoln: University of Nebraska Press, 1999), 53–54; and Martino, *Women at War*, 19–57.
16 B.L. Mirick, *The History of Haverhill, Massachusetts* (Haverhill, MA: A. W. Thayer, 1832), 64.
17 On the development of separate spheres ideology and the cult of domesticity, see Rosemarie Zagarri, *Revolutionary Backlash Women and Politics in the Early American Republic* (Philadelphia: University of Pennsylvania Press, 2007); Linda K. Kerber, *Women of the Republic: Intellect and Ideology in Revolutionary America* (Chapel Hill: University of North Carolina Press, 1980); Carol Berkin, *Revolutionary Mothers: Women in The Struggle for America's Independence* (New York: Knopf, 2005); Norton, *Separated by Their Sex*; and Nancy F. Cott, *The Bonds of Womanhood: "Woman's Sphere" in New England, 1780–1835* (New Haven, CT: Yale University Press, 1977).
18 On the construction of the memory of a maternal martial womanhood and "resolute motherhood" in the early republic, see Martino, *Women at War*, 135–157.
19 Mirick, *The History of Haverhill*, 65.
20 Amy Kaplan, *The Anarchy of Empire in the Making of U.S. Culture* (Cambridge: Harvard University Press, 2005), 25.
21 Fowler, *Woman on the American Frontier*, 81, 527.
22 Boyd Cothran, *Remembering the Modoc War: Redemptive Violence and the Making of American Innocence* (Chapel Hill: University of North Carolina Press, 2014), 19.
23 Cutter, "The Female Indian Killer Memorialized," 20, 26. See also Humphreys, "The Mass Marketing of the Colonial Captive Hannah Duston." On national innocence in the context of settler colonialism more broadly, see Cothran, *Remembering the Modoc War*.
24 On disorderliness, see Little, *Abraham in Arms*, 91–126.
25 Gordon M. Sayre, *The Indian Chief as Tragic Hero: Native Resistance and the Literatures of America, from Moctezuma to Tecumseh* (Chapel Hill: University of North Carolina Press, 2005), 8–10, on curses see ibid., 27–29.
26 Cothran, *Remembering the Modoc War*, 81–105.
27 Sayre, *The Indian Chief as Tragic Hero*, 5–6.
28 Cutter, "The Female Indian Killer Memorialized," 20.
29 John Greenleaf Whittier, "The Mother's Revenge," in *Legends of New-England*, ed. Horace Elisha Scudder (Hartford, CT: Havener and Phelps, 1831), 125–126.
30 Whittier, "The Mother's Revenge."
31 Barbara Claire Freeman, *The Feminine Sublime: Gender and Excess in Women's Fiction* (Berkeley: University of California Press, 1995), 72.
32 Whittier, "The Mother's Revenge," 128. Barbara Cutter rightly observes that Whittier identifies Dustan's actions as the "temporary and natural" response of a woman

whose "natural state" of motherhood "was under attack." Cutter, "The Female Indian Killer Memorialized," 20.
33 Whittier, "The Mother's Revenge," 130.
34 John Greenleaf Whittier, "The Bridal of Pennacook," in *The Complete Poetical Works of John Greenleaf Whittier*, ed. Horace Elisha Scudder (Boston: Houghton, Mifflin & Co., 1904), 29, 34, 36.
35 On Anglo-American disapproval of Native gender roles and families, see Little, *Abraham in Arms*; Kathleen Brown, *Good Wives, Nasty Wenches, and Anxious Patriarchs: Gender, Race, and Power in Colonial Virginia* (Chapel Hill: University of North Carolina Press, 1996).
36 Whittier, "The Bridal of Pennacook," 36–40.
37 For all its significance in American history, the Civil War plays little part in this study. On the remarkable consistency of New England's colonial histories before and after the Civil War (specifically between 1820 and 80), see O'Brien, *Firsting and Lasting*, xx.
38 David D. Hall and Alan Taylor, "Reassessing the Local History of New England," in *New England: A Bibliography of Its History*, ed. Roger Parks (Hanover, NH: University Press of New England, 1989), xxvii.
39 Robert Boodey Caverly, *Battle of the Bush* (Boston: B. B. Russell, 1886).
40 Robert Boodey Caverly, *The Last Night of a Nation*, in *Battle of the Bush* (Boston: B. B. Russell, 1886). The copyright date given for the play (rather than the collection) is 1884.
41 N. S., *The Present State of New-England, with Respect to the Indian War* (London, 1675), 3.
42 Caverly, *Last Night of a Nation*, 55.
43 Caverly, *Last Night of a Nation*, 55. Emphasis is from the original.
44 Caverly, *Last Night of a Nation*, 55, Table of Contents.
45 Caverly, *Battle of the Bush*, Catalog.
46 Robert Boodey Caverly, *King Philip*, in *Battle of the Bush* (Boston: B. B. Russell, 1886), 139.
47 Caverly, *King Philip*, 185.
48 Caverly, *King Philip*, 189.
49 Caverly, *King Philip*, 186.
50 Caverly, *King Philip*, 189–190. This incident likely reflects reports of a third attack on her garrison house, in which Bradley fought alongside her husband Joseph, successfully repelling the assault and avoiding another captivity. George Wingate Chase, *The History of Haverhill, Massachusetts* (Haverhill, MA: George Wingate Chase, 1861), 216.
51 Caverly, *King Philip*, 160.
52 Caverly, *King Philip*, 160.
53 Fowler, *Woman on the American Frontier*, 79–80.
54 Hall and Taylor, "Reassessing the Local History of New England," xxxii.
55 O'Brien, *Firsting and Lasting*, 35.

6

From the Nation to Emancipation: Greek Women Warriors from the Revolution (1820s) to the Civil War (1940s)

Sakis Gekas

Modern Greece does not have an official pantheon of heroes, but it does have a rich tradition of celebrating heroines.[1] This chapter examines historical representations of women warriors in photography, film, and monuments to understand how the image of heroic women fighters has changed during two key moments of Greece's history. This approach differs from European case studies that have looked at individual examples of heroes/heroines or have studied representations of European women warriors in a comparative framework.[2] In the Greek case, representations of women warriors were neither monolithic nor static, but rather changed according to changing social perceptions within Greece about the role of nation, the legacy and meaning of its wars, the state of class conflict, and debates over the policing of gender boundaries and the place of women in Greek society, especially during the Greek Civil War (1946–49) and during times of political liberalization such as in the post-1974 period following the fall of the 1967–74 dictatorship.

Greece emerged as an independent state in 1830 following its War of Independence against the Ottoman Empire in the 1820s. During the Second World War, Greece experienced four years of occupation by the Italians and the Germans, as well as a resistance struggle against the occupiers. Following the German withdrawal in October 1944, a coalition government succumbed to British pressure and insisted that King George II return along with his government in exile. The coalition government failed to elicit the support of the Communist party and other resistance forces in a postwar democratic government. In protest, the Communist party abstained from the elections of March 1946 and the country plunged into a brutal Civil War that lasted for three years. The crushing defeat of the Communist Democratic Army led to the post-Civil War exclusion and persecution of everyone suspected of being a communist sympathizer. The military dictatorship of 1967–74 saw a revival of the "glory" of the 1821 revolution and a proliferation of visual representations of women warriors and heroines; after 1974 the history of heroes and heroines became of secondary importance for historians focusing on the social and economic history of the country.[3] But in the 1980s, parliament formally acknowledged the role of the national resistance as well as the history of the Civil War. As a result, a number of statues and busts in public places signify the commemoration of heroines of the war of independence and fighters in the resistance even in the form of the 'unknown warrior'.[4]

In this chapter, I will argue that the Greek War of Independence (1821–30), the German and Italian occupation during the Second World War (1940–44), and the subsequent Civil War (1946–49) were each specific periods during which women warriors emerged, physically and allegorically, as national heroes; but the visual register and symbolic meaning of these representations of women warriorhood changed over time. In the nineteenth century, they were presented as heroic champions, usually because they fought for the nation in a similar fashion as men did. From the 1940s onward, however, the figure of the woman warrior took on new meaning, especially as Greek women fought, not just against the enemy, but also for gender equality and against patriarchal social norms, as well as national and democratic ideals. While in the nineteenth century some Greek women warriors were celebrated as heroines but also hailed as martyrs, this image changed in the 1940s when political photography depicted women as gallant fighters and used their images for propaganda purposes. The Communist movement and the press sought to establish continuities between the heroes of the Greek War of Independence and the narrative of resistance that developed after the 1940s. The history of representations of women warriors in Greece therefore has been contingent on the prevailing national ideology and the changing political milieu.

Women Warriors in the National Narrative of the Greek Revolution

Beginning in the 1930s, the nationalist dictatorship of Ioannis Metaxas sought to portray Greek history as a story of heroes, heroines, and sacrifice. As part of its campaign of collective memory manipulation, a prominent narrative emerged that celebrated the legendary feats of Greek women warriors during the Greek War of Independence, the nation's foundational moment. For instance, *Souliotes*, men and women from Souli, a rugged, mountainous region close to today's borders with Albania, came to occupy a central place in the Greek national narrative. Stories proliferated about how they fought fiercely for their relative autonomy against Ali Pasha, the Ottoman governor of the Southwestern Balkans whose rule extended from present-day Albania to the Peloponnese. Ali's de facto independence incurred the wrath of Sultan Mahmud II and led to a campaign against Ali Pasha in 1820 and his decapitation, literally, in January 1822, leading ultimately to the beginning of the Greek revolution in March 1821.[5] One story, in particular, emerged during this time and came to form a key moment in the Greek national narrative and the popular imagination: the story of the "death dance" of Zaloggo. This story, which tells of how sixty-three women sacrificed themselves by committing suicide on December 16, 1803, when they jumped off a cliff together with their children rather than surrendering and facing death or humiliation because they would have been enslaved, became legendary. It was told over and over again until it secured a place for the women of Souli in the Greek national historical canon. The women of Souli were probably the first Greek women warriors in the modern era. This does not mean that women in other rural societies in the Greek peninsula, famous for their insubordinate defiance to centralized authority, such as Mani in southern Peloponnese, did not fight side by side with their men; it is just that Souliotisses were remembered for doing so.

As well-remembered as the Souliotisses were, two individual women of the Greek revolution, Laskarina Bouboulina (1771–1825) and Manto Mavrogenous (1796–1848), celebrated as naval commanders, have been lionized above all others in Greek national historical and popular imagination. French and British historians of the Greek War of Independence, in particular, noted the role of these two women in the 1820s war against the Ottoman Empire, often in romanticized ways. But the first book to discuss Manto Mavrogenous's role in the war, published in Paris during the war, did not celebrate her for being a warrior per se but for her contribution to the war effort through her personal fortune.[6] In contrast, Bouboulina, probably the most well-known woman warrior of the revolution, was a fighter. She married twice, first at seventeen to Demetrios Giannouzas and again at the age of thirty to Demetrios Bouboulis. Both husbands were from the island of Spetses, off the east coast of Peloponnese; both were captains and ship-owners;[7] and both were killed by Algerian pirates. The tragic coincidence reveals the extent of fighting in the Mediterranean between traders and pirates, and merchant-captains from the Greek islands were renowned for switching between the two.[8] Bouboulina's deceased husbands left her a large fortune. Following the death of her second husband in 1820, Bouboulina constructed a large armed ship to defend her property. This was the official excuse she gave in requesting a license from the Ottoman administration, although, in reality, she was already preparing for the uprising that would come the following year. The ship, *Agamemnon*, was a corvette, carrying eighteen canons. Built on the eve of the Greek revolution, the ship cost 25,000 tallers and was the largest of three ships Bouboulina built and owned.[9]

The War of Independence did not change gender relations and the figure of the woman warrior-at-sea was exceptional. When the revolution broke out, Bouboulina had already formed her band of Spetsiotes, who she used to call "my brave lads." She armed and fed the crew at her own expense and at a great cost. Indeed, it is estimated that by the end of the second year of the revolution, she had nearly exhausted her fortune. Bouboulina took part in the siege of Ottoman-held forts of Nafplio and Tripoli. Following the takeover of Nafplio on November 30, 1822, Bouboulina settled there. And, a few years later, in recognition of her service to the nation, the revolutionary government granted her a house.

When Britain, France, and Russia, the protecting powers of the nascent Greek state, chose the sixteen-year-old Otto, the second son of the king of Bavaria, to become the first king of Greece in 1831, his father, King Ludwig I, sent the painter Peter Hess with him to Athens to gather materials for pictures of the war of liberation. And, of course, amongst the forty paintings Hess made depicting key figures and events of the Greek revolution up to the coronation of King Otto was a portrait of "Bouboulina." The famous portrait has adorned school books and school classrooms in Greece for decades; it shows an image of a woman directing her ship and her soldiers in battle. French and German romantic painting certainly played a role in the popularity of some women and men warriors of the Greek revolution; this popularity increased with the circulation of stories and images during and immediately after the Revolution, primarily by non-Greek historians, who were quick to provide accounts of the war.[10] In addition to Hess's famous portrait, there is another nineteenth-century image of Bouboulina rendered by an anonymous painter. In this portrait, housed at the National Historical Museum in Athens, Bouboulina appears sterner, fiercer. The image served as the portrait for the 1970 100-drachmas banknote.

Bouboulina was not the only women warrior to be painted during the Greek War of Independence. Images of Manto Mavrogenous also circulated during the conflict. In 1827, Danish painter Adam Friedel included her in his series "Portraits of the Greek revolutionaries." Friedel was one of several romantic painters who created the images of the revolutionaries that still adorn Greek school books and the National History museum. Another, Ary Scheffer, was inspired by Greek themes that extended beyond the revolution; in 1827 Scheffer painted the "sacrifice" of the Souliotisses, a fine example of the romantic liberal art produced in France in the 1820s.[11] After the revolution, one of the early Greek painters, Miniatis, painted a much more warrior-like image of Souliotisses, arms in hand, towering over their men and a dead woman.

In the decades following independence, the Greek state education system praised the two women warriors of the revolution and commemorated the women of Souli for their sacrifice. Sotiria Aliberti's *The Heroines of the Greek Revolution*, the first Greek book that explicitly wrote about heroines and women warriors, was written in the 1890s but not published until 1933. Aliberti was one of the first Greek female

Figure 6.1 Georgios Miniatis. *Souliotisses*. Second half of the nineteenth century, Public Gallery, Corfu.

historians to rise to national prominence through her education and she did so in the last quarter of the nineteenth century, when some social space was opening to Greek educated women. In her stories, her heroines are entangled in events of the Greek revolution. A teacher, philanthropist, and among the first women to promote the cause of women's emancipation in Greece, Sotiria Aliberti wrote her book while teaching at one of the most prestigious schools for Greeks in Istanbul.[12] Her book, therefore, had an explicitly didactic purpose and was partly inspired by the partly historical, partly fictional work by Stefanos Xenos, *The Heroine of the Greek Revolution*.[13] Aliberti's narrative places women firmly at the center of major events before and during the War of Independence. The book popularized the already well-known story of the women of Souli and narrated in detail the role of Bouboulina and Manto Mavrogenous, setting the canon for later narratives.

Since the writings of Aliberti, the women of Souli, Bouboulina, and Manto have all become enshrined within national memory as prominent women warriors. The story of the women of Souli in Zaloggo acquired official status with its inclusion in the commemoration of the events of Greek Revolution on March 25, every year, taught in schools all over Greece even though it happened eighteen years before the revolution. The combination of a strong sense of local identity and kinship, an act of defiance, desperation, and bravery, contributed to the construction of patriotic but passive female identity that was propagated for decades after Greek independence.[14] The national morale of the story is the act of maintaining honor instead of shame. In 1903 Sp. Peresiades turned the mass suicide tragic story into a popular drama that included a song and music for the performance and was labeled as a "dance," which is how it is remembered still today. In 1959, the sacrifice of *Souliotisses* was depicted in Stelios Tatasopoulos's film *Zaloggo; the Castle of Freedom*.[15] This was the first time the Zaloggo story was depicted in film, a romantic nationalist picture that ends with women dancing and singing the famous song and jumping off a steep cliff. The mythologizing of the Greek revolution in film was on. And the elevation of the legend to national story was completed with the erection of a monumental site to honor the women of Souli in 1961 by sculptor N. Zogolopoulos.[16] The six female figures, abstract in form, hold hands and the monument rises up to eighteen meters as it reaches the precipice, becoming more dramatic in its visual impact from a distance as well as from the site.

In 1959, the same year *Zaloggo* was made, the film *Boubloulina* was advertised as the "legendary movie," the "glory of Greek courage." Beyond film, both Bouboulina and Manto are revered in public spaces; the statue of Bouboulina in the Spetses main square confirms her central role in the local and national history and Bouboulina's image has adorned postal stamps and an old fifty drachmas banknote. Manto Mavrogenous has also received by the state the fame and recognition she deserved; a central square in Mykonos was named after her and a statue was erected. There are many streets in Greece bearing her name and she is depicted in commemorative coins, school textbooks, and portraits in class rooms. In 1971, director Kostas Karagiannis made a film, *Manto Mavrogenous*, depicting Manto as a courageous woman who played a pivotal role in the Greek revolution in a world of strong and hot-headed men.

Women Warriors in the 1940s. Women of the Resistance and the Civil War

In times of conventional wars, such as the Balkan Wars of 1912–13, the First World War, and the subsequent Greko-Turkish War (1919–22), there was little room for heroic deeds by women, who were relegated to supporting roles at best. However, when state authority—and established gender norms with it—collapsed under foreign occupation and was subsequently challenged by an armed uprising, women resurfaced as combatants and heroes. American author Henry Miller captured this phenomenon in his classic 1941 travelogue of Greece, *The Colossus of Maroussi*, when he wrote that:

> Everywhere you go in Greece the atmosphere is pregnant with heroic deeds. I am speaking of modern Greece, not ancient Greece. And the women, when you look into the history of this little country, were just as heroic as the men. In fact, I have even a greater respect for the Greek woman than for the Greek man. No wonder Durrell wanted to fight with the Greeks. Who wouldn't prefer to fight besides a Bouboulina, for example, than with a gang of sickly, effeminate recruits, from Oxford or Cambridge?[17]

With Greece about to be enveloped by the storm of invasion, occupation, resistance, and Civil War, Miller was invoking the most famous heroine of the Greek revolution, one who the Greeks themselves still remembered and who they would use in new ways in this new context at the same time as new women warriors would emerge from almost a decade of conflict.

On October 28, 1940, Italy invaded Greece; the Greek army mobilized and achieved the first victory against an Axis power in November 1940, when it stopped the Italians and forced Hitler to draw up plans to invade Yugoslavia and Greece to secure his southern-European—Balkan flank. During the war in the mountains of Pindos between Greece and Albania, women played a new—and decisive—supportive role, for which they have been celebrated as "the women of Pindos." These women assisted men fighting by climbing the mountains to bring ammunition, resources, tending for the wounded and carrying essentials on animals, or even on their own backs. Their role was crucial for morale, demonstrating in practice that the whole nation was behind the defensive effort, with Greek men and women all fighting defending their country against Italian aggression. The story and symbolism became an official monument, erected in Thessaloniki in 1990 and in two additional areas near the border with Albania, one in Pentalofos and a second one in Zagori in 1993.

In April 1941, the German army invaded Greece and, together with the Italian and Bulgarian armies, occupied the country until October 1944. The brutal occupation period saw the plundering of Greek resources and the starvation of up to quarter of a million people during the winter of 1941–42.[18] This disaster led to the emergence of one of the fiercest resistance movements in Nazi-occupied Europe as well as one of the highest death tolls in Europe in reprisals by the occupying forces. The Resistance, as it widely known, saw many heroines, but there were also large numbers of "anonymous" women who took part. However, the memory of the 1940s and the resistance period of 1941–44 is entangled with the much more contested history of the Civil War that followed in 1946–49, which meant that it was not officially acknowledged by the state and made part of the official narrative only in the 1980s. As a result, the story of three of the women who were executed by the occupying forces and their Greek collaborators—Lela Karagianni, Iro Konstantopoulou, Electra Kostopoulou—became known belatedly. There are now many accounts, oral histories, and other works showing that one of the most radical elements in the Resistance movement was the mobilization and participation of women.[19] The shift is also partly due to the rise of public history and historical awareness of the period of the 1940s beyond university departments and academic conferences, where it has enjoyed a popularity for almost twenty years.

The stories of these three heroines reveal the different walks of life they came from but also the representations of women as warriors and heroines in monuments and popular images such as film. Lela Karagianni is probably the best known heroine of the National Resistance against the Nazi Occupation. When the German army invaded Greece in April 1941, thousands of British soldiers did not make the evacuation from the mainland to Crete and then to Egypt. Lela Karagianni quickly found other like-minded patriots and together they helped the soldiers who had been left behind connect with the Resistance and escape from Athens, first to Crete and then to Egypt. Karagianni used her husband's drugstore and perfume store and their house as a hideout for British soldiers seeking to escape, as well as helping numerous starving children in occupied Athens. Karagianni named her resistance organization Bouboulina, after the famous woman-captain. Curiously, this independence hero was also her distant ancestor, although is unclear whether the extraordinary coincidence played a role at all in Karagianni's determination to form and lead a resistance organization. The "Boubloulina" group sabotaged German installations, provided the Allied Middle East Command in Cairo with intelligence, and helped Resistance fighters in any way they could. Karagianni's children also helped, risking their lives and those of their siblings. When arrests started in June 1944 Karagianni was warned that she should go into hiding, but refused and in September 1944, was arrested together with five of her children. She was interrogated and tortured to make her turn in the other members of the organization; she refused to do and on September 8, 1944, only a few weeks before the German army evacuated Athens, on October 12, she was executed, as part of the last major execution of Greek patriots. During her final moments she is supposed to have sung the national hymn and danced the "Zaloggo dance," putting hope into the fearful hearts of her fellow patriots. This is probably more legend than reality but the way the sacrifice of women in the early-nineteenth century was connected in popular imagination with the resistance woman fighter of the 1940s is striking.[20] Karagianni is not nearly as famous as Bouboulina or Manto Mavrogenous because she fought in a different war and because the National Resistance was acknowledged only in 1982; however, she has received state recognition and a street bears her name and in 1963 a small bust was erected near Omonoia Square and Patision avenue, where she lived and run her organization, testament to the acknowledgment of her sacrifice. The legacy of Lela Karagianni and the continuity with the heroines of the Greek revolution is evident in the title of one of her most recent biographies, *The Bouboulina of the Occupation*.[21]

Iro Konstantopoulou joined the resistance youth organization EPON (United Panhellenic Organization of Youth) in 1943 when she was a high school student. In July 1944 she was arrested and tortured. While in jail she became close to Lela Karagianni, who encouraged her through the difficult days. Together with other patriots, the seventeen-year-old Konstantopoulou was executed on September 5, 1944, in Kaisariani, Athens. The report of her death mentioned that, to set an example, she received seventeen bullets, one for each year of her life. A movie about her, *Seventeen Bullets for an Angel* was made in 1981; to emphasize her youth, it depicted her wearing her school uniform. Interestingly, she was awarded the "title" of heroine in 1977, before the political change of 1981, when the Academy of Athens, a traditionally conservative institution, honored her memory for her ultimate sacrifice, acknowledging indirectly the contribution of EPON to the resistance. The official state recognition took place the next year, when the parliament passed the National Resistance Act.[22]

Women in the Resistance came from all walks of life, from the cities to the countryside, from the educated to the illiterate. Some were members of the Communist Party before the War. One of these was Electra Apostolou, who had joined the Communist Party Youth in 1926 and quickly became a leading figure, representing the Greek Communist Party at the Antifascist and Antiwar World Conference of Women in Paris in 1935 and the World Congress of the Communist International Youth in Moscow in 1936. She left home when she was seventeen, abandoning a fairly comfortable bourgeois life. She was imprisoned during the Metaxas dictatorship in 1936 and in 1939, she gave birth to her baby girl in prison. They were both sent to the tiny, isolated island of Anafi (it is still two hours by boat from Santorini), from where she escaped in 1942 to join the Resistance. In 1944, she was arrested, tortured, and executed by the Nazi security forces and Greek collaborators in Athens. Electra Apostolou was a pioneer in the sense that she promoted the aim of establishing equality of women, not only in society and in law but in each family.[23]

The road that Electra Apostolou paved for Greek women would be soon followed by several other women in the Resistance through their participation in the ELAS (Greek People's Liberation Army) and the EPON during the Occupation. The most important representations of these women in arms came political photography, and especially the work of Spyros Meletzis. His work on women fighters of the resistance became known following the photo exhibition at the EAM (National Liberation Front) headquarters in Athens in October 1944, soon after the liberation of the city.[24] One of the women that Meletzis's lens immortalized was Eleni (Titika) Panagiotidou, who he selected as "the face of the Resistance." One of Meletzis's photographs adorned the first stamps that commemorated the Resistance in 1982. Panagiotidou's story became widely known when she died in October 2018.

The carefully staged and posed photograph shows Panagiotidou, ahead of a man; firm, they are both well-dressed, even though she seems to be wearing a slightly lower-quality uniform. The couple are young—probably in their twenties—and the photograph is taken from below, to increase the impact of the image.

Figure 6.2 Spiros Meletzis, "Determined for everything," featuring Eleni Panagiotidou and an unknown soldier from the collection *With the Partisans in the Mountains*, Agrafa: 1944, 242.

Meletzis was not alone in creating heroic representations of female Resistance fighters. Around the same time, in 1944 the images of the Youth Resistance organization EPON represented women equally, next to young men, waving flags, climbing stairs toward the sun, inspiring women during the grim days of the Occupation. The resistance was becoming the great opportunity for emancipation for Greek women.

The last case of group women warriors to be discussed were the women fighters of the Democratic Army (DA), the Communist Party Army that fought against the national government in the Civil War that began in 1946, after the "white terror" in 1945 and early 1946 and the elections of April 1946 from which the Communist Party abstained. Most DA women fighters were teenagers or young women when the ELAS army, the resistance army during the Occupation, allowed women to participate in novel ways. Photographic representations of anonymous women fighters show how images of women fighters were used by newspapers and political organizations to deliver their message. Since the early-twentieth century, photographs of armed men had been prominent and circulated in postcards, newspapers, and other forms of popular literature. They were often accompanied by captions extolling their courage, honor, self-sacrifice, loyalty, solidarity, and bravery. But until the 1940s, photographic representations of women were limited. When photographed at all, they were usually depicted as martyrs or victims of war, not heroic fighters. The Civil War changed that. Photographs of women during the conflict were often based on existing stereotypical postures and ideas about courage, honor, sacrifice. Such images and references are abundant in accounts of the Greek revolution and in the few images that foreign painters produced. The women who fought in the Civil War, however, were not presented as descendants of the Souliotisses and Bouboulina. From an aesthetic point of view the legacy of Meletzis is paramount; shooting from below, depicting women with their heads looking up became the standard image of the struggle of women in the Left. Reactions to those photos ranged from enthusiasm and pride to serious concerns to a moral panic.[25] The Resistance opened a space for women to challenge existing norms and traditional female roles in family and society. In the 1940s and especially during the Civil War, women were present everywhere: enlisting, fighting, nursing, in prison, being executed. Women were not merely "on their men's side" as the traditional narratives from the Greek Revolution suggested, but were fighting equally.

The two sides in the Civil War presented women very differently. On the government side, they were not shown fighting on behalf of the national army, or even assisting them. The government-friendly press published photographs of women that projected family values and virtues, such as raising children, participating in events supporting the army fighting the "insurgents," all neatly dressed, smiling, and full of joy from performing their—gendered—national duty. In contrast, women in the mountains of "free Greece," as it was called, earned their right to political participation by enlisting and becoming party members. This was a space from which, in normal circumstances, they were excluded, both constitutionally and socially.[26]

One of the first representations of women in arms after the liberation was in a left-wing newspaper, *Eleftheri Ellada* [Free Greece], in 1946, before the outbreak of the Civil War. The composition consists of two images that illustrate the transition. The first photograph shows a woman ironing wearing an apron, looking down, occupied in her chores and almost submissive in her traditional role; the second shows women fighting but relaxed, almost casual in their military uniform and bearing arms, two of the three shooting with a machine gun. This is an image that was reproduced in the 2009 film by Pantelis Voulgaris, *Psichi Vathia* [Deep Soul], one of the most emblematic, if stereotypical, depictions of the Civil War in film, that won both acclaim and received severe criticisms for appearing to take equal sides in the conflict—Voulgaris is on the Left—and for striving an as much as possible "realistic" representation of war.

Women organized themselves during the short post-liberation peaceful period. In 1946 the Panhellenic Federation of Women was founded by thirty-five women's groups representing tens of thousands of women. Many women entered the Civil War having organizational experience; some had military experience as well. It is estimated that around 30 percent or 8,000 fighters of the Democratic Army in the Civil War were women, including 688 officers. When British photographer Nancy Crawshaw visited Greece for five months in 1947 between February and July 1947 she went to the stronghold of the partisans in Kastanofito, near the city of Kastoria in North West Greece, where she photographed young women fighters of the Democratic Army. Crawshaw noted that: "the partisan movement attracted many women who believed that they would gain equal rights with men, and who suddenly deviated from the traditional labour they did for centuries, as farmers, confusing in their mind the sound of the machine gun under men's power with true emancipation."[27] Fifty-five DA women were decorated for bravery and 244 for heroic action. Based on autobiographical accounts as well as on reports from other participants, there is no doubt women fighting in the Civil War instilled courage in their comrades in the mountains and shocked and outraged their opponents in the national army. Women were given equal roles in the hierarchy and fought with equal determination for their cause.

Needless to say, many were killed, injured, imprisoned, summarily tried and executed, or exiled and spent decades as political refugees in Communist countries. One of them, Eirini Gini, Slavophone from the village of Xanthogia (Rousilovo) in North Greece, was a teacher before the war and is known for being the first woman that was be executed because of her participation in ELAS, the Resistance army, and for being a member of the Communist Party. Because of the tainted reputation of Slavophone Greeks in the region, who followed the Communist Party line on autonomy for the region of Macedonia, Gini has been elevated to the status of war hero in the Republic of North Macedonia, where there is a statue of her in Bitola. Thus, a Greek warrior who was "expelled" from the national narrative in the post-Civil War period because of her beliefs and her pronounced ethnic identity was "adopted" by a neighboring state for its own nationalist narrative. Another woman, Olga Mastora Psarogianni, who I interviewed in 1999, fought with the communists in the Resistance and the Civil War, was wounded, and found herself in 1948 in Yugoslavia. Olga lost touch with her husband but they reunited in Hungary in 1951; she, however, refused to give up and although she had a two-year-old daughter, she went to Greece for three years to reorganize the clandestine Communist network in the country; if caught, she would have been executed for treason, since the Communist Party was outlawed in 1947. Her story represents a very different kind of fighter, as it transpired in her autobiography after she returned to Greece from Hungary in the late 1980s.[28] Thousands of other women fought in the Civil War and were immortalized in the writings of the Communist Party and many who survived lived in exile for decades in communist East European countries and the Soviet Union before being allowed to repatriate in Greece.

The Democratic Army of the Communist Party managed to print a series of booklets specifically about women and their role in the struggle. Some of those writings propagate the image of women fighter—sometimes called heroine—repeatedly. The publications aimed at condemning the atrocities against women and even teenagers in the villages of Kastoria, a region with many Slavophones who supported the Democratic Army. In 1948 Roula Koukoulou, wife of Communist Party General Secretary Nikos Zachariadis, presented her account of the women fighting on several fronts against the enemy in her speech at the first meeting of the Panhellenic Democratic Union of Women. Despite such declarations, however, it is more likely that the morale of women as well as of all fighters in the Democratic Army gradually deteriorated from 1948 onward, since despair and the appalling living conditions in the mountains took their toll.[29] As the war dragged on, the publications of the Communist Party continued; in March 1949 on the occasion of March 8th, the International Women's Day, the Democratic Army published the booklet "Women Fighters of Freedom," which presented the stories of several women fighters, their struggle and achievements and their determination, against all odds, to defeat the enemy and achieve class victory and female emancipation.[30]

In 1948, as the Communist Party insurgent army was losing ground, the photographs published in government-friendly newspapers such as *Empros* aimed to shock the public and humiliate the captives. The image of a fourteen-year-old girl looking understandably angry was used to "name and shame" her, as her name and village of origin were published; "such a disgrace of a young girl is unimaginable," the caption of the photograph read. The article "revealed" other information about the appearance and characteristics of the women partisan fighters. According to the reporter, partisan women had lost all their femininity and their appearance was "disgusting." Another newspaper, *Ethnos*, published a "report" on the "problem of partisan woman." The newspaper outlined "the problem" focusing—literally—on the female faces while the article fantasized about the alleged sexual promiscuity, the sensuality, and the sexual life of the beautiful women who, it claimed, the partisans had abducted from the virgin Greek countryside and turned them into concubines. The photographs aimed to stress the "point" about the abilities of partisan women to seduce as well as die for their "communist cause."[31]

Conclusion

This chapter has argued that the representations of women warriors as national heroines depended on the context in which their actions took place and on the political conditions that followed. In the period after the 1821 revolution that led to the creation of an independent Greek state, the national narrative elevated few women to the status of national heroines even though they gave their property rather than their lives and did not always take part in the fighting. Bouboulina, the most famous of all, epitomized the woman warrior of the revolution. Other, anonymous, women were portrayed as having chosen to die instead of surrendering to "the Turks" because their story fitted the national imagination. Almost a century and a half later, the Zaloggo legend and the song that accompanied the play acquired a new meaning when it was declared a piece of commemorating the national drama.

In the 1940s, resistance to the German-Italian occupation led to the movement of thousands of women who saw their aspirations for emancipation realized through the ranks of the Resistance organizations. The legacy of the heroic women of the War of Independence resonated in this mew context. Lela Karagianni named the resistance organization she formed "Boubloulina" and women of the Resistance were referred to as the "women of Zaloggo," and the "present-day Souliotisses." Women imprisoned for treason and executed during the Civil War went to their deaths singing the "dance of Zaloggo" song. In the 1940s, however, 120 years on from the Greek Revolution, women were not dancing their way to death jumping off a cliff, but were fighting in uniform with a rifle (or even a machine gun) in their hands, testament to the great strides in women's emancipation that was achieved during the 1940s. Images of those women warriors circulated widely thanks to political photography that popularized and glorified the women of the Resistance. In the 1950s onward popular images in film and monuments of women warriors served the national narrative of the Resistance and representations of women were brought up to date with the place of women in Greek society.

Notes

1. Adrian Shubert, "Women Warriors and National Heroes: Agustina De Aragón and Her Indian Sisters." *Journal of World History* 23, no. 2 (2012): 294. The case of the Greek heroine Bouboulina is similar to Spain's Agustina. There is however a frequent representation of *Hellas*, Greece, as a woman, but not a warrior one.
2. Shubert, "Women Warriors and National Heroes," 279-313. http://www.jstor.org.ezproxy.library.yorku.ca/stable/23320150 (accessed October 13, 2019). The special issue in EHQ concerned mainly with national hero cults. Robert Gerwarth, "Introduction," *European History Quarterly* 39, no. 3 (2009): 381-387. doi: 10.1177/0265691409105058.
3. For a recent account, see Efi Avdela, Thomas Gallant, Nikolaos Papadogiannis, Leda Papastefanaki, and Polymeris Voglis, "The Social History of Modern Greece: A Roundtable," *Social History* 43, no. 1 (2018): 105-125, doi:10.1080/03071022.2018.1394037
4. The issue relates to the place of the history of the 1940s (the period of the Occupation and Resistance 1941-44 and Civil War 1946-49) in Greek society and collective memory and public history (more recently). The literature is quickly expanding as the 1940s are by far the most popular research field for young historians today. For the monuments of the Civil War, see Kostantinos Charamis, "'Nothing and No One Has Been Forgotten': Commemorating Those Who Did Not Give in during the Greek Civil War (1946-1949)," *Cahiers de la Méditerranée*, 70 (2005): 173-193.
5. K.E. Fleming *The Muslim Bonaparte: Diplomacy and Orientalism in Ali Pasha's Greece* (Princeton, NJ: Princeton University Press, 1999).
6. T. Ginouvier. *Mavrogenie, ou, l'heroine de la Grece: nouvelle historique et contemporaine; suivie d'une lettre de l'heroine aux dames parisiennes* (Paris: Chez Delaforest, 1825). The book is a fine example of philhellenic romantic nationalism.
7. The island of Spetses and especially Hydra proved crucial for the Greek revolution because of their ships, wealth, and the men determined and trained to fight on the seas and if need be on land for the cause of independence. Bouboulina's prominence is equivalent to the prominence the island of Spetses played during the war of independence due to its powerful merchant fleet; this role was already acknowledged during and immediately after the revolution; see Alexandra Papadopoulou, "Elliniki naftotopoi sto metehmio. I metavasi ton Spetson apo tin Othomaniki Aftokratoria sto Elliniko Kratos," *Ta Istorika* 67 (2018): 116-141.
8. Leonidas Mylonakis, "Transnational Piracy in the Eastern Mediterranean, 1821-1897," PhD Dissertation, University of San Diego, 2018, https://escholarship.org/uc/item/6k16v5sb (accessed October 13, 2019).

9   Bouboulina she was also partner to several other ships. Sotiria Aliberti, *Ai iroides tis Ellinikis Epanastaseos* (Athinai: Tarousopoulos, 1933), 202. Bouboulina intended to claim some of the ancient Greek glory with the symbolism that the name of the ship carried, of the Mycenian king and leader of the Greek army that conquered Troy.
10 J.L. Comstock, *History of the Greek Revolution: Compiled from Official Documents of the Greek Government: Sketches of the War in Greece by Philip James Green: And the Recent Publications of Mr. Blaquiers, Mr. Humphrey, Mr. Emerson, Count Pecchio, Rt. Hon. Col. Stanhope, The Modern Traveller, and Other Authentic Sources* (William W. Reed & Co., 1789–1858), 1828. Howe, S.G. (1801–1876), *An Historical Sketch of the Greek Revolution*. Second edition. Printed for the author, 1828.
11 Nina M. Athanassoglou-Kallmyer, *French Images from the Greek War of Independence (1821–1830): Art and Politics under the Restoration* (New Haven, CT: Yale University Press, 1989).
12 Efi Kanner, *Emfyles koinonikes diekdikiseis apo tin Othomaniki Aftokratoria stin Ellada kai stin Tourkia. O kosmos mias ellinidas christianis daskalas* [Gender-based social demands from the Ottoman Empire to Greece and Turkey: The world of a Greek Christian female teacher] (Athens: Papazisis, 2012); E. Kanner, "Transcultural Encounters: Discourses on Women's Rights and Feminist Interventions in the Ottoman Empire, Greece and Turkey from the Mid-Nineteenth Century to the Interwar Period," *Journal of Women's History* 28, no. 3 (2016): 66–92.
13 Stefanos Xenos, *I irois tis Ellinikis Epanastaseos iti Skinai en Elladi apo tou etous 1821–1828*, vol. A (London, 1861), 32–44. Xenos's aim was to demonstrate the continuity between ancient Greeks and the heroines of his work.
14 The role of gender in the articulation with class and the nation in Greek history has been highlighted in Efi Avdela and Angelika Psarra. "Engendering 'Greekness': Women's Emancipation and Irredentist Politics in Nineteenth-Century Greece," *Mediterranean Historical Review* 20, no. 1 (2005): 67–79. Another contribution that relates gender and national identity is by Eleni Varikas, "Gender and National Identity in *fin de siècle* Greece," *Gender & History* 5, no. 2 (1993): 269–283.
15 In 1960 Manos Hatzidakis won an Academy Award for the music he wrote for the film *Never on Sunday*, directed by Gilles Dassen and starring Melina Merkouri.
16 The monument was placed on a much higher cliff than the one where women fell off.
17 Henry Miller, *The Colossus of Maroussi* (Harmondsworth, UK: Penguin Books, 1972), 40.
18 Violetta Hionidou, *Famine and Death in Occupied Greece, 1941–1944* (Cambridge: Cambridge University Press, 2006).
19 Polimeris Voglis. *I Adynati Epanastasi* [The Impossible Revolution] (Athens: Alexandreia, 2014), 63. Few studies of the Resistance discuss the role of women specifically. Hagen Fleischer, *Stemma kai svastika. I Ellada tis Katochis kai tis Antistasis 1941–1944* [Crown and swastika. Greece of the Occupation and the Resistance] (Athens: Papazisis, Two volumes 1989–95); Mark Mazower, *Inside Hitler's Greece: The Experience of Occupation, 1941–44* (New Haven: Yale University Press, 1993).
20 Karagianni's son has said in a documentary interview that she only waved goodbye to her children who were imprisoned with her. Film Documentary by Vassilis Loules. "Synantiseis me ti mitera mou: Lela Karagianni" [Encounters with my mother], 2005.
21 Zaousis A. Alexandros, *Λέλα Καραγιάννη, η Μπουμπουλίνα της Κατοχής (1941–44)* [Lela Karagianni, the Bouboulina of the Occupation] (Athens: Ωκεανίδα, 2004).

22 The proliferation of monuments is extraordinary; between 1975 and 2003, there were 345 monuments built for the national resistance and the Civil War in Greece. Charamis, "'Nothing and no one has been forgotten': commemorating those who did not give in during the Greek Civil War (1946–1949)," http://journals.openedition.org/cdlm/915 (accessed October 13, 2019).
23 *Istoria tis Antistasis* [History of the Resistance] 1940–45 (Athens: Avlos, 1979). Iasonas Handrinos, *To timoro heri tou laou. I drasi tou ELAS kai tis OPLA stin katechomeni protevousa 1942–1944* [The people's punishing hand. The action of ELAS and OPLA in the occupied capital] (Athens: Themelio, 2012).
24 The Left in Greece was already acquainted with the photographs of Meletzis when he presented them in the EAM office in 1944 in Athens. Sparos Meletzis, *Fotografia 1923–1991* [Photography], Museum of Modern Art, Vienna 1992, 46.
25 The Spanish Civil War (1936–39) was the first conflict that produced large numbers of photographs of women in arms. Caroline Brothers, *War and Photography. A Cultural History* (London and New York: Routledge, 1997).
26 Tasoula Vervenioti, *I ginaika tis Antistasis: i eisodos ton ginaikon stin politiki* [The women of the Resistance: The entry of women in politics (Athens: Odysseas, 1994).
27 Nina Kassianou, "Gynaikes sta opla" [Women in arms], 157.
28 Olga Mastora-Psarogianni, *Sto drom tou chreous* [On the road of duty] (Athens: Sychroni Epochi, 1995).
29 For the argument on the morale of the Democratic Army fighters, see Nikos Marantzidis, *Dimokratikos Stratos Elladas 1946–1949* [Democratic Army of Greece] (Athens: Alexandria, 2010).
30 *Machitries tis Lefterias* [Women Fighters of Freedom], Eleftheri Ellada [Free Greece], 1949.
31 Newspaper *Ethnos*, September 2, 1948.

Part Three

Gender Fluidity

7

Madeleine de Verchères (1678–1747): Woman Warrior of French Canada

Colin M. Coates

"Each people has had its women warriors," concluded a discussion of the phenomenon of Amazons in a nineteenth-century edition of Voltaire's *Dictionnaire philosophique*, the examples primarily from the Arab and French traditions.[1] The entry could have included an example from France's colony in North America. In the 1690s, Madeleine de Verchères situated the story of her bravery in defending her family's fort in New France in the broad context of the woman warrior tradition.[2] In her own words, she tested the boundaries of gendered military action, thus helping future generations of historical enthusiasts insist on the worthiness of Canadian history through the fact that the country could boast its own woman warrior—at the same time that the actions she claimed to have performed complicated their narratives of gendered historic bravery.

Seven years after the event, Verchères informed Mme de Maurepas, the wife of Louis XIV's Minister of the Marine, that she, at the age of fourteen, had defended her family fort against a Haudenosaunee (Iroquois) attack. In making this claim, Verchères drew on a contemporary act of female heroism in the hopes of receiving a pension. She used imagery and language that evoked the woman warrior tradition in France, in particular pointing to her adoption of male clothing. Her narrative along with some related versions allowed her to become, eventually, the archetypical woman warrior of French Canada, even Canada as a whole. From the mid-nineteenth century, poets, historians, and artists would invoke the better-known figures from French history, Jeanne Hachette (Jeanne Fourquet, the defender of Beauvais in 1472) and Jeanne d'Arc to provide a broader cultural context for her heroism. Her story would inspire a statue erected in her honor in the town of her fame, and a decade later the first feature-length French-Canadian film. Verchères's own story constrained the retellings to some degree, although writers and artists tended to suppress or ignore elements that they found difficult.

This paper is based on material in Colin M. Coates and Cecilia Morgan, *Heroines and History: Representations of Madeleine de Verchères and Laura Secord* (Toronto: University of Toronto Press, 2002).

Verchères's first narrative, dated October 15, 1699, was a short letter petitioning the wife of Minister of the Marine to accord either a pension of 50 écus, like that accorded to wives of officers, or failing that, an ensign-ship for her brother.[3] (The king's officials granted her this pension.) The narrative was direct and clearly organized. After citing the Count de Maurepas's munificence toward the distant colony of New France in the context of the decades of hostility between the French and Haudenosaunee, she provided a summary of her act of bravery. The Haudenosaunee assault on her family's fort, situated eight leagues to the north-east of Montréal, on the south shore of the St Lawrence River, caught the French settlers by surprise. The Indigenous attackers captured about twenty of her compatriots. Verchères herself was four hundred paces outside the fort. She ran quickly back, only barely escaping the clutch of a warrior by unknotting the scarf ("mouchoir de col") around her neck at the precise moment that he had grasped it. Closing the gate behind her, she cried "aux armes" to the inhabitants of the fort, and ignored the women who lamented the loss of their husbands. She mounted the bastion, donned a soldier's hat, and shot the cannon to frighten the Haudenosaunee. This act alerted the neighboring forts to the danger they were facing.

The location of this act of heroism was a fairly new French settlement to the east of Montreal. Although French claims to the St Lawrence Valley dated back to the sixteenth century, royal interest, immigration, and agricultural expansion remained limited until the 1660s. François Jarret de Verchères received seigneurial title to a small fief fronting on the St Lawrence River in 1672, and French settlers began clearing land in Verchères shortly thereafter. Like other settlements in the plains around Montreal, it was located on traditional Indigenous territories and subject to occasional military excursions by the Haudenosaunee, in conflict with the French for many decades. These raids fit into a complicated geopolitics involving opposing alliances, competition for trade and retribution for previous attacks.

With the arrival of the Carignan-Sallières regiment in the colony in 1665, and their eventual encouragement to settle in the colony, many of the men of this generation, like François Jarret de Verchères, were linked to military activity. Between 1663 and 1673, the French government sent over about 800 women from France to help populate the colony and establish an agrarian economy. These women were not usually involved in military feats. Indeed, Verchères recognized how her act of bravery tested gender roles. The third sentence of her letter placed the conundrum at the center of her story: "Although my sex does not allow me to have other inclinations than those expected of me, permit me nonetheless Madame, to tell you that I have the same feelings that draw me to glory as do many men."[4] The wording of Verchères's letter was carefully chosen, with a register of language that one does not normally see in correspondence of the period. Taking on a male guise was not a simple change. Rather, she "metamorphosed" ("je me métamorphosay") in donning the soldier's hat. Verchères also invoked contemporary women who had taken on similar roles in contemporary France. "I know, Madame, that there have been in France persons of my sex in this last war, who placed themselves at the head of their peasants to oppose the invasion of enemies into their province."[5]

This reference was undoubtedly to Philis de la Charce (Charsse), daughter of the Marquis de la Charce in Dauphiné in southwestern France. In the same year as

Verchères's heroic act, Catholic convert Philis de la Charce led a group of provincial inhabitants in opposing a Savoyard invasion during the War of the League of Augsbourg. De la Charce's action was that of a noble, 47-year-old woman. The news of de la Charce's bravery reached the court of Louis XIV, and in the same year, she visited Paris, and importantly—for the comparison to Verchères—was accorded a pension. The court publication, *Le Mercure Galant*, reported a short version of the story in late 1692.[6] Perhaps the notice of this act of heroism took some time to filter to the colony, where communications were restricted to a limited number of sailings and the months during which the St Lawrence River was not frozen.

It is likely that de la Charce's story was the source that inspired Verchères, with the assistance of the colonial official and author Claude Charles Le Roy de Bacqueville de la Potherie, to petition Mme de Maurepas. Bacqueville de la Potherie included two separate versions of the story in the historical account of the colony he later published: both containing much of the same wording as Verchères's missive to Mme de Maurepas. Bacqueville de la Potherie's tomes were not published until 1722, but he probably finished working on the manuscript around 1714. He explicitly claimed that he had assisted Verchères with her letter and he also used the verb "se métamorphoser." There is the possibility of some similarity between the account of de la Charce's actions in the official publication and Bacqueville de la Potherie's first version: (*Mercure de France*: "on ne doit pas passer sous silence le courage & l'intrepidité..."; Bacqueville de la Potherie: "Je ne sçaurois passer sous silence l'action heroïque de Mademoiselle de Vercheres"[7]), although this banal wording cannot provide conclusive proof of the influence.

In any case, Verchères had invoked the woman warrior tradition in justifying her appeal for a pension. She conveyed her transgressive act in donning male apparel, underlining the change with the verb "métamorphosay." She created the framework for portraying gendered aspects of her own heroism: the surrender of her kerchief, an action which protected her virginity; she distinguished herself from the women around her by ignoring their cries and changing her appearance; she represented her region, in this case colonial New France. Like their counterpart in the Dauphiné in metropolitan France, "Canadian women would not lack the passion to reveal their zeal for the glory of the King if they found the occasion to do so."[8] Verchères was a fitting settler colonial addition to the French tradition of woman warriors.

Around twenty-seven years later, Verchères recounted a longer version of the events at the behest of the new governor of the colony.[9] This version highlights Verchères's courage in rallying the French inhabitants of the fort, particularly her younger brothers ("Let us fight to the death. We fight for our country and for our faith. Remember the lessons that my father has given you so often, that gentlemen are born only to offer their blood in the service of God and the king."[10])As she fired muskets and cannons, Verchères's sang-froid distinguished her from other frightened women. She opened the gates three times: to let in livestock returning to the fort; to escort to safety Pierre Fontaine, his frightened Parisian wife Marguérite Antoine, and their two children who had arrived by canoe; and to rescue laundry which had been left by the river. The Haudenosaunee warriors maintained their siege, although they did not undertake any military action after the initial surprise attack, for eight days. To this detailed account,

Verchères added another act of heroism, dated to 1722, when she saved her husband from a confrontation with a small group of visiting Abénaki (normally allies of the French), and was herself rescued by her twelve-year-old son from the Abénaki women who threatened to throw her into the fire. Verchères's account ends with a panegyric praising the new governor, but without a specific request for an increase in her pension.

Six versions of the stories were recorded during Verchères's lifetime: two of them signed by herself, two contained in Bacqueville de la Potherie's books, and brief evocations of the stories appeared in Pierre-François-Xavier de Charlevoix's book, itself in part based on Bacqueville de la Potherie, and a manuscript report of a large-scale survey of the colony. Charlevoix made the clothing change more radical than in the other accounts: "She began by taking off her headdress, tying up her hair, and donning a hat, and a 'justacorps' (jerkin)," and she changed outfits from time to time. The author preceded this story with a similar act of heroism displayed by Verchères's mother in 1690. He referred to the women as "two Amazons."[11]

And then, after her death in 1747, her story was largely forgotten. I located few references to her story in the century or so between her death and rise of the national sentiment in French- and English-speaking Canada in the mid- to late-nineteenth century. These are fairly sparse, and even the man credited with being the first novelist of French Canada, Philippe Aubert de Gaspé, a direct descendant of Madeleine de Verchères, did not know her story very well when he recounted it in an 1866 publication.[12]

But, decades later, in the tradition-building period of the late-nineteenth and early-twentieth centuries, Verchères's story found favor again. In many parts of the Western world, elites identified stories and customs that could embody their nation and champion the nationalist goals to which they aspired. This process was especially relevant in countries whose constitutional frameworks were in the process of redefinition, such as Canada. By referring to traditions and historical precedents, the elites could justify and enhance the aspirations to nationhood. In sub-state groupings like French Canada, which distinguished itself by language and religion from the Anglo-Celtic and predominantly Protestant settlers who surrounded them on the continent, a national hero or heroine provided further justification of their uniqueness. With the establishment of the Dominion of Canada in 1867, French Canadians achieved control, within Quebec province, of key institutions such as education. Quebec would henceforth become the bulwark for French-Catholic culture in North America. Elites cultivated the expression of Québec's Catholic values, and French-Canadian politicians from a range of political positions defended the distinctiveness of the province.

The process of commemoration often involved the discovery or rediscovery of stories that had fallen out of common knowledge, the "invention of tradition" in the famous phrasing of Eric Hobsbawm and Terence Ranger.[13] Researchers identified Verchères's original letters in archives in France and reprinted them for audiences in Canada. The specific details of her narratives undoubtedly appealed to many readers, but it was more the type of story that Verchères personified that mattered. This was a standard national story of a heroic figure ostensibly saving her or his community from an invading or attacking force and thus coming to personify the nation-state. The woman warrior tradition integrated women into such national narratives, allowing them to play an active role in what was usually a male-dominated historical drama. These stories could appeal to a broad cross-section of readers, including school-children being inculcated with respect for the developing nation-state and for some specific virtues. That many of the exponents of these stories actually reflected fairly conservative values themselves provides an ironic commentary on the historical women who tended to recognize that they were taking a novel role.

While there was certainly a range of artistic achievement involved in the construction of the new narratives, in the case of Madeleine de Verchères, most of the retellings of the stories tended to be fairly prosaic in their choice of language and metaphor. On the one hand, the commemorators expressed concern that the stories they had to retell would be forgotten. But at the same time, they tended to repeat the stories that their predecessors had written. Subsequent writers and artists would largely rely on the contemporary narratives, making some choices in resolving anomalies between them. But there was not a great deal of inventiveness in the accounts which appeared in the late-nineteenth and early-twentieth centuries.

One example of a textual recounting of Verchères's story illustrates many features of this commemorative process. In 1898, the poet and doctor William Henry Drummond published a chapbook containing two short poems on French-Canadian themes. Drummond had carved out a reputation as a popular poet during this period. He was particularly well known for his use of a form of vernacular French-Canadian-inflected English. The first poem in the chapbook, "Phil-o-rum's Canoe" reflects that approach, beginning "O ma ole canoe, wat's matter wit' you,/an, w'y was you be so slow?"[14] Drummond initially attempted using this approach in drafting the poem on the heroine's story: "She has pass down by de reever/She has pass down by de shore ...," but he discarded the accented English for the final version.[15]

The published poem, "Madeleine Vercheres," recounted in standard English, with the occasional, easily grasped French word, is a fairly typical retelling of the story. The poet justified his choice of topic, recounted to a child, by the fact that he had not explained it earlier. Rich with references to the weather and associations of the Iroquois with "tigers" and a hurricane which "rends the air," Drummond's poem has Verchères rallying the small group of defenders of the fort. She is outside the fort when the attack occurs, but outruns the attackers, apparently without having to surrender her scarf. (The poet assured the reader twice that Verchères is a "maiden.") Drawing inspiration largely from Verchères's second description of her heroism, Drummond has Verchères and her small band of supporters defending the fort for a week, until French troops finally relieve them. The leader of the troops, "gallant De la Monnière" is stupefied by her courage: "He stood for a moment speechless,/and marvelled at woman's ways." Drummond ends the poem with the pious wish that "we in Canada may never see again/Such cruel wars and massacre.../As our fathers and mothers saw, my child, in the days of the old régime."[16]

In Drummond's version, Verchères's story becomes a heroic tale of the frontier, and it found an enthusiastic readership. Celebrated French-Canadian poet Louis Fréchette expressed his admiration for Drummond's poem: "It is a beautiful effort," Fréchette enthused, inviting his English-speaking colleague to visit him in the village of Varennes, next door to Verchères. The secretary of the governor-general conveyed newly arrived Lord Minto's appreciation for the gift of the chapbook: "His Excellency feels that these poems will give him an insight into French Canadian life which he could not otherwise hope to obtain."[17] Newspapers across North America praised his achievement. The Chicago *Chronicle* wrote of the heroic story, "Dr. Drummond has embalmed it in swift, rhythmic, throbbing, heroic verse that will hardly fail to carry it far into the future."[18] The faithfulness of the retelling to Verchères's second narrative is most striking. There were no inventive deviations from the established narrative. The repetition of the story of heroism was its own gauge of usefulness. Commemorators of Verchères's heroism, both men and women, felt comfortable recounting a tale that, in fact, many had covered before. Prose and poetic accounts begat imitators.

Alongside the many poems, plays, short narratives, and artistic renderings, two particular efforts stand out, if only because of the costs involved: the statue erected in 1913 in the town of Verchères and the film *Madeleine de Verchères*, the first French-Canadian feature-length film, released in 1922.

Statue of a Heroine

Both the statue and the film faced the same problem: how does one visually depict a cross-dressing heroine? In the context of the early twentieth century, the goal must be to emphasize her femininity, at the same time indicating that she could pose enough of a military threat to frighten Haudenausonee warriors. The artist was not constrained by any historical evidence. No contemporary images of Verchères survived—likely none were ever done. Therefore, it was equally valid to depict her a slight blonde girl, or a stockier dark-haired young woman. In theory, she did have to be fourteen years old, although many artists simply avoided the narrative constraints that her age might have placed on their retelling of the story. And a young woman of fourteen could, of course, range a great deal in height.

The artist whose work would eventually take the form of the statue at Verchères, Louis-Philippe Hébert, admitted the difficulty that he faced. Having made some earlier versions, Hébert completed a statuette in her honor in 1908, in a series of figures from the heroic past of Québec. He confided to French journalist Maurice Hodent, how he had struggled to depict her. He wanted to translate her "beautiful energy" but did not want to make her into a "virago."[19] Yet a virago is precisely what she was, and how she had depicted herself. Hébert's statuette did not display an amazon, but rather a pre-pubescent girl, holding a gun almost the same length as her height. The gun pointed toward the ground, an unthreatening gesture. Viewers need have no fear that the gun was pointed toward them. As was fairly typical of the depictions, she wore a kerchief, even though Verchères's own narrative insisted that she had left it in a Haudenosaunee warrior's grasp just prior to entering the fort and grabbing a gun. Therefore, she could not logically have the kerchief and the gun at the same time. Although Verchères left the kerchief behind, it became embedded in the story. It would have been entirely possible to create an image of Verchères without the scarf, but artists preferred to depict it.

To account for her heroic action, Hébert accepted the part of the story where Verchères put on the male hat, and of course he included the rifle as well. Fewer artists

Figure 7.1 Gerald Sinclair Hayward, "Madeleine de Verchères," Library and Archives Canada, Acc. No. 1989-497-1, C-083513.

showed Verchères firing a cannon, a feat what would have undoubtedly been more complicated for a statuette, or possibly even a full-size statue. Verchères's first narrative made no mention of her own firing a rifle, though her second version did. Hébert was happy with the result of his efforts. It captured his intent in rendering visible female heroism: a young woman "transfigured by the idea of saving her people and defending the home of her birth."[20]

Hébert was undoubtedly even happier once his statuette received influential backing. Lord Grey, governor-general of Canada (the English-born representative of the monarch) and a keen promoter of cultural development in Canada, saw in the story of Madeleine de Verchères an opportunity to encourage Canadians' pride in their relatively young country—and to develop a story to bridge the English-French divide that had defined so many of the political arguments in the country. His daughter expressed interest in her heroism, and in 1909 she proposed raising money from schools and women across the country for a 25-foot-high statue in her honor.[21] This campaign never took shape, but it pointed to Grey's fixation with statues.

A couple of years earlier, Lord Grey had himself failed to generate public enthusiasm for a statue at Québec City, the point of disembarkation of many people who arrived in Canada by ship. Grey wanted a "great Goddess of Peace" to greet the ships, rather than the dispiriting view of the provincial gaol.[22] This grand endeavor did not succeed, but the more modest Madeleine de Verchères statue would see the light of day. Not shying away from hyperbole, Grey believed that Verchères's monument would rival the Statue of Liberty, and "give a message to every immigrant going up the St Lawrence to Montreal, of courage, energy, loyalty and patriotism" in contrast to the message of license and individual freedom that the American icon conveyed.[23] He convinced cabinet minister Rodolphe Lemieux to take the statuette to Québec City to a meeting with the premier of the province, place it on the supper table, and ask him to fund a larger-scale statue that would attract attention to the small town of Verchères and "tell the immigrant that the heroic virtues are the bedrock foundation of Canadian greatness."[24] The provincial government was not forthcoming with its support.

But Grey continued his lobbying, and in the run-up to the federal election of 1911, that Sir Wilfrid Laurier's Liberal government would lose, the public works minister William Pugsley committed the $25,000 necessary for the statue.[25] The official provision of funding for the statue was one of the last acts of Laurier's government, signed a few days after their election defeat.[26] Hébert then headed to France to cast the bronze statue. The larger statue would take up the features of the statuette—a young woman, likely somewhat older than fourteen, clutched the long gun, her skirt swirling in the breeze. Again, the gun and the hat were the two hints of the challenge to gender roles. But no one, by design, would mistake Verchères for a man, a feat that she may have managed in 1692, but could not perform in 1913.

It is possible that even without this high-level support for Verchères's story, a statue would have eventually been erected in her honor in the town that shares her family name. The local priest, Abbé François-Alexandre Baillairgé, an indefatigable civic booster, like many priests of the time period, had tried to encourage his parishioners to raise money for the cause. Through tombolas, his work had achieved some success, but Baillairgé was still some way from the amount needed for a statue of the size that eventually was erected. According to the careful accounts that he maintained, he had raised some $2,500 by July 1912.[27] We can't know what design Baillairgé would have preferred, but he was happy enough to welcome the federal government's statue a year later.

By the time the statue was ready, Lord Grey had returned to the UK, and the Laurier Liberals had gone down to defeat, even in the constituency of Verchères. Abbé Baillairgé presided over the inauguration committee, and he was given the opportunity to arrange the unveiling. A special train brought dignitaries from Montréal. Newspaper reporters claimed that some four to five thousand attendees listened through rainy September weather to the series of speeches that celebrated the heroism of the French-Canadian past in the lengthy ceremony that began at 3:00 pm and lasted until fireworks at 7:00 pm. Some speeches acknowledged Lord Grey's initiative in promoting Anglo-French rapprochement, but most emphasized the distinctive French-Canadian character of the event. Son of the parish, House of Commons translator Wilfrid Larose used another French heroine to justify the commemorative event: "If the kingdom of France was delivered and regenerated by Joan of Arc, this colony, then French in its cradle, was exemplified by Madeleine de Verchères." Judge Amédée Geoffrion, a former local member of the provincial legislature, referred to Verchères as a "peasant" (paysanne) who merited being called "la Jeanne Hachette du Canada," after the fifteenth-century defender of Beauvais. Prominent businessman Guillaume-Narcisse Ducharme claimed that "a people which produced children like Madeleine de Verchères had every right to be a considered a brave people."[28]

Figure 7.2 Statue of Madeleine de Verchères, Verchères, QC, about 1925. Archives de la Ville de Montréal, BM42-G0464.

Situated close to the shore and therefore prone to springtime flooding, and erected on a faux stone battlement, unlike the wooden fort that Verchères referred to in her narratives, it was a striking statue, but at one-sixth the height of the larger female figure, it could never be as iconic as its New York City rival (Verchères: 7.2 n; Liberty: 46 m., 93 m. including the pedestal). Sculptor Hébert described it later as a "mute sentinel evoking a past of glory and suffering."[29] It would be difficult for people on board ship to make out the statue as they sailed to Montréal, unless they came very close to shore, as an employee of the federal government board responsible for the statue remarked a few years later.[30] The statue functioned as a relatively minor tourist attraction, the numbers of visitors tracked, possibly fairly haphazardly, by successive caretakers from the 1930s to the 1950s: from about one thousand during the summer months from June to September in the 1930s to about two thousand per summer in the 1950s.[31] It is unlikely that people spent a great deal of time at the statue, and one wonders if many visited more than once. Nonetheless, the promoters had succeeded in part of their goal: they had fixed in metal and stone an image of the youthful heroine of New France. Although the depiction showed her performing an act that might be unexpected of a woman, in that the young woman was wielding, unthreateningly, a rifle, this was in no way an unsettling image. As visitors would gaze on the statue from ground-level, they would first of all see the swirling skirts, before they could make out the maleness of the hat on her head. The statue of Madeleine de Verchères was more woman than warrior. The French journalist who saw the statue in the Paris workshop where it was cast predicted how it would look in situ: "More than one Canadian seeing it as he sails up the St. Lawrence will say that there is no prettier girl under the blue sky."[32]

Film

The first cinematographic portrayal of Verchères's story likewise emphasized her femininity over her military prowess. Filmed and released less than a decade after the erection of the statue, Abbé Baillairgé again played a supporting role in the project. An author on a wide range of topics of contemporary concern, he published a short pamphlet in 1913 retelling the story of Madeleine de Verchères as a fund-raiser for the local museum at Verchères. Although the script-writers for the film did not adhere closely to Baillairgé's version, they did read his pamphlet, and ultimately changed some of the film's inter-titles to provide a greater degree of historical accuracy to the narrative.[33]

We shouldn't exaggerate the historical accuracy of the film, however, nor should we use that as the main measure of the film's value. Even as an example of early silent film, this was a fairly amateurish production, but it was the first feature-length "fiction" film produced by French-Canadians. The film industry had taken off in the early decades of the twentieth century throughout North America, and many social commentators expressed great concern about the immoral stories that found their way to the screen. At the same time, a large public streamed into the cinemas, unperturbed by those concerns. The Catholic Church in Québec was among the strongest opponents of the

Hollywood-dominated industry, often using anti-Semitic arguments. They criticized the frequent depictions of disrespect for authority that created great concern for the church hierarchy.

Given the broader concerns, cultural entrepreneurs saw an opportunity to appeal to a French-Canadian audience while producing films that would not attract the ire of the church. Writer Emma Gendron, photographer Joseph-Arthur Homier, and producer Arthur Larente teamed up in a company reassuringly named Le Bon Cinéma to produce the first French-Canadian feature length "fiction" film. The company saw their role as part of a moral struggle to counter foreign influences. In 1924, they distributed flyers to raise funds by appealing to the morality of their intended product: "it is urgent to begin fighting against the so often inept American cinema and its generally corrupting and fatally anglicizing tendencies with films based on a Catholic perspective, therefore honest and clean, and essentially French-Canadian in direction and tone."[34] But it was difficult to convince Church authorities that a purified cinema was possible. A terrible fire in the Laurier Palace Cinema in Montreal in 1927 that led to the deaths of seventy-eight children heightened anxiety about the dangers posed by film-going and launched the Church's campaign to forbid young people from attending the cinema.[35]

But prior to this disaster, one group tried to find opportunities between the concerns of the Church and the enthusiasms of the public. In 1922, the entrepreneurs behind Le Bon Cinéma turned to an archetypical story of female bravery for their first film. The even more famous known heroics of Adam Dollard des Ormeaux, and his intrepid band of fur traders, who ostensibly held off a Haudenosaunee advance on Montréal in the 1640s, had already received cinematographic treatment by an English-speaking Canadian company in 1912.[36] The same company planned to produce a film version of Verchères's heroism, but this film was never released. Another attempt mentioned in newspapers in 1915 likewise did not appear.[37] The story of Verchères offered a narrative that could have a broad appeal, given the importance of female characters in early films. Film versions of Jeanne d'Arc's story were produced from the earliest days of cinematography, with early shorts produced by 1898 and another by pioneer Georges Méliès in 1899, followed by another four before the French-Canadian film was released in 1922. Jeanne Hachette's heroism would figure in the 1924 film, *Le miracle des loups*.

No version of the *Madeleine de Verchères* film remains, as the highly inflammable nitrate-based copies were destroyed for safety reasons some decades later. But the narrative can be reconstructed from a set of stills registered at the Copyright Office (ultimately finding their way to the British Library) as well as newspaper coverage and other related documentation.[38] Filmed in September 1922, using actors from Montréal and the nearby Mohawk (one of the Haudenosaunee nations) community of Kahnawake, the schedule was compressed because of unfavorable weather conditions. Windy and snowy weather affected the quality of the cinematography. The film debuted in December of the same year. According to the later recollections of Arthur Larente, the film cost $4,412.[39] This represented a smaller investment than the statue, and indeed its impact was more ephemeral. It passed the provincial Bureau of the Censor without issue in one day.[40]

In this version, no one, except for some reason the Haudenosaunee warriors, mistook Verchères for a man. As the story required, Verchères wears a scarf, which

Figure 7.3 Still photograph from scene of motion picture *Madeleine de Verchères*, Library and Archives Canada, PA-028626.

she leaves in the hands of the Haudensonee warrior. Like the statue, the tests to her gendered role entail her donning a male hat and engaging in military activity. She never changes her dress for a jerkin. Going further, the script reasserts her gender by inserting a love story into the narrative. The film opens with scenes of domestic happiness, including a game of blind man's bluff. Verchères, her eyes covered, tags the man who would later become her husband, Pierre-Thomas Tarieu de La Pérade. When Tarieu de la Pérade leaves for Québec along with the other French troops, Madeleine de Verchères hangs a cross around his neck.

Left undefended, the fort is an easy target. The Haudenosaunee attack and lay siege to the fort. This version relies more fully on Verchères's second narrative, as it includes a scene of her rescuing the laundry left drying by the river. She also saves Fontaine when his canoe lands at the shore. When another group of French troops led by lieutenant La Monnerie relieves the fort, she greets them outside, turning over command to him. "You have saved the country," La Monnerie proclaims (wildly exaggerating the military import, if not the heroism, of Verchères's actions). Verchères replies, "I have only performed my duty." La Monnerie rescues the French prisoners, but not before the Indigenous perform an obligatory dance around them. (The producer later complained that the Indigenous actors were rather uncooperative.)[41] The film concludes with prayers of thanksgiving.

Perhaps because of the love interest introduced into the storyline, the main character was visibly older, nineteen-year-old Estelle Bélanger taking the role of Verchères. Tarieu de La Pérade appears older again. Actually, he was fifteen years old in 1692, and he did indeed have a military career. The two did not marry until 1706, some fourteen years after the defense of the fort of Verchères.

By the standards of its comparators in the early decades of filmmaking, this film did not measure up well. Newspaper reviewers and Abbé Baillairgé attempted to be as positive as they could be, wanting to encourage this local initiative. But they could not muster particularly effusive comments. One reviewer writing in the influential newspaper *Le Devoir* concluded: "one could describe it as a man admiring an amateurish portrait of his ancestors. It is more a patriotic than an aesthetic emotion."[42] The Montreal newspaper *La Presse* insisted, "It is also not necessary to focus on certain weaknesses in the film, 'Madeleine de Verchères'. They are excusable in the work of novices."[43] Similarly, Abbé Baillairgé confided to his private journal, after a showing of the film in the parish hall at Verchères, "The film about Madeleine is almost very good overall. Mlle Emma _____ [Gendron] had added many things to the story that could have happened." As a "page in the life of our fathers," the film was most appropriate for schools, convents, and colleges.[44] Criticizing American imports, the influential priest-historian Lionel Groulx recommended the *Madeleine de Verchères* film.[45]

Even the bureaucrats at the National Historic Sites Branch in Ottawa were critical of the film. Although they appreciated the effort, technically, the film was deficient. The filmmakers had not worked out continuities between the scenes. Verchères used her apron as wadding for the cannon, only to have it reappear in a subsequent scene. The clock in the fort was anachronistic, as was the shape of the palisades. They deemed the Haudenosaunee attacks "poor." The script was historically suspect, and to address some of the criticisms, the producers reworked the inter-titles and cut some of the scenes, rendering the narrative even closer to Baillairgé's pamphlet.[46] They admitted to government officials, whom they hoped would provide an audience and commission more historically themed films from them, that it was "far from being perfect."[47]

As an historic source, the film, like the statue, reflects the problem with depicting a woman warrior in the early twentieth century. Although there is no record of the intertitles that were used, it does not appear that the film made its own overt references to Jeanne d'Arc, or Jeanne Hachette, and of course Philis de la Charce, until the recent film dedicated to her story[48] did not achieve much fame outside of the Dauphiné. Perhaps surprisingly, newspaper reviews of the film did not make the link between Verchères and other French heroines, although some of them referred to the famous order issued in the bloody Battle of Verdun in 1916: "They shall not pass."[49] Nonetheless, as the film illustrates, in the early twentieth century, women warriors were à la mode. Jeanne d'Arc, the archetypical French woman warrior, would be canonized in 1920. Madeleine de Verchères, through her actions and her letters, provided a Canadian counterpart to Jeanne d'Arc and thus validated the history of peoples, both French-Canadian and Canadian in general, who looked to the past to justify their existence in the present.

Conclusion

As a reflection of the broader cultural figure of woman warriors, Madeleine de Verchères's narrative was worth commemoration. But women warriors remained an ambiguous historical category. When a people could point to a heroic female figure in their collective past, it ensured their place among the nations of the world. Governor-General Lord Grey proudly referred to Verchères as "the Joan of Arc of Canada," as did many other writers. Creating nationalist heroes out of people who could be compared to heroes in other national contexts was a common practice in the period of nation-building and invention of national traditions that swept the Western world, at least, in the late-nineteenth and early-twentieth centuries. And it was especially significant for countries that tried to assert their validity internally and externally, whether this was French Canada specifically or Canada as a whole.

The woman warrior tradition offered a place for women in that grand national narrative, but the unruly women had to be contained. The return to the domestic sphere had to form an essential part of the story, and Canadian depictions of Verchères emphasized her femininity. While historical commentators in the grand period of national history in the late-nineteenth and early-twentieth centuries could celebrate figures like Madeleine de Verchères, they also had to curb the potentially subversive elements of the story. A woman warrior must, emphatically, remain a woman.

Notes

1. "Chaque peuple a eu des guerrières." Voltaire, *Dictionnaire philosophique*, tome premier, A.-App. (Paris: Édition Touquet, 1822), 202.
2. On the woman warrior tradition, see, among many other titles, Rudolf M. Dekker and Lotte C. van de Pol, "Republican Heroines: Cross-Dressing Women in the French Revolutionary Armies," *History of European Ideas* 10, no. 3 (1989): 353–363; Louise Anne May, "Worthy Warriors and Unruly Amazons: Sino-Western Historical Accounts and Imaginative Images of Women in Battle" (PhD thesis, University of British Columbia, 1985); Megan McLaughlin, "The Woman Warrior: Gender, Warfare and Society in Medieval Europe," *Women's Studies* 17, nos. 3–4 (January 1990): 193–209; Marina Warner, *Joan of Arc: The Image of Female Heroism* (New York: Alfred A. Knopf, 1981); Julie Wheelwright, *Amazons and Military Maids: Women Who Dressed as Men in the Pursuit of Life, Liberty and Happiness* (London: Pandora, 1989).
3. This narrative is contained in the colonial archives, a microfilm copy of which is located at the Library and Archives Canada (LAC), Archives des Colonies, série F3, vol. 4, folio 341. A complete transcription of the letter is in Diane Gervais and Serge Lusignan, "De Jeanne d'Arc à Madeleine de Verchères: La femme guerrière dans la société d'ancien régime," *Revue d'histoire de l'Amérique française* 53, no. 2 (automne 1999): 198–199.
4. "Quoyque mon sexe ne me permette par d'avoir d'autre inclinations que celles qu'il exige de moy, cependant permettez moy, madame, de vous dire que j'ay des santiman qui me portent à la gloire comme a bien des hommes."

5 "Je scay, Madame, qu'il y a eu en France des personnes de mon sexe dans cette derniere guerre, qui se sont mises a la teste de leurs paisant pour s'opposer à l'invasion des ennemis qui entroient dans leurs province."
6 *Le Mercure Galant*, 30 septembre 1692, pp. 330–331. A statue to de la Charce was erected in 1911 in Grenoble, and the film *Mademoiselle de La Charce* was released in 2016.
7 Ibid., p. 330; Claude Charles Le Roy de Bacqueville de la Potherie, *Histoire de l'Amérique septentrionale* (Paris: chez Nyons fils, 1723), vol. I: 328. Bacqueville de la Potherie recounted Verchères's story twice in the same book, vol. I: 324–328 and vol. III: 152–154. The phrasing in the second is similar: "L'action de Mademoiselle de Vercheres (Fille d'un Officier qui a cinquante ans de service) me paroît trop heroïque pour la passer sous silence," vol. III: 152. In my longer analysis of the time between the event and Vercheres's letter to Mme de Maurepas, I discuss the possible influence of internal colonial politics. The earliest surviving account of Vercheres's story is her own letter to Mme de Maurepas in 1699. The contemporary annual reports from 1692 to 1693 mention an attack on Verchères fort, but fail to mention the young woman's courage. Coates and Morgan, *Heroines and History*, 33–36.
8 "Les Canadiennes n'auroient pas moins de passion de faire eclatter leurs zele pour la gloire du roy si elles en trouvoient l'occasion."
9 The document is not dated, but it mentions Governor Charles de Beauharnois de la Boische, who arrived in New France in 1726. A complete transcription of the letter is in Gervais and Lusignan, "De Jeanne d'Arc à Madeleine de Verchères," 199–205.
10 "Battons-nous jusqu'à la mort. Nous combattons pour notre patrie et pour la religion. Souvenez-vous des leçons que mon pére vous a si souvent donné, que des gentils hommes ne sont nés que pour verser leur sang pour le service de Dieu et du Roy."
11 Pierre-François-Xavier de Charlevoix, *Journal de Voyage fait par Ordre du Roi dans l'Amérique septentrionnale* (Paris: chez Rollin fils, 1744): "Elle commença par ôter sa Coëffure, elle noua ses Cheveux, prit un Chapeau, & un Juste-au-Corps ..."; "deux Amazones," 125, 124. Gédéon de Catalogne's reference is very brief in "Report on the Seigniories and Settlements ...," in *Documents Relating to the Seigniorial Tenure in Canada, 1598–1914*, ed. William Bennett Munro (Toronto: Champlain Society, 1908), 114.
12 He probably followed Bacqueville de la Potherie's account. Philippe Aubert de Gaspé, *Mémoires* (Montréal: Fides, 1966 [1866]), 402–404.
13 Eric Hobsbawm and Terence Ranger, eds., *The Invention of Tradition* (Cambridge: Cambridge University Press, 1983).
14 William Henry Drummond, *Phil-o-rum's Canoe and Madeleine Vercheres* (New York: G. P. Putnam's Sons, 1898), vi.
15 McGill University, Osler Library, William Henry Drummond Collection, Acc. 439, Box 1, 5.7, Draft of poems.
16 Drummond, *Phil-o-rum's Canoe and Madeleine Vercheres*, x–xvi.
17 McGill University, Osler Library, William Henry Drummond Collection, Acc. 439, Box 9, 17.2/35, L.G. Drummond to W.H. Drummond, December 28, 1898.
18 Clipping from the Chicago *Chronicle*, January 30, 1899, ibid., Box 11, 32.3/6.1–6.7.
19 David Gauthier maintains that Hébert was influenced by Frédéric Auguste Bartholdi's "Liberté éclairant le monde" (the Statue of Liberty), "La représentation de l'Autochtone dans l'œuvre du sculpteur Louis-Philippe Hébert (1850–1917): figure d'altérité au service d'une idéologie nationale" (Mémoire de maîtrise, Université de Montréal, 2007), 62–63.

20 "transfigurée par l'idée de sauver les siens et de défendre la maison natale" (quoted in M. Hodent, "Philippe Hébert, le maître de la sculpture canadienne", *La Canadienne* (septembre 1913): 164).
21 LAC, Grey of Howick Papers, MG27 II B2, vol. 28, Grey to Lord Strathcona, March 11, 1909, 7411.
22 LAC, MG27 II B2, vol. 28, drawer 4, file 2, Lord Grey to Mr. MacBride, Premier of British Columbia, November 9, 1907.
23 LAC, Parks Canada, RG84, vol. 1255, File HS-7-25, vol. 1, Lord Grey to Mr. Pugsley, November 2, 1910.
24 LAC, Fonds Rodolphe Lemieux, MG27 II D10, vol. 10, Lord Grey to Lemieux, May 16, 1910, 11878.
25 Canada. House of Commons. *Debates*, February 9, 1912, columns 2824–2825. Pugsley himself referred to a commitment of $15,000 for the statue, but Hansard records $25,000.
26 LAC, Parks Canada, RG84, vol. 1255, File HS-7-25, vol. 1, J.B. Hunter to W.W. Cory, April 30, 1923.
27 Archives nationales du Québec à Trois-Rivières (ANQ-TR), Fonds de la famille Baillargé (*sic*), Livre de compte personnel, 96–99.
28 Larose: "Si le royaume de France fut délivré et réhabilité par Jeanne d'Arc, cette colonie, alors française à son berceau, fut illustrée par Madeleine de Verchères." Ducharme: "un peuple produisant des enfants comme Madeleine de Verchères ne peut être qu'un peuple de braves." *Le Devoir* (22 septembre 1913): 4–6; *La Presse* (22 septembre 1913): 8.
29 "Sentinelle muette évocatrice de tout un passé de gloire et de souffrance," LAC, RB84, vol. 1255, HS-7-25, vol. 4, part 3, "Note on the Heroine of Verchères," Philippe Hébert, mars 1916.
30 LAC, RG84, vol. 1255, HS-7-25, vol. 2, part 1, A. Pinard to J.B. Harkin, December 10, 1923.
31 LAC, RG84, vol. 1255, File HS-7-25, vol. 3, pt 1 to vol. 4, pt 3, reports by the caretaker.
32 "Plus d'un canadien qui l'apercevra en remontant le fleuve dira qu'il n'est pas de plus jolie fille sous le ciel bleu." Hodent, "Philippe Hébert," 164.
33 ANQ-TR, P29, Fonds de la famille Baillairgé, S.T. Grenier to Rev Fred A Baillairgé, 7 février 1924.
34 "il est urgent d'entrer en lutte avec le cinéma américain inepte trop souvent, à tendances généralement corruptrices, fatalement anglifiantes, par un cinéma catholique de mentalité, donc honnête et propre, et essentiellement canadien-français de direction et de ton." Archives de la Chancellerie de Montréal, dossier Cinéma (campagnes de censure), 773–80, 924-1b, "Des noms qui sont une solide garantie d'habile et honnête administration," investment flyer for Le Cinéma Canadien.
35 Magna Fahrni, "Glimpsing Working-Class Childhood through the Laurier Palace Fire of 1927," *Journal of the History of Childhood and Youth* 8, no. 3 (Fall 2015): 426–450.
36 On the commemoration of the story of Dollard des Ormeaux in this period, see Patrice Groulx, "Dollard des Ormeaux," in *Symbols of Canada*, ed. Michael Dawson, Catherine Gidney, and Donald Wright (Toronto: Between the Lines Press, 2018), 138–147.
37 This background history is covered in Pierre Véronneau, "Les nombreux visages de Madeleine de Verchères," website: Vues québécoises: Histoire et cinéma,

24 octobre, 2008, http://cinemaquebec.blogspot.com/2008/10/les-nombreux-visages-de-madeleine-de.html, consulted August 5, 2018.
38   In addition to stills and newspaper reviews, I have used the following sources to reconstruct the script: notes in the personal journal of abbé Baillairgé, company correspondence with the National Parks Board, and records of an interview with the producer.
39   LAC, National Film Archives, Hye Bossin Collection, reel 2, file 27, no. 27, Arthur Larente to Hye Bossin, May 31, 1964.
40   Archives nationales du Québec à Montréal, Régie du cinéma, E188, Direction du classement des films, box 62, November–December 1922, December 6, 1922.
41   LAC, Records of early Canadian cinema, Interview with Arthur Larent, 1973.
42   "On dirait un homme qui admire le portrait de ses ancêtres—brossé par un artiste d'occasion. C'est une émotion plus patriotique qu'esthéstique." *Le Devoir*, 11 décembre, 1922, 2.
43   "Il ne faut pas tenir compte aussi de certaines faiblesses du film de 'Madeleine de Verchères.' Elles sont bien excusables dans une œuvre de début." *La Presse*, 16 décembre, 1922, 43.
44   "Le film de Madeleine est Presque Tres Bien dans son ensemble. Melle Emma ____ a ajouté plusieurs choses très vraisemblables."; "une page de la vie de nos pères." ANQ-TR, P29, Journal of F.-A. Baillairgé, 22 décembre, 1922.
45   Abbé Lionel Groulx, "Notre avenir politique" *L'Action Française*, IX, 5 (décembre 1922): 348.
46   LAC, RG84, vol. 1255, File HS-7-25, vol. 2, part 1, memorandum on Madeleine de Vercheres film, J. Pinard, October 12, 1923; ibid., S.T. Grenier to J.B. Harkin, January 7, 1924.
47   LAC, RG84, vol. 1255, File HS-7-25, vol. 2, part 1, S.T. Grenier to J.B. Harkin, October 5, 1923.
48   "Mademoiselle de la Charce," Lionel Baillemont, director, 2016, 1h 40 min.
49   "Ils ne passeront pas." *Le Devoir*, 5 décembre, 1922, 2.

8

Jeanne d'Arc, Arab Hero: Warrior Women, Gender Confusion, and Feminine Political Authority in the Arab-Ottoman *Fin de Siècle*

Marilyn Booth

> From history, we learn how many a secluded chamber yields a slayer-lion, and how many a hennaed hand receives the reins of kingdoms....
> ... How many they are: intrepid women like Jeanne d'Arc, philosophers like Hypatia, poets like al-Khansa'. How many thousands of women have made intellect their leader [*amir*], striving their backer [*zahir*], piety their general [*qa'id*], probity their scout [*ra'id*], and contemplation their companion [*jalis*].[1]

In the opening article of the first Arabophone women's magazine, *al-Fatat* (The young woman, Alexandria, Egypt, 1892-94), founder-editor Hind Nawfal (1860-1920) invoked a parade of formidable women in history as she addressed herself, letter-like, to women (*sayyidati,* "my ladies"). It was an era when middle- and upper-strata women across the world were working to garner expanded educational, political, economic, and legal rights for females. Like late-nineteenth-century rosters of "women worthies" elsewhere, Nawfal's list intimated a sense of gender-specific community across place and time—in this case, from medieval France to classical Greece to ancient Arabia, and—in her first-person address—to contemporary Egyptian and Ottoman Arabophone women.

Nawfal's metaphor—slayer-lion issuing from the boudoir—was readable within a topos familiar to her presumed readership: the male leaders that fledgling modern nation-states needed were products of strong, devoted mothers. "Lion" was an ancient Arabic descriptive for male warrior-heroes. But maybe the lions were women. Implicitly at least, Nawfal's figuration mobilized the venerable presence in Arabic popular epic and Arab-Islamic history of warrior women, juxtaposed to women as rulers everywhere and to Jeanne d'Arc as that singular, malleable figure of female heroism, celebrated across so many sites and eras. Nawfal capped this obeisance to history's female heroes by hailing a veritable surge of unnamed women, marshaling their ideal and operative character traits, personifying these feminine qualities as male figures associated with political leadership and military campaigns: "masculine" positions fulfilled by heroic traits of a platoon of leaderly women.

The women's magazines following *al-Fatat* deployed famous-woman biography to encourage shifting roles for female readers while embedding new feminine lives in constructions of "tradition." From ancient to contemporary, profiles of women across the world—especially those with origins around the eastern Mediterranean—might inspire readers. This genre paralleled the "greatest-events-and-most-famous-men" feature fronting the mainstream magazine *al-Hilal,* its first issue appearing two months before *al-Fatat*'s. Perhaps it was a riposte to another popular miscellany, *al-Muqtataf,* where women's biographies and writings mostly appeared in the "Home management" section toward the back of the magazine.[2]

In 1895, Jeanne d'Arc was amongst *al-Hilal*'s "most-famous-men." Surfacing occasionally in "malestream" magazines, Jeanne made more appearances in the women's press. At least sixteen biographies of her appeared in Egypt-based women's magazines before 1940 (at least nine in the 1920s), others in Ottoman Syria. My study some years ago of "the Egyptian Jeanne d'Arc" as an evolving localized figure within Egyptian anti-colonial modernizing nationalist formulations, crystallizing in post-First World War activism, demonstrates how such representations were saturated in local, time-salient gendered expectations. Biographies of the 1920s emphasized Jeanne's modesty, religiosity, and devotion to homeland as earlier Arabic biographies had, but they gave increased emphasis to her domestic proclivities and salt-of-the-earth peasant-ness. They celebrated her resistance—anachronistically—to "the British."[3] And they suggested that her religiosity was transferable across faiths. In other words, she was a perfect 1920s Egyptian female patriot and representative of the nation, akin to the peasant woman in Muhammad Mukhtar's famous statue "Egypt's renaissance." As in France, so in Egypt: the imagined Jeanne played a unificatory-nationalist role, subtending (or uneasily repressing) political differences—although in France "she" also carried the banner for competing political agendas.[4] Decades later—following the Free Officers' revolution, Egypt's allegiance to the non-aligned movement, and the emergence of organized Palestinian resistance—Jeanne appeared in 1950s–60s collections titled "Fighting women" and "Woman in theatres of struggle," an anti-colonial fighter.[5]

Jeanne's earliest sightings in Arabic, 1879–96, were not so marked by imperatives of anticolonial nationalism. In Egypt, nationalist sentiment and organizations emerged gradually after the shock of the 1882 British invasion; Ottoman Syrians' trajectory was differently complex. *Fin-de-siècle* contestations around gender and nature, women's rights, and women's political demands shaped figurations of Jeanne and other "fighting women" in biographies by early female intellectuals with feminist inclinations, and equally, those produced by male intellectuals interested in harnessing women's capacities for modernizing societies. In the 1880s–90s, it was less nationalism than questions of gender identity in and for modernity that seem to infuse these texts, consonant with debates on "the woman question" (and more silently, "the man question"). One cannot predict perspectives from gender of authorship, but divergences among these texts suggest an alignment of gender and political perspective. This becomes particularly evident when the imagined Jeanne d'Arc joins her warrior-women sisters-in-arms as they were portrayed in the same and contemporaneous publications. For these were populated by a cosmopolitan panoply of female figures. Jeanne was not the only militarily engaged female hero in the public eye.

The woman warrior was an available heritage figure, indigenous and borrowed, multiple of guise, appropriated by early proponents of expanded lives for women *and* by their opponents. Medieval and early modern anecdotes of historically attested Arab-Islamic and pre-Islamic warrior women resurfaced in *fin-de-siecle* publications. In cafes and homes, people listened to popular epics from a vernacular tradition celebrating legendary pre- and early-Islamic-period women combatants. Overlaying this was an emerging practice of women's history writing as a polemic enmeshed in "the woman question" as a local debate responding to critiques from Europe. Finally, as women's activisms elsewhere generated nervousness in the Arabic press (via articles translated from Europe and North America), representations of women's struggles for national-political rights were framed in narratives of politically aspirational women as violent and uncontrollable, desexed or resexed as men manqué. These narratives depended on cross-societal, masculinist, often misogynist, narratives of "modern Amazons."

Nineteenth-century Arabic biographies of Jeanne d'Arc, this chapter argues, index a relatively fluid, ambiguous and differentiated (but forceful) discourse on gender and political voice that would later become more fixed. A certain gender fluidity in 1890s portrayals of Jeanne, even those that do not foreground story elements such as cross-dressing and military leadership, is evident, later giving way to a national-feminine iconography set within a rather rigid role-separation schema where women are aligned with "home." But the Jeanne representations accrue additional nuance when read alongside other woman-warrior images contributing to a contentious public conversation on gender-rights in modern nation-formation. Because any woman-warrior figure—against the backdrop of patriarchal gender expectations—tends to highlight issues of gender ambiguity or fluidity, the woman-warrior imagined serves as one notation of how open to question a community's gender conventions are at a given moment. Competing representations of woman-warrior figures suggest a multiplicity of perspectives and attempts to manage them.

Reading Jeanne d'Arc

In the nineteenth century, Arabophone elites' earlier intellectual activity became a concerted knowledge movement. While manuscript production continued, government-initiated and private ventures energized newspaper, magazine, and book publishing. Bringing their senses of heritage into conversation with how-to debates on attaining suitable indigenous modernity, intellectuals reassessed the premodern Arabic literary-intellectual-scientific heritage—repackaging medieval and later works for new audiences—as they evaluated challenges posed by Europe's politico-economic power and social models. Like nineteenth-century thinkers elsewhere, many vigorously supported educating girls—for intelligent motherhood—but differed on what girls must study, raising the specter of challenges to patriarchal authority in and beyond the family. Recognizing and re-voicing European stadial histories that linked societies' advancement to women's status, they argued over how to define and enact that status.

Integral to this intellectual-political activism were new modes of history writing incorporating older historiographical models (biographical dictionaries, dynastic chronicles) and creatively appropriating histories composed elsewhere. Ancient histories of the eastern Mediterranean region offered senses of heritage and authority to colonized or peripheralized elites. World histories and the European-format encyclopaedia, plus prolific borrowing of material from European periodicals, opened up the world to local readers, as works by Arab travel writers had done for centuries. These contemporary works did not necessarily espouse stadial history, but they collated material that nourished it.

In such works, Jeanne d'Arc made her first Arabic appearances. Syrian-Armenian writer-translator Yuhanna Abkariyus (c. 1832–89), from a Beirut-based intellectual family, published a massive world history, *Qatf al-zuhur fi ta'rikh al-duhur* (The plucking of blooms on the history of the ages, 1875, 1885); it would become a source for the first full-length Arabic biographical dictionary of world women. Abkariyus included a few warrior queens and fighting women: Semiramis, Zenobia, Boudicca. His sketch of Jeanne d'Arc was admiring. In a time of turmoil for France, "in 1429 a star of good fortune shone on the horizon: there appeared the maiden Jeanne d'Arc, daughter of a peasant, exhibiting piety and godliness." Neither cross-dressing nor visions (highly visible, problematic story elements in so many Jeanne representations across time and space) figured in this abbreviated, circumspect account. When her father refused her permission to leave home, Jeanne went "secretly to [Dauphin] Charles...her intrepidity astonished him." She led the small army he provided "with courage that men's zeal and bravery fall short of," soon "breaking" the English hold. "She did things so astounding to the English that they thought she must be a sorceress." Her trial and murder were "blameworthy deeds, a grievous sight to make bodies shiver."[6]

Jeanne appeared in the encyclopaedia project spearheaded by educator, translator-publisher, and all-round culture activist Butrus al-Bustani (1819–83). He realized this ambitious work with help from family members (possibly including his well-educated writer daughters). Eleven volumes came out (nine in Beirut 1876–87, two in Cairo 1898–1900), though the project remained unfinished.[7] Jeanne's entry in volume six (1882) was more detailed than Abkariyus's narrative. Did the language imply Protestant skepticism on visions? People of her region were strong believers in *khurafat* (legends), this text noted. Jeanne

> was full of imaginings, very pious.... She loved stories of the Virgin and was familiar with the widespread tale that a virgin would save France.... She believed in supernatural apparitions, talking about voices she heard and visions she saw. A few years later, it came into her head that she was called to save her country and crown its king. The Burgundians' descent on her village strengthened her belief in the soundness of her musings.[8]

This narrative mentions that Jeanne wore armor and carried a sword but is silent on cross-dressing. Jeanne is portrayed less as leader than as helper. (For example: she stayed on after Charles's crowning only because his male deputy feared losing Jeanne's aura of authority.) The sympathetic focus is on her steadfast and consistent performance in the interrogation she faced from malevolent foes, *lahutiyyin muhtalin*, "cunning, deceitful theologians."⁹ (Harsh language on French men of religion is interesting given al-Bustani's difficult personal history with Maronite clergy and Protestant missionaries.) "Feminine" characteristics predominate: "The life history of that maiden was beyond reproach; she was renowned for purity, chastity, and modesty. No spilt blood stained her hand; her majestic bearing evinced gentle graciousness. Her awe-inspiring gravity was a source of wonder to all... these [qualities] restrained her soldiers from brutish behaviour."¹⁰ Jeanne becomes a "civilizing" figure, exemplary for "feminine" refinement as constructed in the era's discourse, which posed these qualities as natural and timeless attributes and yet also as outcomes of a trained domesticity that could and ought to "tame" men's behavior. Aspects of her history that have generated unease in many contexts are not completely absent, but they are not foregrounded, and this treatment minimizes gender ambiguity: Jeanne's ability to lead is predicated not on a fighter image but on that of a morally and socially impeccable, and subordinate, feminine presence.

Abkariyus's and al-Bustani's texts maintain gender-differentiated boundaries of aspiration, attribute, and efficacy. Two biographies in works authored by women position Jeanne somewhat differently. By 1879, Maryam al-Nahhas Nawfal (1856–88), mother of *al-Fatat*'s editor Hind Nawfal and her writer-sister Sara, had composed a biographical dictionary of women, *Ma'rid al-hasna' fi tarajim mashahir al-nisa'* (The fine woman's exhibition: Biographies of famous women). Only a sixteen-page prototype appeared; what was apparently a two-volume work was otherwise lost. The prototype comprised lives of Çeşm-i Âfet Hanım (d. 1907)—third consort of Khedive Isma'il (reg. 1863–79) and sponsor of Egypt's first state girls' school, who underwrote publication—and "a few famous women," as Zaynab Fawwaz (c. 1850–1914) would describe it in her profile of al-Nahhas several years later. In February 1879, the author's husband Nasim Nawfal published an announcement of the project, with a plea for subscribers to facilitate publication. Among the subjects, he said, were "women famous... for storming barricades."¹¹ The prototype included Jeanne d'Arc—at least five pages out of sixteen—followed by Catherine of Russia and ancient Arab poet Layla bt. Hudhayfa b. Shaddad.¹²

Ma'rid portrays Jeanne as emotionally cognizant—as a very young girl—of the political scene (framed in national terms) as it impinged on her environs: "She was frightened at what she saw of France's disturbed state and the disasters befalling it; her heart broke and her alarm grew when the English drew near her birthplace."¹³ Jeanne would linger in a nearby wood to "contemplate her homeland's misery, her head hanging low, held-back tears now flowing... hot with the flame of fervour and regret." Recalling the prophecy about France's salvation, "she wondered: could she be that girl?" Then, "she recalled what she had heard about women famed for courage and bold initiative, saving their country from ruin, like the women of Bohemia who bore weapons and defended the homeland's honour." The text lists several names.¹⁴ Al-Nahhas, biographer of women, assumes Jeanne's political precocity and worldly knowledge; but would this unlettered peasant girl have heard of the women who became known as Bohemia's Amazons?¹⁵ In *her* time, al-Nahhas might well have heard of the Bohemian Amazons. Ensconced in Czech nationalist and feminist imaginaries by the 1880s, these figures drew fascinated attention from writers across Europe. As Jitka Malečková has shown, moving "from mythology to national myth," their story became one of fighting for the homeland rather than (as in some, myth-laden, versions) for reinstitution of lost female rule.[16] Al-Nahhas's proto-nationalist interpretation highlighted Jeanne as putative "reader" of famous-woman history, inspired by exemplary forebears. Jeanne becomes the perfect nineteenth-century female-feminist patriotic pupil. But she is less demure or devout than she is politically astute and ambitious.

The narrative is vivid and fiction-like, describing Jeanne's first visions with none of al-Bustani's cautious tone. It puts Jeanne in the ranks of young patriots seeking to serve militarily through less exalted channels: "she showed her desire to enlist with the volunteers, to go to the battlefield, to join the French army. Her father grew troubled and angry.... He decided to marry her off." Thus, Jeanne's first test occurred in response to her attempted transgression of gender norms. She surmounted it. Then, won over, "the captain" at Vaucouleurs gave her "a burnished sword ... the garb of war ... and a mount." By the time she reached Charles, "news of Jeanne soared near and far." The people of Orléans begged Charles to send her to them in their hour of need. Only then did she have a royal audience: "she entered wearing the uniform of war and armed: eighteen years of age, her darkish complexion imbued with a pale glow, her eyes black with a magical quality, her form slender, her face displaying knight-like bravery and intelligence." With this gender-bending appearance, even at her first audience she found adversaries insinuating her guidance was satanic. The text injects a note of feminine solidarity: the bishops whom Charles assigned to interrogate her feared to disturb the royal house: "they knew Jeanne had gained the liking of the queen and queen mother."[17] There is evidence that Charles's mother-in-law, Yolande of Aragon, supported Jeanne. But likely this was because Yolande opposed the pro-Burgundian views of Charles's advisors: she saw Jeanne as possibly useful.[18] It was not a question of gender solidarity, as al-Nahhas's text intimates. This narrative also sketches Jeanne as popular hero. Facing interrogation, "she buckled on her sword, donned the garb of war and struggle, mounted her noble steed, and grasped the lance of St Catharine."[19] This biography emphasizes her determination despite the clerical establishment's harsh opposition, and highlights her cross-dressed visibility as unproblematic for her staunch followers, specified as the civilian masses and soldiers.

Sketching Jeanne as an unconventional, politically motivated, and astute girl, al-Nahhas's biography showed her growing into a decision-maker and risk-taker (under divine guidance). The coronation at Rheims "was a momentous occasion heralding joy and delight: by Jeanne's courage and bold initiative the French ably chased the English from the territory they had seized."[20] Historically, the English were not expelled from France in Jeanne's lifetime. It is unclear whether the biography's declaration refers to the initial, partial rout, or implies credit to Jeanne posthumously for final victory. The vocabulary naming Jeanne's powers—"courage and bold initiative," terms associated with masculine heroism—echoes precisely the text's earlier characterization of the Bohemian women whom Jeanne recalled. Al-Nahhas gestured to a connected history of women's active defense of nation or homeland, a theme more explicit and yet less military in later, "high-nationalist" Arabic biographies of Jeanne.

The narrative's intense emotional register culminates in direct address, linking the biographical subject to young unmarried female readers, and all to patriotism: "Woe is you, O Jeanne, do you die hands fettered after your sword touched the head of the enemies [?] ... O maidens, do you bewail a young woman who dies for love of her homeland, having prepared for it victory's crown even as it wilfully neglected to save her from destruction [?]."[21] This biography's tone and terms of address diverge from al-Bustani's and Abkariyus's, a difference of genre *and* apparent communicative intent. Perhaps al-Nahhas had read the English translation of Dumas père's 1842 romance *Jehanne, la pucelle*.[22] (None of the texts discussed here name their source, and none acknowledge each other.)

About fifteen years later, readers could peruse Zaynab Fawwaz's biography of Jeanne in her sweeping biographical dictionary of women, published by Egypt's prestigious government press. This work comprised 453 biographies, the majority Arab and/or Muslim while incorporating women of Europe, North America, and South and East Asia. Fawwaz's massive project drew from al-Bustani and Abkariyus, but was rooted in the Arabic-Islamic tradition of biographical dictionaries. These had emerged as a technology for remembering and ascertaining the veracity of the Prophet's contemporaries and later followers, sources on his deeds and words. The genre ramified, embracing variously defined groups of eminent individuals. Fawwaz signaled her focus—women—and her generic intertext in her title, *al-Durr al-manthur fi tabaqat rabbat al-khudur* (Pearls scattered on the classes [*tabaqat*] of ladies of cloistered spaces). *Tabaqat* was shorthand for the genre tradition, wherein women had appeared but never as *sole* focus. Quite a few of Fawwaz's subjects were warrior women.

Fawwaz's biography of Jeanne d'Arc differed from al-Nahhas's in structure, language, and tone, perhaps a product of diverse educations. Al-Nahhas's training in Anglophone missionary schools in Ottoman Syria gave her facility in English and familiarity with European literature. Fawwaz was tutored in Arabic letters and Islamic texts in her native south Lebanon and later in Egypt, her adopted home. She was familiar with historical chronicles, premodern Arabic *adab* works, and biographical compendia featuring anecdotes of literati, scholars, and other eminent women and men. The profile's debt to Arabic biographical style is evident, beginning with a physical description that verifies a moral countenance. Jeanne was

> pure of complexion, slender of build, with deep-black eyes; coal-black hair fell over her shoulders. Visible on her graceful countenance were the qualities of bashfulness, sweetness and humility, while her features gave indication of strength of purpose, far-reaching goals, and cool self-possession.[23]

The latter comprised heroic traits ascribed conventionally to male leaders (and in Fawwaz's volume, to females). My earlier study on Jeanne d'Arc saw this constellation of attributes as describing "the core of the ideal woman: bashful and resolute, sweet and self-possessed, mild and highly ambitious.... Such a mix of traits was crucial for the dynamic yet self-effacing female presence that many in Egypt... saw as underwriting their vision of national progress."[24] But the self-effacing aspect becomes more dominant in the 1920s biographies. It is there in Fawwaz's characterization, but not in the way she describes Jeanne's life and actions. When Fawwaz wrote, opposition to the British occupation was gathering. Yet there was not the programmatic energy that would emerge slowly beginning from 1906. Her Jeanne was less a national/ist image than a feminine-heroic one: Jeanne as consistently ambitious, visible, and strong. "She was always mounting a stallion[25] and racing ahead, without saddle or bridle, in boldness and valour."[26] Again, these terms—*jara'a, furusiyya*—conventionally labeled the heroic masculine.

In February 1921, *al-Mashriq* (founded 1898), a Catholic journal produced by the University of St Joseph in Beirut, republished Fawwaz's biography in celebration of Jeanne d'Arc's recent canonization. Its title highlighted Fawwaz's identity and mentioned Jeanne's sainthood anachronistically (Jeanne was not yet beatified, let alone sainted, when Fawwaz wrote): "A Muslim woman writer honours the saint Jeanne d'Arc." The introduction celebrated ecumenical and geographical communion: both Egypt and Syria could celebrate Fawwaz as their own. The biography, said this Catholic journal, displayed Fawwaz's "balanced judgment."[27] Reproducing the biography "word-for-word" to honor both St Jeanne and Zaynab Fawwaz, the magazine announced, however, that it would "indicate what was missing due to lack of knowledge and incorrect wording." Its third footnote read: "She did not mount a horse before going to war." Presenting Jeanne as a fearless, skilled horse-rider early in life, Fawwaz's text was challenged by this 1921 republication as historically inaccurate.[28] Consistently, the magazine's footnotes played down Fawwaz's emphasis on Jeanne as bold, athletic, and visible.

Fawwaz narrates another sign of precocity: "aged five," the girl began "having religious apprehensions." *Al-Mashriq* corrected this to age thirteen and assured readers that Jeanne's *hawajis* "were not fancies but rather, noble sentiments of piety."[29] The magazine glossed Fawwaz's wording: Jeanne's "asserting [*za'ima*] that angels and saints appeared to her" signified "confirming," not the more doubtful (or rebellious) "asserting." This annotated version promoted Jeanne's veracity and her visions' religious validity, taming the tenor of forceful individuality that surfaced matter-of-factly in Fawwaz's narrative. Jeanne left her village for a nearby town, working for a widow who ran a *pensione*, said Fawwaz, "striving loyally, showing initiative in her work and chastity in her behaviour, and then returning to her father."[30] It may be over-reading to link this to Fawwaz's feminist politics, but in the early 1890s she was publishing essays upholding singlehood for women over painful marriages, advocating women's right and need to be prepared for waged work, and declaring their upright conduct as workers despite men's bad behaviors.[31] *Al-Mashriq* again "corrected." Rather than "returning to her father," Jeanne "returned with her family to Domrémy": thus, the new version implied, she was not a working girl on her own.[32] The magazine's version seemed to negotiate prevailing understandings of respectability for 1920s Arab middle-strata girls.

Preparing to meet the heir apparent, Jeanne "donned a knight's garb," said Fawwaz. *Al-Mashriq* added a footnote: "She did so on the basis of inspiration from her visions."[33] Fawwaz, in other words, did not explain or justify Jeanne's equestrian skills, assumption of masculine attire, or act of leaving town alone. For the Catholic magazine thirty years later, without an external guiding force none of it could have happened. Divergence between Fawwaz's text and its later "reproduction" displays a process of narrative containment, consistent with prevailing nineteenth-century woman-warrior representations across cultures.³⁴

In an interesting spin, Fawwaz localized Jeanne's story linguistically, borrowing from the Qur'anic story of Yusuf (Joseph) to ventriloquize English soldiers' reaction as, witnessing Jeanne's triumphs, "terror possessed [their] hearts." The words Fawwaz put in the mouths of the English echoed words said (in the Qur'an) by women assembled in the Egyptian official's home after his wife tried to seduce Yusuf: "He cannot be a human being, he is none other than a noble angel." In the Qur'an, the hostess commanded his detention, Joseph declared incarceration preferable to the women's demands, "and God...deflected their wiles from him."³⁵ For the stunned English in Fawwaz's narrative, Jeanne "cannot be a human being, she is none other than a noble angel." This intertextual element introduces a note of gender inversion, linking Jeanne to Yusuf's purity, and the English military men to the dishonorable women's "wiles." *Al-Mashriq* noted the quotation's source without comment. (Fawwaz had not: presumably her envisioned readers would recognize the phrase.)

Of these nineteenth-century Arabic Jeanne narratives, Fawwaz's is the most challenging to conventional visions of gender roles and relations. Fawwaz's Jeanne is an astute leader, agential negotiator, and skilled horsewoman. This contrasts both with earlier biographies and with 1920s portraits highlighting Jeanne's self-effacing sacrifice for France, portrayed as a colonized nation.³⁶ Indeed, *al-Mashriq*'s 1921 version made this move. Fawwaz described Jeanne's confusion and dismay at Charles's refusal to let her return home; if this led to failure, still the text emphasized heroic qualities: only now did Jeanne's "zeal and bravery" desert her. *Al-Mashriq* glossed Fawwaz's narrative as "something of an exaggeration," observing that "St Jeanne d'Arc...remained in confusion and doubt for a time, but preferred in the end to sacrifice herself in service to her homeland, her death spurring France's complete and final victory over England."³⁷ Even as *al-Mashriq* praised Fawwaz for writing "judiciously" about the Catholic saint, it muted her emphasis on Jeanne's fearless *furusiyya*—her knightly qualities—while highlighting religiosity and divine guidance.³⁸

Malestreaming Jeanne

Just before *Pearls Scattered* was fully printed, Jeanne d'Arc appeared in *al-Hilal*'s biographical series. In volume four (1895–96), the front-page "greatest-events-and-most-famous-men" feature mostly offered biographies of deceased nineteenth-century men: newspaper editor Salim Taqla; Louis Pasteur; Joseph Garibaldi; Victor Emmanuel II; former Ottoman governor Rustum Pasha; Butrus al-Bustani; and Gordon Pasha. Lebanon-based missionary educator Cornelius van Dyck was still alive but died soon after. Molière, Dante, Martin Luther, and Socrates also appeared. And there was Jeanne d'Arc. In its heading, was *al-Hilal* using *rijal* (men) in its extended and notionally ungendered sense of "trustworthy sources" or (as *rijalat*) "distinguished people"? This seems unlikely. Jeanne was one of the men, and it was an awkward fit.

True to the magazine's practice, the issue's first page featured an image (Figure 8.1).³⁹ Unlike Fawwaz's description of Jeanne, *al-Hilal*'s visual Jeanne had short "pageboy"

Figure 8.1 Frontispiece to *al-Hilal* 4: 4 (October 15, 1895). Private collection of M. Booth.

hair; a lowered, averted gaze; short tunic or skirt; and upper body armor. Her crossed arms hugged a sword; another hung at her side. The tip of the sword between her arms fell in front of a second element, a vague shape with hands or gloves distinct. This image must have been derived ultimately from a popular statue by Marie d'Orléans (1813–39) of Jeanne d'Arc at prayer. This work was the model for numerous images of Jeanne across Europe, particularly as it could be "read" in various ways, as Nora Heimann has demonstrated (Figure 8.2).[40] Likely, *al-Hilal*'s image came from a secondary rendering;

Figure 8.2 Jeanne D'Arc, from the statue by Princess Marie of Orléans in the gallery at Versailles, from Mark Twain, *Personal Recollections of Joan of Arc*. Courtesy of Wikimedia Commons.

it suggests simplicity and modesty but not necessarily prayer, though readers might see her crossed hands as a sign of Christian piety. As Heimann observes of the original sculpture, what was absent was as significant as what was there: it did not include "overt references to religious convention" or royalist overtones.[41]

Likely, *al-Hilal*'s editor Jurji Zaydan (1861–1914) came upon this image and took the opportunity to use it, rather than selecting it deliberately from among several. It was not a simple matter to acquire usable images. (At one point, Zaydan pleaded with readers to send images of individuals he wanted to feature—giving a list—but for whom he had no portrait.[42]) Even so, it seems he found a particularly appropriate image that could speak meaningfully to various audiences.

As Heimann notes, the statue "acknowledges Joan's controversial cross-dressing, but it de-emphasizes it by placing a modest skirt over the armor covering the Maid's lower limbs."[43] In other ways, the image might appear as androgynous to *al-Hilal*'s readers. This not only complicated the "most-famous-men" rubric but presaged ambiguities in the text that followed. But in its very gesture to gender ambiguity, this text placed "heroic qualities" firmly and exclusively in the grip of men as a gender category with fixed attributes.

> She is the maiden who has bewildered historians over how to name and label her. For she united the gentle sex's delicacy with the courageous audacity of great military leaders and the wisdom of great philosophers. She came close to matching the prophets.[44]

This contrasts with al-Nahhas's and Fawwaz's deployments of such traits as ungendered designations of excellence garnering authority.

Matching the prophets? In an Islamicate context, though controversially, some have regarded Mary, mother of Jesus, as holding *almost* prophet status.[45] For any other woman to approach this seems unthinkable, as much for Christians such as Zaydan as for Muslims. In her milieu, Jeanne was associated with (suspect) female visionaries, and with Mary through the mystery play genre. In the 1890s, for a non-Catholic Christian (Zaydan, whose magazine advanced a modernizing, science-oriented ethos), was "near-prophetic" status the least troubling way to explain divine visions? The text mentioned "portents" around Jeanne's birth in a secular vein, as instantiating legendary elements common in life histories of the great:

> Rarely do you read biographies of great men, especially military leaders and heroes, without encountering events said to have occurred before or in the hour of their births.... There might be no truth in those narratives beyond some phrase spoken by the mother, female neighbour, or midwife, but not often does one attend a birth without hearing the like.[46]

Framing a female hero's birth, this statement placed her in the category of male heroes and set this category apart from women as sources of dubious information. The comment feminized "legendary" tales to dismiss them, articulating an oft-expressed view: "women's gossip" (like *khurafat*) was antithetical to true and serious history (not to mention sober nationhood). In other ways, too, this biography bore the imprint of Zaydan's didactic mission. The novelist in him emerged in the profile's vivid detail, while the popular-didactic historian surfaced in its digressive mini-narratives on French and English history. Like the historical novels Zaydan wrote, this text presented history through a female figure at once centered (as narrative protagonist) and decentered (as agent of political action). This was the volume in which Zaydan published serially his novel *Armanusa al-misriyya* (Armanusa the Egyptian), recounting the Muslims' conquest of Egypt, with the eponymous female protagonist at the center of the story but not in the thick of battle.[47]

Al-Hilal's Jeanne introduced the emergent domesticity discourse which justified girls' schooling as necessary to efficient home management and child-raising. Partly adapted from Victorian England and contemporary France, grafted onto local notions of a gendered division of labor, this ideal of family management (incorporating companionate monogamous marriage) became a consistent theme in later Jeanne biographies and—complicatedly—throughout Arabic women's biography. It shaped Jeanne's desire to return home, resituating herself in the bosom of family. But for *al-Hilal*, it explained Jeanne's oddity, a girl kept *apart* from the domestic hearth with consequences for her understanding of the world: "She spent her childhood with her godmother, because her mother was occupied with homemaking tasks and managing her husband's and children's needs."[48] The godmother reprised the talkative women mentioned earlier and the pervasive "old women" ['aja'iz] of oral lore. Her intervention explained Jeanne's imaginings:

> In leisure hours this godmother related stories told by that age's old women, of spirits and demons and sainted ancestors who were (they claimed) in perpetual communication with the living. Because the girl was keenly intelligent and sensitive, she was affected ... whenever her godmother ceased talking she asked for more ... and so she grew up believing in spirits.[49]

This passage echoes homilies in the press on the dangers of storytelling (and novel-reading) for the impressionable young (especially females), and the corruption of children's minds by "ignorant" mothers relating tales of the supernatural.

Marie d'Orléan's statue included a plinth on which a helmet and gloves lay. In *al-Hilal*, the treatment of the plinth seems possibly ambiguous in its lack of detail, possibly readable as a shadowy second figure, crouching and half-hidden, perhaps even reaching for the sword-tip. Might readers interpret this as a gesture to the narrative's supernatural aspect, a hint that Jeanne's agency was not her own? For al-Nahhas, Jeanne was the perfect pupil, the unconventional girl with female military heroes, the precocious politico. For Zaydan, she was the impressionable victim of women's tales. Perhaps men like al-Bustani and Zaydan, whose writings advanced ecumenical, even secular, notions of modern society, including support for girls' education, had to explain away the visions. Fawwaz and al-Nahhas appeared not to have such anxieties.

Jeanne sallied forth in *al-Hilal* as determined and courageous. But unlike al-Nahhas and Fawwaz, *al-Hilal* related this demeanor as almost a sex change, Jeanne as gender anomaly rather than exemplar of feminine aspirations and abilities: "Where she had been soft of voice, gentle in speech, fond of solitude, she now became a person of ringing tones, so bold she spoke to others only through commands and prohibitions." In Arabic, "to command and prohibit," a common idiom with religious overtones, is the province of political-religious leaders (presumed male). This biography offered leaderly details. But even as Jeanne appeared in her actions as bold and self-sufficient (staunching her wounds), she "screamed like a child at the pain." Later, "her eyes gushed tears" at Charles's "cowardice"; celebrating the coronation, she addressed him "with tears in her eyes." The biography does recount that in prison, alive on condition that she not resume wearing masculine garb, "she felt abased and servile, and could not bear staying in prison; they found her once again in men's clothes."[50] *Al-Hilal*'s awkwardness in situating Jeanne—man or woman?—bespoke debates transpiring in its own pages.

Women Warriors around Ancient Arabia and Elsewhere

Although neither later women's magazines nor *al-Hilal* featured many women of war and politics, Fawwaz's biographical dictionary did. There, Jeanne was in the company of warrior leaders including those of ancient Semitic or other "nearby" communities. Fawwaz profiled women who led resistance to foreign (including colonial) rule, defended against invaders, or kept invaders distant through vigorous state management. Boudicca, "mother of the tribe of Briton-land,"[51] Cleopatra VII, and Zenobia of Palmyra resisted the Romans; Bilqis, queen of Sheba, resisted Solomon and his army. There were ancient Assyrian ruler Semiramis; Amalekite ruler al-Zabba' Na'ila bt "Umar, a quasi-legendary figure who triumphed militarily—and revenged her father's death—through ruses; Daluka bt. Zabba," queen of ancient Egypt's Copts; Sajah, tribal leader and rival prophet resisting Muhammad's leadership in seventh-century Arabia; Azarmidokht and Purandokht, daughters of Khosrow II and, briefly, Sasanian monarchs in the seventh century as the Persians attempted to repel the Arab Muslims' eastward march; tenth-century St Olga of Kiev, presented as assiduous war strategist; Dahya, medieval Berber sorcerer-queen and indefatigable military leader; Khadija of the Maldives (reg. 1347–80?); Khamani bt Ardeshir b. Bahman, consummate strategist, mother of Darius I; Habus al-Shihabi of nineteenth-century Lebanon, scion of a ruling house, a politician in her own right; and Agustina de Aragón, known as "the virgin of Zaragoza" (1786–1857), an epithet and a story that echo Jeanne's proto-nationalist construction. At a crucial moment in her city's resistance to French invasion in 1808, Agustina stepped up to fire a cannon, rallying her compatriots. As portrayed in *Pearls Scattered*, these women combined smart strategy—reversing misogynistic notions of "women's wiles" famously named in Yusuf's story and deployed in Arabo-Islamic polemics to demolish women's public activities[52]—with strength and loyalty to family and "nation." Often they succeeded *to* leadership because of kinship ties to men. But they succeeded *in* leadership through their own abilities. The probably legendary Daluka marshaled Egypt's women—including female sorcerer Budur—to protect the realm, following the drowning of Pharaoh and his male warriors, explained the biography. If women leaders fill a vacuum in times of need, Daluka's abilities and loyalty—and her political success—contrast with the corrupt, tyrannical, and drowned ruler she replaced.[53]

Comparing Fawwaz's warrior-women lives to those in al-Bustani's encyclopaedia and Abkariyus's history elicits striking divergences. The encyclopaedia entry on Semiramis (in volume 10, published after the patriarch's death) foregrounded her as legend, calling her heroic story dubious; deeds attributed to her were the accomplishments of "her husband and other kings." But Fawwaz treated her as an indisputably historical personage responsible for her kingdom's flourishing as well as military ventures. Her people so venerated her that they deified her, "asserting that... she assumed a dove's likeness." Expressing scepticism on this point, Fawwaz maintained focus on the *historical* personage: "Whatever the case, she was the pride of women of the ancient era, the light in that age's lamp." Here is a different deployment of gender fluidity: confirming women's equal abilities, rejecting arguments that their feats were due to men. This reasserted the feminine, but as infinitely capacious in the subject's roles as individual hero and categorical force. Fawwaz named not "her folk"— the adoring subjects—but rather "ancient women," constructed as collective subject through Semiramis's example.

Fawwaz's biography appeared before al-Bustani's but after Abkariyus's history: the latter's chapter on Semiramis was likely one of her sources. Yet Fawwaz eschewed Abkariyus's emphasis on motives: "Covetous greed aroused her," he observed, "inciting her to seize all other kingdoms in the world." This divergence is noticeable because otherwise, Fawwaz's narrative of Semiramis's major campaign is almost identical to Abkariyus's. Unlike Fawwaz, Abkariyus did not find Semiramis worthy of pride: "Thus ended the life of this great queen whose assiduous habit was to raid and mount wars out of greed for conquests and booty, rather than spending her time effectively managing her kingdom and making a success of her nation."[54] Fawwaz's highlighting of Semiramis's military skill and peacetime leadership abilities shows more affinity with Christine de Pisan's life of Semiramis in her *Book of the City of Ladies* (1405). Neither sexualized Semiramis as had Dante and Boccaccio.[55]

Fawwaz's biography of Agustina followed al-Bustani's word-for-word: "When the French besieged the city... she participated in the defence, growing famous for the courage she displayed. She was called 'Artillerywoman' because she wrested a fuse from an artilleryman's hand... and fired cannon at the besiegers. As reward for her service... she was given leadership of a soldiers' unit and several medals." Here ended al-Bustani's text. Fawwaz added: "She persisted in battle, gaining victory time after time... against the French."[56] This text highlights Agustina's military heroism less as spectacular single event, more as vocation. In Spain she was constructed as national/ist hero. But there is no indication of continued military leadership, though Agustina probably provided continued support work, and her act of heroism earned her an honorary military title and pension.[57]

This array of historical and quasi-legendary figures was clustered around the Mediterranean, including the lands of Fawwaz's first readers. Amongst women warriors and consistent with the volume's overall population, Europeans were there but at the margins. Warrior women "locals" spanned wide swathes of territory and time. They intermingled alphabetically with writers, scientists, religious figures, royal consorts, and queens known for statecraft (Militarily significant figures could include Byzantine queens, Elizabeth I, Catherine II, Victoria, Isabella I—all *Pearls*.) Fawwaz's profiles emphasized the "warrior" role as one in a range of public feminine heroic positions. She assumed the historical existence of subjects rather than dismissing them as mythological. Historically attested or not, figures such as Semiramis—and Jeanne d'Arc—had been mythologized through myriad retellings. Distancing such figures from "History" meant dismissing their historical existence, hence significance. Mythologizing the female heroic was one means to discount the transgressive figure of the woman warrior.[58] For Fawwaz, these figures were neither dismissible nor confinable within gender-conventional attributes. Concerning women's potential, they were not necessarily extraordinary. In company with Semiramis, they were "the pride of women." And they were real.

Fighting for the Faith

Although Jeanne's gender normativity varies strikingly across these early Arabic biographies, her religious devotion is consistently attested as a praiseworthy feminine trait, albeit with intimations of "womanly superstition" in the biographies published by al-Bustani and Zaydan. The mix of religious and battlefield zeal aligns biographies of Jeanne with those of another group among Fawwaz's *Pearls,* early Muslim women famous for battlefield roles alongside the Prophet, often downplayed by later pre-modern treatments.[59] These early Muslim women fit into a longer tradition of local female warriors, pre-Islamic Arabian Bedouin adept at poetry and the sword, creating continuity if not direct linearity with stories of the region's ancient warrior-queens. Warrior-poet Khawla bt al-Azwar and others, profiled by Fawwaz and in women's magazines, exemplified women's dedication and capability as workers for the nation or the community of Muslims. The second issue of the long-running women's magazine *Fatat al-sharq* (Young woman of the East, est. 1906) featured 'Ikrisha bt. al-Atrush, 'who united attributes of courage and of literary refinement. She attended the Battle of Siffin and, standing between the two ranged sides, made eloquent speeches provoking the army of 'Ali b. Abi Talib, may God be pleased with him."[60] Ancient Bedouin women on the battlefield were a usefully vague referent. Were they fighting? Or playing crucial support roles: shaming men into action and challenging foes, binding wounds, carrying water, composing elegies for martyred male family members? This ambiguity echoed the gender equivocacy of profiles of Jeanne. Like her, they were profiled as "national" heroines in the post-1918 intensification of nationalist anti-colonial sentiment. In the 1890s as in the 1920s, their images might be problematic for proponents of gender-differential clarity, who magnified women's domestic roles and expressed anxiety at women's political-professional demands. Correspondingly, they were useful to those like Fawwaz who argued the opposite.

Another familiar warrior-woman category was problematic for congruent reasons: heroes of Arabic popular epic. As a genre associated with unlettered audiences and vernacular oral recitation, which publishers repackaged in cheap, poorly printed booklets, epic and its spin-off tales were a cultural institution not celebrated by modernizing intellectuals. If the performance tradition foregrounded male storytellers, in reform discourse epics and other vernacular narratives were sources of legend fuelling "old wives' tales," akin to the women's talk inferred in *al-Hilal*'s Jeanne biography. But epic was also associated with the feminine through women-warrior characters. These works "usually present legendary versions of historical events, and famous historical figures play leading roles in them," Remke Kruk notes, while they combine the fantastic and the mundane—like some *Pearls* sketches.[61] They draw on legend-wrapped tales of ancient Arab and early Muslim exploits, and a tradition of female knights in Arabic, Persian, Urdu, and Turkish literatures, although no archaeological record exists of indigenous female warriors.[62] Kruk finds epic's female heroes paralleling legendary ("historicized" through biography) tales of Semiramis or Zenobia, indeed themes present in retellings of Jeanne d'Arc's life: her childhood penchant for horseback riding, her cross-dressing.

Associated with "the popular" and the imaginary, perhaps epic's warrior women were too challenging as gender-bending female icons to figure in turn-of-the-century discussions of gender right. They inhabited an archive of performance and memory: did their shadows haunt the modern production of women-warrior biography for a target audience of middle-class readers? Woman-warrior figures did not appear in Arabic novels (though a few female characters found themselves obliged to take up arms in self-defense), even though the early Arabic novel did borrow motifs, character types, and plot lines from oral storytelling.

Enter the Amazons

As Kruk notes, Arabic epic's warrior women are not the Amazon figures of ancient Greek myth and modern imaginaries. Their stories accord with some classical historical descriptions of fighting women within Scythian and adjacent semi-nomadic groups. The epics' women did not live in female-dominated societies or woman-only communities. They were not hostile to men: they married, and supported sons and daughters both. They partook in Muslim piety and fought for Islam. They did not challenge patriarchal structures.[63] Occasional motifs suggest that artists and audiences crafting epics over time were familiar with Amazon legends coming into Arabic through translations of Hippocrates and Alexander narratives.[64] A tenth-century Arabic medical work described Amazons to illustrate the category of "masculinized woman":

> those whose femininity has coarsened and whose nature has become close to that of men who are brave, gallant, and courageous. There are thus those who set out to encounter valiant warriors and to venture into battle, in the way that 'A'isha (with whom God may be pleased) did on the Day of the Camel, and like some of the women of the Turks and the daughters of the Byzantines do.[65]

That this writer associated the Prophet's spouse 'A'isha with the Amazons, for her presence and role at a battle over leadership succession, disapproved of then and later, presages the term's modern re-emergence as women were seeking political subjecthood. In Arabic, by the 1890s Amazons had made their way into public discourse—but not as role models. The "Amazons" entry in al-Bustani's encyclopaedia had been succinct and neutral: some writers insisted on their historicity, others disputed it. They "made great conquests, to the border of Assyria, penetrating Africa and Europe, building cities in Asia Minor." The entry mentioned famous Amazon queens associated with the Greeks; African Amazons; Scythians; and the eighth-century Bohemians "whose existence is not doubted; under Libussa and Vlasta's leadership they sowed terror…and were destroyed only with great effort."[66] These were the Bohemians al-Nahhas cited as inspiring young Jeanne. Al-Bustani did not draw on Czech nationalist reconstructions of these figures as embodying "gender harmony." But neither did he highlight competing images coupling "Amazons" with "cruel man-haters."[67]

Da'irat al-ma'arif was not the Amazons' first nineteenth-century Arabic staging ground. In September 1871, the Egyptian government's school magazine, *Rawdat al-madaris al-Misriyya* (Egyptian Schools' Garden, established 1870 in Cairo) published "The Amazon Army." Reprising Greek and Roman reports on ancient Amazons of the Caucasus and Spanish explorers' narratives on Central American Amazons, the essay incorporated a report on the "Amazon army" of Dahomey translated from the London magazine *Gentleman's Journal*. It translated the *Journal*'s "news hook," too: Women "of strong minds" who were "demanding insistently to have men's pleasures, their work, their management [roles] and responsibilities" needed reminding that this might mean facing "sailors' storms and battleground fighting." (The Arabic translation modulates the original's racialist language somewhat.[68]) Dahomey's fighting women are described as fierce, cruel warriors, "surpassing men in their savage deeds"; they are portrayed as frenzied and half-mad rather than as skilled soldiers.[69] The Arabic article's opening echoed the translated text's warning:

> Often, we have recollected the Amazonian army, especially when it reached us that the women of France were enlisting as soldiers during the siege of Paris to help defend their homeland and cherished possessions. No doubt everyone who has heard about the Amazon Army of France wants more information about that army which inspired France's women to take it as a model in the recent war—especially since many women in the world, notably those famous for strong minds and bold initiative, are trying these days to obtain all the rights men have, and are demanding that female rulers, ministers and judges be drawn from among them. Obviously, women who demand to enjoy the rights of men are obligated to bear the [same] hardships and troubles they [masc.] bear, and must prepare themselves accordingly.[70]

The spectral "Amazon" was shaped by contemporary gender politics.

There is another layer to the story: *Rawdat al-madaris* took this article—including the embedded translation—from Butrus al-Bustani's serial *al-Jinan* (established 1870), published in Beirut.[71] Here, "Amazons" did not enjoy the neutral tone of the encyclopedia. *Al-Jinan* reported on France's new "Amazon Army of the Seine" in December 1870, based on an article in *The Times*. Women were enlisting in droves, ready to defend their capital city.[72] The description gestured to gender-specific expectations that drew on familiar narrative conventions, emphasizing these female soldiers' envisioned role as carers for (male) soldiers. There was a "gender-segregated" aspect to the venture: the women's uniforms and light rifles were to be funded by donations from wealthy women aware that "the enemy would plunder their finery and jewels, were they to conquer Paris."[73] In this circulating discourse on "Amazons"—the *Gentleman's Journal* and *Times* in London, the *Times* correspondent in Paris, *al-Jinan* in Beirut and *Rawdat al-madaris* in Cairo—prescriptions and proscriptions for women, their own public aspirations, male commentators' palpable anxiety about women's political demands, and spectral histories were intertwined. Was *al-Jinan* trying for the jocular in its conclusion? "What is hoped is that our women do not come to the point of forming an Amazon army like America's Amazons, lest we lose them; nor like Asia's Amazons, lest they kill males; nor like Africa's Amazons, lest they lose their gentle kindness."[74] But, it added, women *were* to acquire "energy and initiative" such that they might no longer frighten their children with tales of the bogeyman and the ghoul. Did *al-Jinan*'s hopes (repeated in *Rawdat al-madaris*) dispel the "ghoul" of the Amazon by evoking the figure of the mother?

A decade later, male Beiruti intellectuals were deploying the Amazon image even more antagonistically, as Arab women were becoming increasingly publicly vocal on political rights.

Amazons and Electoral Politics

Although the 1890s were not the first decade witnessing Arabic-language debates on masculinity and femininity, or "nature" and social roles, arguably issues of gender and authority came to the fore then. As girls' schooling became more acceptable, worries about elite girls' and women's reading, public presence, and alleged consumption practices mounted (discursively, at least) with the growing presence in Arab cities of Europeans and their social, commercial, and cultural institutions (including "French novels" in translation). Anger at European imperial reach in Africa and Asia, especially Britain's ongoing occupation of Egypt and French colonial rule in northern Africa, gave a political edge to local attacks on European "habits" and their moral and material costs. The proliferation of daily newspapers, magazines, and translated-adapted texts in print fostered discursive intimacy with European practices and concepts. Nowhere was this dialectic of familiarity/anxiety more evident than in debates and commentaries on gender and authority in modern societies.

Local outlooks and imported ideas converged in discursive constructions of domesticity as a privileged space of trained activity. This would (proponents believed) enhance women's and girls' senses of their appointed role in society and underwrite support for girls' schooling, while maintaining females' association with home and discouraging aspirations for public careers. But where did one draw boundaries? What of women who did not adhere to domestically bounded lives, whether out of choice or necessity? (As Fawwaz and others pointed out, Egyptian peasant and urban working-class women did not have the luxury of staying home.) As Fawwaz worked on her compendium, argument raged on gender's significance to modernity. Just months before *al-Hilal* published its biography of Jeanne d'Arc, it hosted a venomous debate, involving over a dozen contributors, women and men, on the question: "whether women can demand all the rights of men." This continued for six months, centered on issues of gender and nature, invoking European polemicists—a reminder that local commentators drew on all manner of European thought, against as much as for more flexible understandings of gender's meanings for social organization. Readers responded to a physician in the Egyptian Delta city of Damietta who attacked women's intellectual and professional aspirations by drawing on European medical authority, brain "science," and theories of animal evolution. Women, he argued (drawing also on medieval European and Arabophone writings), were inferior mentally and physically, capable only of housework and childcare, a difference decreed by science, history, and God. Two months after Jeanne's biography appeared, this doctor, Amin al-Khuri, sent *al-Hilal* a similarly dismissive response to the question: "whether learning or wealth most improves a woman's status." Other contributors opted for "learning" (the era's keyword). Khuri went for "wealth," saying female "learning" could never amount to much. Implicitly, he excluded women as unable even to comment.[75] Not long before, Fawwaz had engaged in a debate on similar themes via a different approach: Qur'an interpretation and (mis)uses of historically accreted exegesis.[76] This followed her earlier debate with Lebanese writer Hanna Kurani over "women and nature": Kurani espoused the "liberal" view that women equaled men in capacity but their capabilities were altogether different. But, like many in Europe, she saw gendered spheres as divinely decreed. Did Fawwaz's biography of Jeanne, with her divine sanction for military leadership, possibly contest that view? Or was the extraordinary Jeanne, as *al-Hilal* suggested, temporarily an honorary male, graced by feminine traits? What sparked the women's debate was Kurani's critique of British suffragists. Fawwaz responded furiously, taking a social-constructionist position and praising British women for far-sighted activism.[77]

The Beirut newspaper *Lisan al-hal* agreed with Kurani, said Fawwaz. This prominent paper's management (founder-editor Khalil Sarkis and managing-editor nephew Salim Sarkis) was bothered enough to publish a long editorial citing previous articles supporting "women's work," although, it clarified, the newspaper did not support women's political rights. Many amongst the liberal male intelligentsia liked to champion "women's rights," but only insofar as these did not challenge family or state authority, vested in men. Why, an article in *Lisan al-hal* commenting on the *al-Hilal* debate mused, did women waste their time writing on political rights? Couldn't they write about childrearing instead? Biographies of politically and militarily active women had to contend with this constrictive discursive environment.

Consider *Lisan al-hal*'s representation of European and North American women's political activities. Its coverage of the 1892 women's suffrage bill, narrowly defeated in Parliament following William Gladstone's strong opposition, disparaged the suffragists' campaign as it told readers its sole concern was "moderation." Its anxiety about women's political aspirations was already clear in frequent reporting on women's activisms, the tone varying from guarded approval to patronizing scepticism to outright disgust. In 1890, an article on "American Amazons" conveyed "wonder" that a group of New York women sought martial arts training. The headline was footnoted: "Amazon is a name applied to mannish women who fought in ancient times." This gloss distanced fighting women even as the article brought "Amazons" into contemporary life. The opening linked "Amazons" to contemporary gender politics as a widespread preoccupation.

> In the port city [Beirut], they confer on whether civilisation rests more on men or women. Amongst one who's positive and one who's negative, one passive and another combative, and yet another demanding women's rights, a group of men exists who recognize no womanly distinctions except childbirth, child-raising, and keeping an orderly home. Today we give them news that newspapers in the West are transmitting in astonishment—and we are first among them in that.[78]

The article exploited gender stereotypes to belittle women: surely it was only a matter of their attraction to wearing military uniform. Amidst furious debates over gender equality in the United States, *Lisan* speculated, men were asserting superior strength; women felt obliged to respond. Hence the New York "Amazons"—explicable only in responsive mode, whether to fashion or to men.

The article ended on a note of uneasy mockery. "It is hoped that when these new armies enter the battlefield, they will restrain themselves from killing and spilling blood viciously, as related by the historian Herodotus about Scythian Amazons, who on the battlefield proved harsher-hearted than men." The tone[79] turned darker: colonialist racism and misogyny conjoined in a derisively worded image of "women's armies" defending the continent from native inhabitants: "How sad if we see more than one Amazon skull—known to more than one hairdresser in New York City—becoming décor elements in Indian huts." Contempt, stereotype, and patronizing "humor" laced the finale.

> If the American Amazons intend to stand firm and serious they must prepare to suffer.... we don't doubt women's ability to appear strong and active, judging by the energy we have seen... in their daily transactions with husbands, brothers, and male relatives. It is not farfetched that they could resemble Napoleon or Wellington, knowing how to direct their strengths at the enemy's weaknesses. Anyway, America's men might be very willing to accept their mothers-in-law entering the battlefield.[80]

If this was gender fluidity or gender inversion, it was only sarcastically so, and served rather to solidify the cross-societal misogynistic notion of "women's wiles." The tone echoed the anxieties voiced in the 1870s "Amazon" articles about women's evident martial proclivities, the fear that they would lose their "gentle kindness."

Other articles on women's crime and violence used the label "Amazon."[81] Referring to readers' disbelieving reaction to the American Amazon story, one addressed recent election of an all-woman slate to the town council, Edgarton, Kansas. "Woman and her political position" linked women's "militancy" to political aspirations. Seemingly approving male politicians' support for political women (based on women's "propensities": "cleanliness and peace"), it concluded that women's political work was doomed to failure. "A bad outcome for women's rights: recently in Edgarton, Kansas, women were elected.... They exerted themselves to the extent of their ability to undertake action and reform, but their actions were criticized widely and forcefully. They all resigned, returning to the domestic domain."[82] *Da'irat al-majlis al-baytiyya* ("sphere of the domestic council") satirically echoed the municipal council, *al-majlis al-baladi*. If the women's adversaries were implicitly to blame, so, it seems, were the women.

In *Lisan al-hal,* stories of women asserting physical strength and acting violently and criminally commingled with reports of women's political aspirations. Circulating amongst newspapers—from London or Chicago to Beirut, between Beirut, Cairo, and Istanbul—stories created a multi-sited echo chamber of masculine authority. Reporting on a young German woman arrested for spying near Paris, *Lisan* reproduced the comment of its source, Egyptian paper *al-Mahrusa*: "It is astounding how far the Europeans have gone: their women are not satisfied vying with men in knowledge's arena, refined comportment's racecourse: no, they must compete in the military and political arts."[83]

As images in London's *Punch* and, earlier, in the *Gentleman's Journal,* suggest, "Amazon" was a powerful symbolic figure for a scare-discourse on suffragists. Amazons stood for women *en masse*, women fighting for women, emasculating men. Warriors by "profession" rather than for temporary missions, they challenged a patriarchal discourse dependent on "nature's" order. For medieval Europeans, Carol Myscofski notes, Amazons "represented the wholly other...warriors possessed of corrupted femininity and disharmonious attributes."[84] This captures *Lisan al-hal*'s "Amazons."

A potent trans-cultural image, Amazons in *Punch, Lisan al-hal,* and other organs calibrated men's anxieties about women's political demands as threatening the gender-hierarchical social order. For their image challenged assumptions about patriarchal authority as "natural." Unlike Jeanne d'Arc, these women could not be made palatable as figures acting in obedience to a higher (male or male-god) authority. No wonder German feminists' writings were embracing Amazon collectivities.[85] In Arabic, al-Nahhas's reference to Bohemian queens as Jeanne's model appears the sole positive reference. Was it deliberate that she did not call them "Amazons"? For Amazons as a named category were deployed to discredit feminists. And if, in men's imaginings across centuries, Amazons were "peripheral to the world of whoever [was] writing about them,"[86] finding Amazons in North America or France was alarmingly close, in a world where some local women supported their Western peers' activism, even if they carefully distantiated their praise.

Lisan al-hal welcomed female writers. This was significant, a strong signal that women were entirely capable of participating as serious commentators on public issues of the day. But the lines were firmly drawn. *Lisan al-hal* drew on the language of rights in a particular way: "Female writers of Syria...have risen, demanding women's rights in the sphere of knowledge: if she cannot bear a sword, she can carry a pen." Calling women's intellectual work central to the "knowledge awakening," *Lisan* circumscribed acceptable "women's rights" to the right to write, and took credit for generating women's presence in public discourse.[87] It encouraged men to "aid" daughters, wives, and female relations to "cultivate their minds...and push them to action." Again, the newspaper congratulated itself: "*Lisan al-hal* has...fervently encouraged them, opening its doors to accept their writings, now seeing the fruit of its efforts ripening." No wonder women like Fawwaz, Nawfal, and al-Nahhas emphasized women's own efforts—women's heroism, whether on horseback or as journalists, in a fluid space not fully defined by biology. *Lisan al-hal*'s self-congratulatory article appeared one day after defeat of the woman suffrage bill in London, a defeat *Lisan* approved.[88]

From Gender Fluidity to Invisibility: Jeanne d'Arc's Ottoman Sex-change

"Jeanne" as Arab hero entailed multiple displacements. Yet, her un-Amazon-like story as well as her remoteness—in time, place, and cultural milieu—facilitated her adoption as a locally intelligible hero, while the emerging ideological commitments of anticolonial nationalism would mute that distance, as Jeanne in Arabic became the thoroughly feminized embodiment of nationalist womanhood: the refined peasant, the home-oriented woman dedicated to the nation, the faith-and-nation anti-"British" helpmeet. Earlier, in the 1890s, gender fluidity appeared more prominent. Al-Nahhas's heroic history maven and Fawwaz's forthright horsewoman contrasted with Zaydan's gender "confusion," al-Bustani's implied discomfort with visionary leadership, and *Lisan al-hal*'s prescriptively gendered spheres. The profiles appearing in men's publications tended to mute gender fluidity or turn it into a more troubling gender confusion or ambiguity. In Jeanne's very oddity, and her linkage to feminine and unlettered "superstitions" (antithesis of modernity), and differently in *Lisan*'s Amazon figures, these narratives reaffirmed patriarchal normativity. In contrast, Fawwaz and al-Nahhas embraced Jeanne as a figure whose fluidity was a chosen strategy, a prooftext for feminine agency. They showed Jeanne and other warrior women as working on behalf of female and popular collectivities, whereas for the texts produced by men, Jeanne's acts were singular. As Jeanne became increasingly nationalized, that she fought for a national collectivity and not for herself or for women meant that her "excesses" in gender terms could be reassigned as self-sacrificial crisis strategy. This was not the case for Amazons and other women warriors. For those advocating women's activisms on behalf of women, the woman warrior might stand as a challenge to masculine patriarchal authority.

Halfway between the nineteenth-century profiles and the 1920s nationalized and domesticated Jeanne, two texts demonstrate her multifaceted symbolic utility across spatial-cultural divides. In March 1910, *Fatat al-sharq* published a translated notice: "View of the Chinese on Jeanne d'Arc." Someone with a French name "about to join missionaries in China" had sent it to a French journal dedicated to Jeanne d'Arc. He described asking a senior Chinese bureaucrat "what his people thought of France's saviour." The bureaucrat remarked on "Europeans' prevalent view" of China as "steeped in the darkness of savagery." But, he added, "knowledge is spreading fast." If "the common people" were ignorant of Jeanne d'Arc, "eminent people have studied European history." Not only did they know of Jeanne: "our scholars, even if most follow Confucius, believe in her divine mission.... The barbarians of the Far East prove they have more finely-tuned views than some European historians." The speaker mentioned an ancient Chinese monarch, Wu Zetian [*Wutsutin*] (624–705), as a powerful female ruler, but privileged Jeanne as "image of your nation, in her live belief, dignity and courage." Knowledge of Jeanne d'Arc becomes a marker of worldliness, of belonging to a "global" modern elite, as she is "back-translated" from Chinese to French, and thence to Arabic. Yet this particular instance of the international circulation of Jeanne's symbolic potential effaces her gendered embodiment and social embeddedness. The same was happening more locally.

In 1908, the triumph of Ottoman constitutionalist opposition to Sultan Abdülhamid II's repressive regime was greeted euphorically throughout Ottoman Arab territories. Among peons to the constitution was a poem appearing in *Fatat al-sharq*, edited in Egypt by a female Syrian émigré. "That's Jeanne d'Arc, but she's Turkish," authored by a male poet in Alexandria, appeared initially to celebrate female heroism.[89]

> A splendid maiden of the Turks has rescued us
>   Through exertions of the free and the army
> She has pulled us out from the darkness into light,
>   And we've become like the peoples of strength

The poem recounted a history of tyranny ending in emergence of a "party" (*hizb*) demanding "justice, brotherhood, equality." It criticized political division based in religious difference. "Do not say you are Muslim or Christian/for in rights we are all equal." The reformist party, Young Turkey (Turkiya al-fatat; in English, usually Young Turks), is gendered feminine through the grammatically feminine place name, modified by the adjective "young" a homonym of the magazine's title rubric for young womanhood. This party is "the Turkish Jeanne d'Arc." But the "maiden" party is a brotherhood, eclipsing the femaleness of *al-fatat*. The poem lauds by name individual male heroes of the constitutional movement, with no mention of women amongst this hailed collectivity. This Turkish (in Arabic) Jeanne d'Arc is a man amongst men; or, the allegorical sign of fraternity.

This translation of Jeanne's image into a collective brotherhood—exploiting the historical figure of Jeanne to efface it, rendering Jeanne's history as myth and re-gendering it male—appeared a decade before the surge of Egyptian-nationalist-inflected Arabic Jeanne biographies. Literarily obscure, it speaks volumes on gender's political uses in an emerging nationalist agenda that invoked Woman as sign and helpmeet of modernity while silencing women as political agents. Instantiating a familiar phenomenon in liberationist nationalisms the world over, this figuration of gender "fluidity" used feminine heroism in the name of masculinized activism, even as women in Egypt and across the Ottoman Empire were articulating gender-specific demands in a language of rights.

Read in these overlapping contexts, did Jeanne d'Arc in Arabic moderate and manage the emergent-envisioned nation's need for active women against the frightening specters of militant or demandingly political women that the local press "retweeted" translationally as it reported on world events? Perhaps Jeanne in Arabic also tamed warrior-woman images from classical Arabic biographical sources and heroic epics of quasi-legendary ancient conflicts. Jeanne was not the Amazon haunting men's writings on women, rights, and politics. Yet "Jeanne" as female action hero, in the known lineaments of her life, remained inevitably an ambivalently posed figure, embodying gender transgression *as well as* acceptable "feminine" sacrifice for the homeland. Arabic treatments of her history, posed amidst those of other sword-wielding women, testify to the difficult salience of feminine public articulation and political work in a transitional-colonial modernity: what would it mean for the men?

Notes

1 [Hind Nawfal], "Iydah wa-iltimas wa-istismah," *al-Fatat* 1, no. 1 (November 20, 1892): 1–7. Translations mine unless noted otherwise. I use "Jeanne d'Arc" rather than the English equivalent because this is how it was transcribed in Arabic: *jān dārk*.
2 Marilyn Booth, *May Her Likes Be Multiplied: Biography and Gender Politics in Egypt* (Berkeley: University of California Press, 2001).
3 Marilyn Booth, "Jeanne d'Arc, Egyptian Nationalist: Community, Identity, and Difference," chap. 6 in Booth, *May*, 233–269; Marilyn Booth, "The Egyptian Lives of Jeanne d'Arc," in *Remaking Women: Feminism and Modernity in the Middle East*, ed. Lila Abu Lughod (Princeton: Princeton University Press, 1998), 171–211. On her family's socio-economic status, e.g., Jan van Herwaarden, "The Appearance of Joan of Arc," in *Joan of Arc: Reality and Myth*, ed. D.A. Berents et al. (Rotterdam: Erasmus University; Hilversum Verloren, 1994), 19–70; 35; Larissa Juliet Taylor, *The Virgin Warrior: The Life and Death of Joan of Arc* (New Haven: Yale University Press, 2009), 5–6. I mention more of the many sources on Jeanne in *May*.
4 Venita Datta, *Heroes and Legends of Fin-de-Siècle France: Gender, Politics, and National Identity* (Cambridge: Cambridge University Press, 2011), 31; chap. 4.
5 Sufi 'Abdallah, *Nisa' muharabat* (Cairo: Dar al-ma'arif, 1951); Fa'iza 'Abd al-Majid, *al-Mar'a fi mayadin al-kifah* (Cairo: Wizarat al-thaqafa, 1967).
6 Yuhanna Abkariyus, *Qatf al-zuhur fi ta'rikh al-duhur*. Second edition (Beirut: n.p., 1885), 451–452. On the sorcery issue, Marina Warner, *Joan of Arc: The Image of Female Heroism* (New York: Knopf, 1981); "Joan of Arc: A Gender Myth," in D.A. Berents et al., *Joan of Arc*, 97–117.
7 On these works, Marilyn Booth, *Classes of Ladies of Cloistered Spaces: Writing Feminist History through Biography in* Fin-de-Siècle *Egypt* (Edinburgh: Edinburgh University Press, 2015), chap. 5.
8 Butrus al-Bustani, *Kitab da'irat al-ma'arif* 6 (Beirut: Matba'at al-ma'arif, 1882), 360–361.
9 al-Bustani, *Da'ira* 6, 360.
10 al-Bustani, *Da'ira* 6, 361.
11 Nasim Nawfal, "I'lan," *Misr* 2, no. 47 (May 24, 1879): 4, quoted in Booth, *May,* 2–3. Fawwaz detailed al-Nahhas's project and setbacks (Booth, *May,* 317–18n7).
12 Maryam ibnat Jabra'il Nasrallah al-Nahhas qarinat Nasim Nawfal al-Tarabulusiyya al-Suriyya, *Mithal li-kitab Ma'rid al-hasna' fi tarajim mashahir al-nisa'* (Alexandria: Matba'at Jaridat Misr, April 25, 1879). The copy I used lacked pp. 2–6; Jeanne's entry must begin on p. 6.
13 al-Nahhas, *Mithal*, 7.
14 The copy is torn; partial names remain.

15 It may be that Amazon lore was more known and accepted than contemporary scholars have generally thought—but in medieval France? Adrienne Mayor, *The Amazons: Lives and Legends of Warrior Women Across the Ancient World* (Princeton: Princeton University Press, 2014).
16 Jitka Malečková, "Nationalizing Women and Engendering the Nation: The Czech National Movement," in *Gendered Nations: Nationalisms and Gender Order in the Long Nineteenth Century*, ed. Ida Blow, Karen Hagemann, and Catherine Hall (Oxford: Berg, 2000): 293–310.
17 Quotations: al-Nahhas, *Mithal*, 7, 8, 8, 8.
18 Taylor, *The Virgin Warrior*, 35, 204; van Herwaarden, "The Appearance," 30–32, 40.
19 Al-Nahhas, *Mithal*, 9.
20 al-Nahhas, *Mithal*, 10.
21 al-Nahhas, *Mithal*, 11.
22 Alexandre Dumas [père], *Jehanne, la Pucelle, 1429–1431* (Paris: Magen et Comon, 1842). In English, Alexander Dumas, *Joan the Heroic Maiden*, trans. Louisa C. Ingersoll (New York and Philadelphia: E. Ferrett and co., 1846). After writing this, I was interested to see that Nora Heimann mentions this work as an important contributor to Jeanne's shifting image in nineteenth-century France, toward the spiritual. Nora M. Heimann, "The Princess and the Maid of Orléans: Sculpting Spirituality During the July Monarchy," in *Joan of Arc and Spirituality*, ed. Bonnie Wheeler and Ann Astell (ser. The New Middle Ages, New York: Palgrave/St. Martin's Press, 2003), 229–247; 238.
23 Zaynab bt. 'Ali b. Husayn, b. 'Ubayd Allah b. Hasan b. Ibrahim b. Muhammad b. Yusuf Fawwaz al-'Amili, *al-Durr al-manthur fi tabaqat rabbat al-khudur* (Bulaq/Misr: al-Matba'a al-kubra al-amiriyya, 1312 AH), 122. On the publishing history, Booth, *Classes*, chap. 1.
24 Booth, *May*, 242.
25 *al-faras* can be masculine or feminine: Fawwaz refers to it using the masculine pronoun.
26 Fawwaz, *al-Durr*, 122.
27 "Ikram katiba muslima lil-qadisa Jan dark," *al-Mashriq* 19, no. 2 (February 1921): 108–114; 108.
28 "Ikram," 108fn3. Footnotes correct spellings of personal and place names—errors perhaps due to Fawwaz (or typesetters) not knowing French.
29 "Ikram," 109fn2.
30 Fawwaz, *al-Durr*, 122.
31 See my forthcoming book, *Feminist Thinking in Fin-de-siècle Egypt: The Career and Community of Zaynab Fawwaz*.
32 "Ikram," 109fn6.
33 Fawwaz, *al-Durr*, 122; "Ikram," 109fn8.
34 Helen Watanabe-O'Kelly, *Beauty or Beast? The Woman Warrior in the German Imagination from the Renaissance to the Present* (Oxford: Oxford University Press, 2010); Warner, *Joan of Arc*.
35 Fawwaz, *al-Durr*, 123; "Ikram," 110fn3, noting a grammatical discrepancy; Fawwaz changes masculine [*hadha… huwa*] to feminine [*hadhihi… hiya*]. Q 12: 31. This passage is one source of notions of "women's cunning" (here, "treachery"), *kayd al-nisa'*, which Fawwaz interrogates in biographies.
36 Booth, *May*, chap. 6.
37 Fawwaz, *al-Durr*, 123–124; "Ikram," 112fn2.

38 Fawwaz: as fire consumed Jeanne, "she implored, in a voice to make her enemies weep." *Al-Mashriq* adds: "She was imploring the Messiah, asking that his sacred cross be placed before her, and she died saying: O Jesus. The flame did not touch her heart." Fawwaz, *al-Durr*, 124; "Ikram," 113fn7.
39 "Jan dark: Fatat Urliyan," *al-Hilal* 4, no. 4 (October 15, 1895): 121–128; 4: 5 (November 1, 1895): 166–168; 121.
40 Nora M. Heimann, *Joan of Arc in French Art and Culture (1700–1855): From Satire to Sanctity* (Aldershot, Hampshire: Ashgate, 2005), 130–154; Heimann, "The Princess," 229–247. And personal communication from Nora Heimann (January 2019), for whose comments I am grateful, while remaining responsible for my own interpretation.
41 Heimann, "The Princess," 236.
42 [Jurji Zaydan], "Iltimas min hadarat al-qurra'", *al-Hilal* 4, no. 11 (February 1, 1896): 407.
43 Heimann, "The Princess," 229–230.
44 "Jan dark," *al-Hilal*, 122.
45 Barbara Stowasser, *Women in the Qur'an, Traditions, and Interpretation* (Oxford: Oxford University Press, 1994), 69, 77–78.
46 "Jan dark," *al-Hilal*, 122.
47 Marilyn Booth, "Fiction's Imaginative Archive and the Newspaper's Local Scandals: The Case of Nineteenth-Century Egypt," in *Archive Stories: Facts, Fictions, and the Writing of History*, ed. Antoinette Burton (Durham: Duke University Press, 2006), 274–295.
48 "Jan dark," *al-Hilal*, 122.
49 "Jan dark," *al-Hilal*, 122.
50 Quotations: "Jan dark," *al-Hilal*, 124, 127, 128, 128, 168.
51 Fawwaz, *al-Durr*, 102, taken largely from Butrus al-Bustani, *Kitab da'irat al-ma'arif*, vol. 5 (Beirut: Matba'at al-ma'arif, 1881), 658–659. See Booth, *Classes*, 409n97. In a coeval historical (publication) parallel, Charlotte Stopes began her 1894 pamphlet for British suffrage societies with "ancient British women," specifically Boudicca. Charlotte Carmichael Stopes, *British Freewomen: Their Historical Privileges* (London: Swan Sonnenschein, 1894). Jane Rendall, "Recovering Lost Political Cultures: British Feminisms, 1860–1900," in *Women's Emancipation Movements in the Nineteenth Century: A European Perspective*, ed. Sylvia Paletschek and Bianka Pietrow-Ennker (Stanford, CA: Stanford University Press, 2004), 33–52; 33.
52 Booth, *Classes*, 55–56, 84–85, 170, 183–185.
53 Fawwaz, *al-Durr*, 191–192; Booth, *Classes*, 83–84. The sorcerer is also a *Pearl* (*al-Durr*, 90).
54 Quotations: Booth, *Classes*, 187–190. Fawwaz, *al-Durr*, 251–252; al-Bustani [family], *Kitab da'irat al-ma'arif*, vol. 10 (Cairo: Matba'at al-Hilal, 1898), 90–91; Abkariyus, *Qatf al-zuhur*, 18–20.
55 Watanabe-O'Kelly, *Beauty*, 194.
56 Butrus al-Bustani, *Kitab da'irat al-ma'arif* vol. 3 (Beirut: Matba'at al-ma'arif, 1878), 792; Fawwaz, *al-Durr*, 42; Booth, *Classes*, 178.
57 Adrian Shubert, "Women Warriors and National Heroes: Agustina de Aragón and Her Indian Sisters," *Journal of World History* 23, no. 2 (June 2012): 279–313, esp. 297–301.
58 Watanabe-O'Kelly emphasizes the mythological as prevalent until the modern era in German warrior-women representations. What work does the history/myth divide itself accomplish?

59 Asma Afsaruddin, "Early Women Exemplars and the Construction of Gendered Space: (Re-) Defining Feminine Moral Excellence," in *Harem Histories: Envisioning Places and Living Spaces*, ed. Marilyn Booth (Durham, NC: Duke University Press, 2010), 23–48.
60 Tammam al-Karkabi, "Shahirat al-nisa': 'Ikrisha ibnat al-'Atrush," *Fatat al-sharq* 1, no. 2 (November 15, 1906): 47–48; 47. A pioneer of under-the-masthead female biography, *Fatat al-sharq* would publish biographies of Jeanne d'Arc and Zaynab Fawwaz.
61 Remke Kruk, *The Warrior Women of Islam: Female Empowerment in Arabic Popular Literature* (London: I.B. Taurus, 2014), 2.
62 Kruk, *The Warrior Women*, 15–16. See also Mayor, *The Amazons*.
63 Kruk, *The Warrior Women*, 19–20.
64 Kruk, *The Warrior Women*, 16–17.
65 Kruk, *The Warrior Women*, 16, quoting al-Marwazi's *Kitab taba'i' al-hayawan*.
66 Butrus al-Bustani, *Kitab da'irat al-ma'arif*, vol. 4 (Beirut: Matba'at al-ma'arif, 1880), 349–350.
67 Malečková, "Nationalizing Women," 304.
68 My quotes are from my back-translation from the Arabic rather than from the original. The translation of Amazons from English to Arabic is fascinating but cannot be pursued further here. "Female Warriors of Dahomey," *The Gentleman's Journal* (November 1, 1869): 77–78.
69 "Jaysh al-Amazun," *Rawdat al-madaris al-misriyya* 2, no. 11 (15 Jumada II 1288 [September 1, 1871]): 6–10; 8, 9.
70 "Jaysh al-Amazun," 6.
71 "Jaysh al-Amazun," *al-Jinan* 2, no. 14 (July 5, 1871): 474–477.
72 "Jaysh al-Amazun al-Sayni," *al-Jinan* 1, no. 23 (December 1, 1870): 711–712.
73 "Jaysh al-Amazun al-Sayni," 712.
74 "Jaysh al-Amazun," 477.
75 The debate occurred largely in *al-Hilal* January, June 1894. Fruma Zachs and Sharon Halevi, *Gendering Culture in Greater Syria: Intellectuals and Ideology in the Late Ottoman Period* (London: I.B. Tauris, 2015), 32–33; Marilyn Booth, "'This tyranny you have called nature': Misogyny and/as Science," unpublished ms. Later intervention: Amin Khuri, 'Bab al-murasalat: Hal ta'lu manzilat al-mar'a bi-l-'ilm akthar am bi-l-mal," *al-Hilal* 4, no. 13 (March 1, 1896): 489–494.
76 Discussed in my forthcoming book on Fawwaz.
77 Marilyn Booth, "Peripheral Visions: Translational Polemics and Feminist Arguments in Colonial Egypt," in *Edinburgh Companion to the Postcolonial Middle East*, ed. Anna Ball and Karim Mattar (Edinburgh: Edinburgh University Press, 2018), 183–212.
78 "al-Amazuniyyat al-amrikiyyat," *Lisan al-hal* 13, no. 1236 (May 29, 1890): 2. Source given as the *Telegraph*.
79 It is unclear what is quoted and what is *Lisan al-hal*'s commentary.
80 "al-Amazuniyyat al-amrikiyyat," 2. Source given as the *Telegraph*.
81 E.g., "Amazuniyya nimsawiyya," *Lisan al-hal* 13, no. 1238 (June 5, 1890): 3. Space limitations preclude elaboration.
82 "al-Mar'a wa-markazuha al-siyasi," *Lisan al-hal* 13, no. 1238 (June 5, 1890): 2.
83 "Jasusa alamaniyya," *Lisan al-hal* 14, no. 1277 (October 20, 1890): 3.

84 Carole A. Myscofski, *Amazons, Wives, Nuns and Witches: Women and the Catholic Church in Colonial Brazil 1500–1822* (Austin: University of Texas Press, 2013), 36. Also Watanabe-O'Kelly, *Beauty*, 38–42.
85 Watanabe-O'Kelly, *Beauty*, 39, chap. 8.
86 Watanabe-O'Kelly, *Beauty*, 39.
87 "Khatir mulahiz," *Lisan al-hal* 14, no. 1391 (March 14, 1892): 3. Unusually, a notice at the bottom of page 1 highlights this: "On the issue's last page is the article 'Khatir mulahiz', so consult it." See also "Majallat al-fatat," *Lisan al-hal* 14, no. 1404 (May 2, 1892): 1.
88 Booth, "Peripheral."
89 As'ad Efendi Arqash, "Tilka Jan Dark innama Turkiyya," *Fatat al-sharq* 3, no. 1 (October 1908): 22–24.

9

Gender and Transgender in the Mexican Revolution: The Shifting Memory of Amelio Robles

Gabriela Cano

Today, in the village of Xochipala, in the state of Guerrero, there is a school named in honor of Coronel Amelio Robles, a prominent man and landowner, and a hero of the Mexican Revolution. In that same village in southern Mexico there is a community museum honoring the life of woman warrior Coronela Amelia Robles, displaying some of the furniture and personal effects of this former Zapatista fighter. Standing a few hundred yards apart, these two sites of memory represent a profound transformation and a paradox, for Amelio Robles and Amelia Robles were not brother and sister, father and daughter, or husband and wife, but one and the same person.

This chapter examines the conflicting perceptions of Robles's gender identity in the context of twentieth-century Mexican history. Amelio Robles was a transgender man during most of his life. His gender transition—from female to male—began after he joined the Mexican revolution as a rank-and-file fighter when he was twenty-five years old. The violence and confusion of the armed struggle were propitious for his transition. Robles's personal qualities as a guerrilla fighter were valued and accepted as expressions of his masculinity by his comrades in the army. In the post-revolutionary era, masculine identity was generally accepted and Robles was respected as a local male hero of the Mexican revolution—being honored as an official war veteran by the highest military authorities. People in his community bore no doubts of his masculine gender, and he eventually became a respected elder in his town. The local school was given his name in order to honor his contributions to education in his village and the revolutionary movement in the state of Guerrero. Robles's family, neighbors, friends, and comrades-at-arms generally accepted his masculinity.

Moreover, identification papers and photographs gave his maleness an official status. However, the social perception of Amelio Robles's identity shifted radically after his death at the age of ninety-four. Local state agents imposed a feminine gender on Robles's identity. A narrative that blatantly contradicted Robles's masculinity took over in Guerrero's memory of the Mexican Revolution, where Robles became a symbol of the nationalist woman fighter and was included as such in the local historical narrative.

Figure 9.1 Colonel Robles School, Xochipala, Guerrero in 2012. Photo credit: Gabriela Cano

Figure 9.2 Community Museum Coronela Amelia Robles in 2012. Photo credit: Gabriela Cano

The binary gender of the nationalist celebration of the Mexican revolution that paved the way for the official acceptance of Robles's masculinity would eventually deny his transition and push him back into a female identity.

The following pages tell the microhistory of Robles and the paradoxical workings of the memory of the Mexican Revolution that led to the coexistence of a school honoring the male revolutionary and a community museum celebrating the woman fighter in the same town. A short biography of Amelio Robles highlights his gender transition in the context of the Mexican revolution. However, the focus of the chapter is not the details of his life but the effects of gender and nationalism in the social memory of Robles's identity.

Amelio Robles took part in the 1910 Mexican Revolution as a combatant in Emiliano Zapata's popular agrarian movement, or Zapatismo. Zapata led a rebellion of "campesinos" (poor peasants), who took up arms to defend their communal lands against the encroachments of privately owned haciendas (estates). Zapatismo emerged in the state of Morelos, where the movement was centered, and also had followers in some parts of Guerrero, Puebla, and the State of Mexico, in the country's central southern region. Robles's home town of Xochipala, in Guerrero, joined the movement with a twofold aim: to defend its traditional boundaries from the expansion of the nearby hacienda and to resolve a long-running agrarian dispute with the neighboring town.[1]

Zapatista fighters were not a professional army but rather a "people in arms"; that is, they were guerrilla groups who served a community leader skilled in combat.[2] War brought about the destruction of social and cultural structures and broke up traditional social hierarchies. Social instability, evacuations, and constant danger were all part of daily life for the Zapatista guerrillas. This created a favorable atmosphere for Amelio Robles to adopt the desired masculine identity and appearance: the young rural woman became a male Zapatista guerrilla. Robles adopted the wardrobe, pose, and body movements typical of men in the early-twentieth century from his rural part of Mexico.[3] The transition also entitled a name change: from the feminine Amelia, which was given to her at birth, to the masculine grammatical Amelio. (In Spanish, an "o" or "a" at the end of a word indicates grammatical gender; in proper names, the feminine or masculine grammatical gender usually corresponds to the person's gender identity.[4])

Unlike other women who only passed themselves off as men while participating in the armed uprising, Amelio Robles sustained his masculinity throughout his long life and even into old age. He died at the age of ninety-four, and those who knew him in his final years remember him as an elderly man respected by his family and the wider community alike. Neighbors and family members addressed him in the masculine grammatical gender as "Mr.," "colonel," "uncle," or "great uncle" and treated him with the respect typically accorded to an elder property owner and household head. Robles's standing in his community stemmed largely from his involvement in the Mexican Revolution.

The Mexican Revolution was an internal war and political process that overturned an oligarchical state and established a new government apparatus. The new state heard the voices of workers and peasants. The broad cross-section of society— middle class, peasants, workers, and landowners—entering the armed struggle made the Mexican Revolution a multi-layered and contradictory military and political process. Nationalism also came to the fore at various junctures. And to complicate matters further, the struggle not only pitted rebels against the former government and those supporting a continuation of the regime; there was also infighting among the revolutionary forces due to socio-economic, cultural, and regional differences, together of course with ideological and political reasons.

The term "Mexican Revolution" has several meanings and its polysemy is relevant to the story of Robles's transgendering and the conflicting perceptions of his masculinity. The term refers not only to the armed struggle, but it is also used to encompass a long process of state-building and legislation. That process includes the Constitution of 1917, which mandated a major agrarian reform, the protection of workers' rights and the imposition of restrictions on the Catholic Church. It also includes other laws such as the Law of Family Relations (1917), the Civil Code (1928), and the Federal Labor Law (1932). The Constitution of 1917 is often invoked to mark the end of the Mexican Revolution but this is a contentious issue. Some scholars argue that the Constitution of 1917 was a turning point, but not its end. Rather, they point to 1940, the final year of President's Lázaro Cárdenas term, and a period of extensive agrarian distribution, support of workers' demands, and nationalist economic policies as the conclusion of the revolutionary process. Furthermore, regime political actors, such as the single political party that ruled from 1929 until 2000, maintained that the ongoing, institutional process of the revolution continued until the 1970s or even later. The word "revolution" or one of its derivations has remained in the ruling party's name over the decades. In 1929, it was called the National Revolutionary Party, and in 1938 it took on a corporate structure and adopted the name Party of the Mexican Revolution. In 1946, its name changed to Party of the Institutional Revolution, as it continues to be known today.

There are at least three meanings of the term Mexican Revolution: first, as a war and political conflict; second, as the drafting and institutional enforcement of laws and reforms; and third, as rhetorical trope. The revolution was a symbol of nationalism, social justice, and secularization. Throughout the century, both governments and opponents invoked the Mexican Revolution as a source of their legitimacy. More importantly for our purposes, the Mexican Revolution celebrated binary gender as the epitome of national identity: the male peasant, worker, and soldier, and the sexually available young woman and mother.

Each of these meanings is relevant to the story of Amelio Robles. The process of his gender transition took place during the armed struggle and his constructed masculinity was accepted in the midst of the instability and violence of the war. The form of masculinity that Robles enacted had great value in the battlefields: he was courageous and capable of responding violently and immediately to aggression. His able handling of horses and firearms made him an outstanding warrior. His excessive indulgence in alcohol and his reputation for being a womanizer affirmed his masculine identity. The political influence his comrades exercised at the local and federal levels was crucial for the acceptance of his masculinity by state institutions. Finally, the later erasure of his masculinity and the imposition of female identity to his name and biographical narrative were also part of the discourse of the Mexican Revolution.

The Zapatista guerrillas fought to keep collective ownership of the land and to defend their traditional boundaries and autonomy in the face of the gathering forces of modernization and long-running agrarian disputes. They were notably active in demanding agrarian reforms, and although their political faction was eventually defeated, the revolutionary government dictated policies of land distribution and collective ownership and constructed poor peasants as one of the central symbols of the Mexican Revolution.[5] The majority of Zapata's followers were poor peasants, but some small-scale landowners also joined the movement. This was the case of Robles and other Zapatistas from the state of Guerrero who joined the Mexican revolution either to support their communities in a border dispute with a neighboring town or to open political spaces for themselves in otherwise closed systems. The Robles family owned several hectares of land for crops and cattle. We can only speculate on the reasons Amelia Robles joined the Zapatistas. However, the opportunity to leave her small village and adopt a life in which appreciation for her horsemanship and use of firearms was prevalent undoubtedly counted as one of her motives.

The passing of the Constitution of 1917 gave the country a regulatory framework but did not halt the armed conflict. The restoration of peace came about only in 1920, and it was only then that a new state apparatus could be built. Robles fought with the Zapatistas for a few years, from 1913 or 1914 until 1920 when he left the revolutionary guerrillas and enlisted as a soldier in the federal army along with other Zapatista chiefs and comrades. Zapatismo was eroding, and Emiliano Zapata was assassinated in 1919.

In the army, Robles fought with the forces of Álvaro Obregón, the military leader from northern Mexico who supported agrarian reforms as part of the revolutionary coalition, and who was elected president of Mexico in 1920.[6] Between 1923 and 1924, Robles was deployed first to the states of Tabasco and Chiapas and later to the state of Hidalgo, where he fought against the Adolfo de la Huerta rebellion, which unsuccessfully tried to oust Álvaro Obregón's government. After sustaining a bullet wound in Hidalgo, Robles was discharged from the Mexican Army and returned to Guerrero as a civilian. During his active service in the Mexican Army, Amelio Robles fought under Adrián Castrejón, a former Zapatista leader. Castrejón had a very successful political and military career, became governor of the state of Guerrero, and remained in office from 1929 to 1933. Castrejón's influence in the army and his home state's local political networks proved decisive for Robles's masculine identity to be accepted by the local organizations and institutions of the post-revolutionary government.

In the late-twentieth century, the inhabitants of Xochipala, particularly the youngest members of the community, were unaware of Robles's gender transition. They had no inkling that "El Coronel" (The Colonel in the masculine grammatical gender) had lived his childhood and adolescence as a girl and then a young woman, conforming to the gender identity assigned to him at birth and in line with the social norms of the time. Only the older generation knew his background and, on the whole, they respected him and did not confront him (at least not in his presence). His change of identity was not an issue discussed in his family, and even his nieces were oblivious to it despite living under the same roof. Robles's masculinity was taken as a matter of fact. This acceptance was confirmed by Amelio Robles's marital relationship with Ángela Torres, openly acknowledged by the local community. It was further reinforced by the fact that the authorities gave a public school established in 1966 the name Coronel Amelio Robles,[7] in recognition of his donation of the school grounds to the community.[8] This act of largesse reaffirmed the former Zapatista guerrilla fighter's reputation as a benefactor of the people and promoter of non-religious public education—a top priority for the revolutionary and post-revolutionary governments, as the basis for society's secularization and modernization.

The federal government also legitimized Amelio Robles's masculine identity when the Ministry of National Defense (Secretaría de la Defensa Nacional, SDN) recognized him in 1974 as an official veteran of the Revolution, using the masculine grammatical gender.[9] This recognition and the corresponding paperwork referred to his masculine identity and name, which Robles used himself when signing letters and petitions addressed to the military authorities and veteran organizations, which responded in kind.[10] He received official acceptance of his masculine identity both locally and elsewhere in Mexico thanks to his bonds of friendship with his comrades-at-arms. Some of these men climbed the rungs of the military and political ladders and recognized Robles's masculinity. Such was the case of the above-mentioned Adrián Castrejón who addressed Amelio Robles in the masculine in official correspondence, although he still occasionally poked fun at Robles by calling him "La Coronela" in the feminine grammatical gender.

The Shifting Memory

Paradoxically, the Mexican Revolution undermined the officially recognized masculine status sustained by Amelio Robles for over fifty years: "La Coronela" Robles emerged as an icon of the armed movement within Guerrero's cultural apparatus, upon which Amelio Robles had no influence. This feminization of Robles was a long process that began in the mid-twentieth century and took place in parallel with the reconfirmation of Amelio Robles's masculine identity among his close circle of family and friends, in local organizations, veteran associations, and the SDN. The process ran in parallel but opposite directions: on the one hand, Amelio Robles's masculine status gained legitimacy on a regular basis in his official correspondence, social and family life; on the other hand, he was assigned a feminine nationalist identity, transforming the Zapatista into an iconic revolutionary Mexican woman.

After his death in 1984, Robles's feminization went even further. His tombstone at the local cemetery bears an inscription: "Coronela Amelia Robles," in the feminine grammatical gender. Some years later, in 1989, a small community museum was set up in Xochipala to celebrate Coronela Robles's female identity and her contributions to the Mexican Revolution. The objects on display include some of Robles's personal effects and pieces of furniture belonging to his period as a Zapatista. Following the standard nationalist historical narrative that begins with the country's pre-Hispanic past, the museum's collection also includes objects found in the nearby archeological site of La Organera.

It is indeed a paradox that the Mexican Revolution facilitated Amelio Robles's gender transition and the post-revolutionary authorities legitimized his masculinity while commemorations of these complex political and military processes have promoted his femininity. The volatility of war and the Zapatista movement's popular roots enabled Robles's masculinization; this identity had the official backing of the post-revolutionary state thanks to the influence of some former Zapatista commanders and Robles's comrades-at-arms who went on to occupy high-ranking political and military positions. At the same time, the commemorative discourse of the Mexican Revolution celebrated the nationalist feminine images of "Adelita" and "Coronela." Local commemorative histories of the state of Guerrero included mentions of the woman colonel or "coronela," and Amelio Robles was powerless to oppose the feminization of his identity. As previously mentioned, the contradiction remains evident in Xochipala today where the Coronel Amelio Robles school still exists, bearing its original name, while a few blocks away, the Coronela Amelia Robles community museum receives local visitors as well as those who come from farther afield.

The most famous transgendering process in the military history of Latin America is that of the nun Catalina de Erauso or La Monja Alférez, the so-called Nun Lieutenant. In the seventeenth century, she took up a masculine identity and appearance to enlist in the service of the Spanish Empire.[11] Despite fighting and being treated as a man in the army, evidence suggests that she identified as a female and used a feminine narrative voice in the memoirs she wrote toward the end of her life. Unlike Robles, Alferez's transition was not permanent, and it appears not to entail a gender transition.

The story of Enriqueta Faber, who lived in Cuba with a male identity in the nineteenth century after fighting with Napoleon's army in France, bears certain similarities with the story of Amelio Robles. Faber's masculinity was recognized in the armed forces and in identity documents. He was even able to marry a woman and became a medical doctor. Although Faber's story has been told by several authors and versions of his life vary in some aspects, his image remained unchanged as part of a larger nationalist narrative or political discourse.[12]

Amelio Robles's permanent gender transition is relevant because it illuminates the social construction and performativity of gender identities. However, the most relevant part of his story is the perceptions of his gender identity. Such conflicting perceptions are embedded in gendered discourses of the Mexican Revolution, which were very powerful in the twentieth century.

Masculinization

Many women adopted masculine identities and wardrobes as a strategy to survive the dangers of the Revolutionary War. Passing as men gave them advantages in the army: access to weapons and horses, promotions through the military ranks, and better pay. Encarnación "La Chona" Mares wore men's clothing and pretended to be a man when joining Venustiano Carranza's forces. She aimed to avoid becoming a "soldadera" (camp follower), women who followed armies in the rear and, in the absence of professional groups, took charge of essential duties such as supplying the troops and cooking for them. They also often provided the men sexual and emotional company. The services of these "camp followers" were seldom appreciated; on the contrary, the women were frequently mistreated and abused.[13] Angelina Jiménez experienced something similar to Encarnación "La Chona" Mares, as Jiménez was given a military rank in the Constitutionalist Army, in which she fought with a masculine appearance and identity. The conclusion of the armed struggle marked the end of the masculine period for them both, and each subsequently lived traditional lives as women and mothers, forming families and taking on domestic tasks.[14]

The same happened in the case of Coronela María de la Luz Espinosa, who came from the Costa Chica region of Guerrero and, similar to Amelio Robles, fought for the Liberation Army of the South.[15] Espinosa adopted a masculine identity and appearance when she took up arms, but later abandoned this masculinity to live a domestic life, looking after her children and husband in post-revolutionary Mexico. Some parallels exist between the lives of Robles and Espinosa: both took on masculine identities, fought as guerilla combatants, and received a military rank. However, their respective processes of masculinization contrasted significantly in other respects. Espinosa's decision was pragmatic and temporary without implying a subjective transformation. In contrast, Robles underwent an internal transformation and repositioned himself within the social gender order. His masculine identity, which outlasted his period as an active soldier, expressed a vital impulse, a desire to act, be treated, and be recognized as a man in every aspect of life.

Not all female Zapatista leaders who had men under their command and had a military rank took on a masculine identity. For example, Coronela Rosa Bobadilla, viuda de Casas who was notoriously tough and disciplinarian with her troops, maintained her feminine identity. The same was true of other women, such as the Zapatista colonels called "La China" and Pepita Neri, both of whom had a reputation for being extremely violent.[16] Their stories reveal that, because the Zapatistas lacked the traditional hierarchy and discipline of a regular army, women could take up arms and assume leadership roles without having to shed their female identities.

The specific circumstances under which Amelia Robles first took up arms are unknown due to a lack of supporting documentation. Oral accounts do exist, however. The memories of those who lived through the Mexican Revolution in Xochipala and their descendants were recorded more than fifty years after the events, and are inevitably colored by the expectations and emotions of the present. They, nonetheless, offer some clues (though subjective) about the circumstances under which Robles joined the Zapatistas and about how Amelia Robles, who was twenty-five years old at the time, still maintained the feminine identity assigned to her at birth.[17]

According to these accounts, joining the guerrilla forces offered Robles a way out of the dissatisfaction she felt with the isolation of her home town and a chance to escape her step-father, hence it was a decision born of personal necessity rather than ideological beliefs. The problems with her step-father had worsened since he abused her sexually. These oral histories also consistently indicate that at the time of joining the Zapatistas Robles was in a relationship with Leonor López, a young woman also from Xochipala; however, no subsequent information is available about López.

Amelio Robles's gender transition is another gap in the story, but we can reasonably suppose that it happened gradually over the years of the Zapatista struggle and that his masculinity became firmly established after he enlisted in the Mexican Army, which was out of bounds for women. In 1915, Venustiano Carranza decided to cancel the military appointments of "señoras y señoritas," (ladies and young ladies), thus excluding female soldiers from the ranks.[18] This ruling did not apply to the Zapatistas who were not part of the national armed forces. At the time of joining the Mexican Army, Amelio Robles's masculine status was already firmly entrenched and recognized by her leaders and fellow comrades-at-arms. Otherwise, she would not have been accepted as a soldier or allowed to fight.

Several official documents confirm his masculine identity. Such is case with the credentials and identity documents from the local organizations he joined: the Socialist Party of Guerrero (Partido Socialista de Guerrero) in 1934, which was affiliated with the National Revolutionary Party, was appointed a delegate in Xochipala for the League of Agrarian Communities (Liga Central de Comunidades Agrarias) in 1945, and became a member of the local Ranchers' Association (Asociación Ganadera) in 1956. These were all corporate organizations recognized and authorized by the State and subject to controls over the selection of their leaders and the formulation of their demands.[19]

The veteran organizations that emerged from the late 1930s also accredited his masculine status.[20] In 1948, Amelio Robles joined the National Confederation of Veterans of the Revolution (Confederación Nacional de Veteranos) and then the Zapatista Front (Frente Zapatista), led by Adrián Castrejón, who appointed Robles as Secretary of Military Action in 1950. The front represented campesinos who requested land, jobs, and pensions, among other benefits. Despite being a landowner,

Robles requested a military pension for his participation in Zapatismo, but he only received a one-off payment from the government during Luis Echeverría's presidency.

The weightiest official document attesting to Robles's masculine identity is perhaps the birth certificate issued by the Civil Registry Office of Zumpango del Río, which records the birth of the boy Amelio Malaquías Robles Ávila. The particulars given—the name of the father and mother, date and place of birth—are the same as the handwritten entry in the original birth registration held by the same office.[21] It is impossible to know whether the apocryphal document was issued in Guerrero or Mexico City, or to know the identity of the intermediaries, but its existence is evidence of the networks of power and the complicities in which Amelio Robles was involved. However, these networks were powerless to prevent the feminization of his identity that had already begun as part of the commemorative discourse of the Mexican Revolution in Guerrero.

Feminization

The Guerrero state government and the federal ministry of education (Secretaría de Educación Pública) published the *El Coyote: Corrido de la Revolución* (1951) (The Coyote: A Ballad of the Revolution) by a rural teacher called Celedonio Serrano Martínez.[22] This book was produced to commemorate the centenary of the official founding of this Mexican state in 1949. Written in verse, The Coyote narrates the story of the Mexican revolution in the state of Guerrero. The literary style emulates the popular ballads penned by troubadours during the period of armed struggle. These popular songs recounted the latest news and also past events about which people were already familiar. Listening to a troubadour sing a ballad in fairs or local fiestas was a form of entertainment. The verses and the music of the popular ballads were a means of cultural expression and gave cohesion to particular social groups or regions.[23] Often the verses of the ballads were printed on colorful rice paper sheets, expanding circulation to a larger public. The broadsheets were a form of penny-press that offered entertainment and information at low cost. Engraved primitive illustrations were often printed on the flyers making them more attractive to semi-literate audiences. José Guadalupe Posada was the author of the prints that depicted scenes of life of the popular classes. In the post-revolutionary period Posada's style was celebrated as an expression of the culture of the Mexican revolution and national identity, and some artists emulated his primitive style.

The verses in The Coyote are similar in their form to the popular ballads of the revolutionary period, with the notable difference that they were not authored by popular poets but by a teacher from Veracruz, Celedonio Serrano Martínez. The format was also different, as the Coyote was a large, heavy coffee table book with more than three hundred pages of bond paper. The sheer size of the publication made reading it a difficult task if a table and chair were not available. Its large, heavy format belongs in a library or school and not a fair. Its purpose was not entertainment but educational and was not aimed at a popular audience.

The account of the Mexican Revolution presented by Serrano Martínez celebrates Zapatismo, focusing on the developments of the movement in Guerrero and naming revolutionaries who fought in this Mexican Republic state, giving Guerrero the same standing Morelos had in the history of Zapatismo. By the time the book came out in 1951, the revolutionary movement had become a dominant rhetorical trope in Mexican political discourse. Many saw the Mexican revolution as the foundation of nationalism, progress, and social justice.

The volume's illustrations include engravings and vignettes of Gabriel Fernandez Ledesma and Francisco Moreno Capdeville in a style that was used by artists of the League of Revolutionary Writers and Artists (Liga de Escritores y Artistas Revolucionarios) and the Popular Graphic Work Shop (Taller de la Gráfica Popular). The League and the Workshop were groups of artists committed to protesting against oppression and capitalism with their art. One of their aims was to stir people into action. Their political position was not an obstacle for accepting government commissions especially after the Second World War when their anti-fascist art flourished.

El Coyote dedicates a couple of verses to "La Coronela Amelia Robles" (Colonel Amelia Robles, in the feminine grammatical gender), praising her bravery and femininity. The prints also add to her feminine identity through visible features including her plaits and long dress, as well as the depiction of her mounted side-saddle on horseback. The long skirt and hair are typical of Mexican female national identity. The "China Poblana" (the Chinese woman from Puebla) and "Adelita" (Adelita is the diminutive for the female proper name Adela) sported that style, both figures who were part of the nationalist imagery popularized during the post-revolutionary period and both of whom are depicted as young, attractive, and sexually available rural women.[24] They fulfill the role of being the complement of Mexican masculine identity. The "China Poblana" has oriental features and plays the role as the partner of the virile, horse-riding charro, a horseman with a typical outfit and sombrero that became a symbol of national identity. The "Adelita" accompanies the revolutionary soldier, and her feminine traits—beauty, youth, and submissiveness—are the polar opposite and complement of the brave soldier possessing the power to make women submissive.

The revolutionary "Adelita" was an idealization of the camp followers, women who traveled with armies moving across northern Mexico, cooking for them and offering sexual and emotional company as well as rudimentary first aid. The woman colonel or "Coronela" on the other hand was a reference to armed women in command of troops, idealized as figures who enjoyed a certain degree of personal autonomy and freedom, and yet, by remaining within the parameters of complementarity between masculine and feminine identities, did not threaten any upset to the established social gender order. Despite the contrasts between the two idealizations, the "Adelita" and female colonel figures, they also shared some characteristics: they were abstractions without any individualizing features since they lacked any specific geographical origin, political affiliation, or ideological position. As abstract entities, they could be adapted to suit accounts of the revolutionary movement in the country's different regions and political factions. As such, they were idealizations that rendered invisible poor women who had names and surnames and individual stories—some submissive, others violent—and, like the menfolk, also joined the uprisings with the hope of changing their situation.

The image of the female colonel was less widespread as compared to the "Adelita," who became an icon of women's participation in the Mexican Revolution. In spite of the differences between them, both figures can be seen as "invented traditions" establishing a feminine nationalist and revolutionary identity.[25] Invented traditions, according to Eric Hobsbawm, are discourses that serve the political needs of the present and give legitimacy to certain practices by assigning them a link to a past that never existed and are usually invented. The feminized figure of "Coronela Robles" (Colonel Robles, in the feminine gender) adds legitimacy to the revolutionary participation of women who had a certain control over their situation and did not question the binary nature of gender, according to which feminine and masculine identities and social spaces are immutable and clearly contrasting human expressions. At the same time, Coronel Amelia Robles, in the feminine gender, was accepted in the Guerrero region because she offered the local history of the Mexican Revolution a person of comparable status to the famous "Adelita." The anonymous author of a popular ballad from the 1970s on this topic supports this point: in the north, there was "Adelita," and in the south, the warrior Amelia Robles.[26]

The "Adelita" and the image of the female colonel entered the spotlight in the 1930s as part of the heated public debate on women's suffrage, and came to form a part of the commemorative discourse of the Mexican Revolution. Those defending women's right to vote praised these figures to draw attention to women's participation in the Revolution, and recognized the female combatants' political usefulness as a way of countering the argument put forward by those opposing the women's vote, namely that women were more Catholic and conservative than men, and that giving them citizenship rights risked jeopardizing the country's stability and reform process.[27] The activists supporting women's suffrage used various ways to refer to their commitment to the armed struggle and revolutionary causes. Refugio García, secretary of an organization supporting The Sole Front of Women's Rights (Frente Único Pro-Derechos de las Mujeres), insisted that the smell of gunpowder remained fresh in the hair of women who had participated actively in the armed uprising; other suffragettes drew up a list of the names and details of specific women who fought in the Revolution. Amalia de Castillo Ledón, President of the Mexican Athenium of Women (Ateneo Mexicano de Mujeres), was also active in this regard, arguing that providing equal suffrage for women and opening government positions to women with suitable training was not at odds with a clear and unequivocal feminine identity, as expressed in their attire and refined manners. This was a modern form of feminism, which Castillo Ledón called "feminine feminism," a means of including women in public affairs but without undermining the feminine and masculine characteristics that society attributed to women and men respectively. Castillo Ledón considered that the definition of such feminism needed to move away from "primitive feminism," led by the movement's precursors—feminist women who encountered such a hostile reaction that they were forced "to look butch and dress accordingly, ignoring their hygiene, forgetting their manners and the psychology of their gender."[28] The times had changed, and it was no longer necessary to resort to such excesses.

The female colonel was a figure embodying a feminist spirit who demanded equal civil rights for women, while at the same time placating those opposed to women's suffrage. Objectors to the suffragette movement were concerned that women would become masculinized as a result of their involvement in public affairs; they feared that women would become so absorbed in public life that they would neglect their domestic duties and reject the typically feminine and maternal characteristics attributed by society to women, leading inexorably to their masculinization. This was not a newfound anxiety but one that had already surfaced in the early twentieth century when women began to enter the workplace; the public debate over women's suffrage simply brought the issue back into focus. Those defending women's right to vote needed to overcome people's fear on two counts: women's assumed political conservatism, and the loss or impairment of feminine identity defined by maternity and dedication to the home. These fears delayed universal women's suffrage for a further fifteen years, and it was only granted in 1953.

Furthermore, a modern dance performance called La Coronela, premiered in 1940 with choreography by US dancer Waldeen and musical score by Silvestre Revueltas and Blas Galindo, revived the revolutionary, nationalist, and armed feminine revolutionary figure. The iconography that provided inspiration for the choreography was José Guadalupe Posada's print of a woman wearing a rebozo (typical shawl) and holding a rifle propped up against her side. The show's choreographer and creator imagined the female colonel as a full expression of Mexican nationalism and feminine identity.[29]

Gabriel Fernández Ledesma, whose prints were used to illustrate the aforementioned book called El Coyote, also worked on the set design of this dance performance. This artist, founder of the League of Revolutionary Artists and Writers (Liga de Autores y Escritores Revolucionarios, LEAR), conceived art as an expression of nationalism and social commitment. The Coronela Robles print shares at least two of the elements found in Posada's work: the rifle and the long dress, graphic components more closely linked to the conventional artistic image of the "Coronela" than to Amelio Robles's actual appearance. The female colonel was also the subject of the novel by Francisco Rojas González, *La Negra Angustias* (1944), awarded the National Literature Prize in the same year of its publication. Matilde Landete made a film of the book, in an adaptation critics have described as feminist both for its intention to show women's contributions to the revolutionary movement and for the obstacles facing them.

Gertrude Duby Blom, a journalist and photographer exiled in Mexico in the early 1940s, also produced a feminized representation of Robles. Duby visited and interviewed Robles in Xochipala as part of a feature to raise the profile of Zapatista women of the Mexican Revolution.[30] In this text—which remained unpublished for decades—the journalist finds that Amelio Robles's masculinity is not an expression of identity but a strategy to cope with an environment hostile to women's attempts to participate in politics. In this respect, she agrees with Amalia de Castillo Ledón's explanation of the motives for their precursors in adopting a manly aspect and identity and moving away from the "the psychology of their gender."[31] However, both aspects of Robles overlook her gender transition and Amelio Robles's satisfaction with his masculine body image and movement, something he nurtured carefully and effectively with his choice of wardrobe, gestures, and behavior of a rural Mexican man.

The feminine image of the "Coronela" Amelia Robles was affirmed with the establishment of the community house-museum in Xochipala, which bears his name in its feminine form, and is also supported by the National Institute of Anthropology and History (Instituto Nacional de Antropología e Historia, INAH). The Robles family backed the museum project, providing furniture and personal effects of the former Zapatista fighter and for some years provided a large room in their house to serve as a community museum (it later was moved to the premises of the former school building). Amelio Robles's gender identity began to change after his death, and his great-nephew and nieces were pleased at the attention lavished on the female Colonel Robles as an iconic historical figure. The change in attitude can be understood due to the attention received by Coronela Robles in the press and from the state government.

The initiative came from a group that supports the elderly and was backed by the Guerrero State Government's Ministry for Women, a government office established in 1987 (two years before the museum opened its doors). Its objectives included supporting gender equality, promoting women's participation in the economy, and combating violence against women. This ministry's support to the museum can be explained because honoring outstanding women from the past by publishing their names and photographs is a manner of celebrating women's contributions to society in the present.[32]

A large-sized color portrait, kept by the family and now one of the main exhibits in the museum, characterizes Robles as an iconic woman of the revolution. Also on display are identity documents showing Robles in his masculine identity, effectively leaving visitors to reach their own conclusions. Perhaps many people simply do not care much about the matter and readily accept Robles's feminine identity as expressed in the museum's name.

Amelio Robles's memory found a new vocation as a symbol of sexual diversity in the twenty-first century, particularly following the centenary celebrations commemorating the 1910 Revolution.[33] Activists began to work in different parts of the country: in Monterrey, in the state of Nuevo León, one group set up the Amelio Robles prize for people, organizations, and companies that support equal rights for gays, lesbians, and transgender people; a profile in the LGTBI Obituary mentions his identity change and a transmasculine activist called Mario Sánchez sung a ballad he wrote himself, dedicated to Robles, while wearing a rainbow flag across his chest, reminiscent of the cartridge belts worn by the fighters and seen in the numerous portraits in the revolutionary period. The spirit of acceptance of sexual diversity is also found in a leaflet about his life. This text—published in Chilpancingo, the capital of the state of Guerrero—mentions his "difference" and, while remaining ambiguous, appears understanding of his gender transition.[34] However, both cases focus on his participation in the Zapatista uprising, overlooking the fact that this was just one period in his military career and that his active service in the Mexican Army was instrumental for him to have received official recognition of his masculinity.

Conclusion

Transgender identities are authentic and consist of the same elements as feminine and masculine identities assigned at birth: people's body language and gestures, social behavior and subjective constitutions. The history of Amelio Robles is interesting partly because it reveals the social construction of gender and also because it allows us to explore different perspectives on this matter.

The history of Amelio Robles's permanent transgender identity is a unique case because of the existence of documents and photographs that register both his masculinity and the construction of a feminine icon that uses his name and biography. The first group of archival material accounts for his legal and subjective masculinity and his bodily performativity, while the second group of materials register those who blatantly erase or try to erase his acquired maleness. Amelio Robles's military career first as a guerilla fighter and then as a rank-and-file federal soldier was otherwise short and unremarkable: if not for his transgender identity, perhaps his life story may never have become known outside the circle of his family and local community, as his military career was not particularly outstanding and resembled that of thousands of other Zapatista combatants.

In terms of documentation, no equivalent to the story of Robles exists in the history of Mexico or Latin America. Its significance, however, is not to be found in his transgender identity but the conflictive perceptions of his masculinity. The exaltation of his often violent masculinity in the troubled and unstable days of war, shifted to the institutional oral and written recognition of his masculine identity. The local and national political connections that protected Amelio Robles in the post-revolutionary recognition could not avoid the feminization of his image and name in nationalist cultural expressions underpinned by the invented traditions of the "Adelita" and female colonel figures. These identities were bolstered by a feminism that demanded equal and universal suffrage for women from the 1930s, and their femininity provided a means of confronting people's anxiety about women's supposed masculinization resulting from their possible inclusion in public life. Therefore, the regional commemorative history of the state of Guerrero helped attribute feminine qualities to Robles by using him to give the state a certain standing in the history of the revolutionary movement.

Notes

1 Edith Pérez Abarca, *Amelia Robles. Revolucionaria Zapatista del Sur* (Chilpancingo, Guerrero: PACMYC 2000, CONACULTA, Unidad Regional Guerrero de Culturas Populares e Indígenas, Instituto Guerrerense de la Cultura, 2007), 15–16.
2 Rosa E. King, *Tempest over Mexico* (Boston: Little, Brown, and Company, 1940).
3 Gabriela Cano, "Unconcealable Realities of Desire: Amelio Robles (Transgender) Masculinity in the Mexican Revolution," in *Sex in Revolution. Gender, Politics and Power in Modern Mexico*, ed. Jocelyn Olcott, Mary Kay Vaughan, and Gabriela Cano (Durham: Duke University Press, 2006), 35–56.
4 I use the feminine name Amelia, and the feminine pronouns she and her when referring to Robles in early life when she lived as a girl and as a woman and reserve the masculine name, Amelio, for the longer period when he was recognized as man.
5 Publications on Zapatismo in the Mexican Revolution include: John Womack, *Zapata and the Mexican Revolution* (New York: Vintage Books, 1968); Samuel Brunk, *Emiliano Zapata: Revolution and Betrayal in Mexico* (Alburquerque: University of

New Mexico Press, 1995); Felipe Arturo Avila Espinosa, *Los orígenes del zapatismo* (Mexico City: UNAM, 2001). The meanings and representations of Zapatismo following Emiliano Zapata's death are the subject of Samuel Brunk's book, Samuel Brunk, *The Posthumous Career of Emiliano Zapata: Myth, Memory and Mexico's Twentieth Century* (Austin: University of Texas Press, 2008).

6   Robles's military career is documented by paperwork contained in two personal files: AHSDN, Cancelados, 112– 119, Amelio Robles Avila, AHTF, INAH, Gro.-06.
7   Pérez Abarca, *Amelia Robles*, 17.
8   The date is recorded on one of the walls of the former school building where the Amelia Robles house-museum is now located.
9   Gabriela Cano, "La masculinidad trans de Amelio Robles en sus documentos de archivo," in *Sociología y género: estudios en torno a performances, violencias y temporalidades*, ed. Karine Tinat and Arturo Alvarado (Mexico City: El Colegio de México, El Colegio de México, Centro de Estudios Sociológicos, 2017), 27–49.
10 AHSDN, Cancelados, 112–119, Amelio Robles Avila.
11 Catalina de Erauso, *Lieutenant Nun: Memoir of a Basque Leutenant Nun* (Boston: Beacon Press, 1996).
12 James Pancrazio, "Reescritura, invención y plagio: Enriqueta Faber y la escritura del travestismo" in http://www.habanaelegante.com/Fall_Winter_2014/Invitation_Pancrazio.html, viewed on February 1, 2019.
13 AHSDN, Cancelados, María Encarnación Mares, C-801 and AHTF, INAH, Gro-06.
14 Esther R. Pérez, James Kallas, and Nina Kallas, *Those Years of the Revolution, 1910– 1920* (San José, California: Aztlan Today, 1974), 37–39.
15 Anna Macías, *Against All Odds: The Feminist Movement in Mexico* (Westport, CT: Greenwood Press, 1982), 42–43.
16 Womack, *Zapata and the Mexican Revolution*, 170.
17 Pérez Abarca, *Amelia Robles*, 31–34.
18 *El Constitucionalista, Diario oficial del gobierno constitucionalista de la República mexicana*, March 29, 1916, 1.
19 AHTF, INAH, Gro.-06.
20 A portrayal of these veteran organizations can be found in Martha Eva Rocha Islas, *Los rostros de la rebeldía. Veteranas de la Revolución Mexicana, 1910–1939* (Mexico City: Secretaría de Cultura, Instituto Nacional de Estudios Históricos de las Revoluciones de México, Instituto Nacional de Antropología e Historia, 2016), 43–101.
21 AHSDN, Certified copy of birth certificate, April 8, 1957.
22 Celedonio Serrano Martínez, *El coyote. Corrido de la revolución* (Mexico City: Secretaría de Educación Pública, 1951).
23 These notes on the *corrido* are based on the work of Guillermo Bonfil Batalla "Trovas y trovadores de la región Amecameca-Cuautla," Guillermo Bonfil Batalla, Teresa Rojas Rabiela, and Ricardo Pérez Monfort, *Corridos, trovas y bolas de la región de Amecameca-Cuautla. Colección de don Miguelito Salomón* (Mexico City: Fondo de Cultura Económica, 2018), 11–32.
24 I have referred to Adriana Zavala's analysis on the "Adelita" in her study of La China Poblana in Adriana Zavala, *Becoming Modern, Becoming Tradition: Women, Gender and Representation in Mexican Art* (University Park: The Pennsylvania State University Press, 2009).

25 Eric Hobsbawm, "Introduction: Inventing Traditions," in *The Invention of Tradition*, ed. Eric Hobsbawm and Terence Ranger (Cambridge: Cambridge University Press, 1983), 1–14.
26 JCR, "Corrido a la sin par guerrillera, Amelia Robles," Zumpango del Río, 1959.
27 Gabriela Cano, "The Feminist Debate in Mexico," in *The Routledge History of Latin American Culture*, ed. Carlos Manuel Salomón (New York: Routledge, 2018), 284–297.
28 Amalia de Castillo Ledón, "Memorial del Departamento de la Mujer," in Gabriela Cano (sel.) *Amalia de Castillo Ledón. Mujer de Letras, mujer de poder. Antología* (Mexico City: Consejo Nacional de la Cultura y las Artes, 2011), 99.
29 César Delgado Martínez, *Waldeen. La Coronela de la danza mexicana* (Mexico City: Escenología, AC, 2000).
30 Gabriela Cano, "Gertrude Duby y la historia de las mujeres zapatistas de la Revolución Mexicana," *Estudios Sociológicos* 28, no. 83 (May–August 2010): 579–597.
31 Amalia de Castillo Ledón, "Discurso ante la Columna de la Independencia," in Gabriela Cano (sel.) *Amalia de Castillo Ledón. Mujer de Letras, mujer de poder. Antología* (Mexico City: Consejo Nacional de la Cultura y las Artes, 2011), 93.
32 "Secretaría de la Mujer. Catorce años de lucha a favor de la mujer guerrerense, 1987–2001," *Mujeres del Sur. Bienestar, política, salud y cultura en Guerrero*, April–June 2011, 4–6.
33 In an article illustrated with photographs, aimed at a general readership, and subsequently in a chapter published in English in 2006 which was later translated into Spanish in 2009, I proposed that Robles should be seen as a transgender person rather than as a revolutionary woman. This work attracted media attention in 2010 during the celebrations marking the centenary of the Mexican Revolution. Gabriela Cano, "La íntima felicidad del coronel Robles," *Equis*, no. 14 (June 1999), 25–34 and Cano, "Unconcealable Realities", 35–56. In a previous publication, I wrote a short article on Robles, whom I saw as a revolutionary woman: Gabriela Cano, "El coronel Robles: una combatiente zapatista," *Fem.*, April 1988, 22–24.
34 Pérez Abarca, *Amelia Robles*, 22.

Part Four

Survivors

10

"Amazons" in the Pantheon? Women Warriors, Nationalism, and Hero Cults in Nineteenth- and Twentieth-Century Chile and Peru

Gabriel Cid

War, Nationalism, and Female Heroism in Nineteenth-Century Latin America

This chapter addresses the paradox of female access to the national pantheon of heroes in two Latin American national cases: nineteenth-century wars fought in Chile and Peru and their aftermaths. Warfare was endemic in Latin America during the nineteenth century and played a decisive role in the process of nation-state formation. Conflict played a role not only in setting borders and in the concentration or fragmentation of state power, but also in the creation of narratives about the nation-state.[1] Since the break with the Spanish Monarchy, the war and the army have had a key impact in the region. Power disputes, competition over economic resources, provincial and ethnic conflicts, as well as the definition of the limits between the new states, made military violence an everyday factor on the continent. A widespread consensus exists regarding the deep-seated links between warfare and nationalism, as military confrontations between communities tend to accentuate their contrasts and differences, often linking them to real or imagined ethnic factors. Memory policies, with their associated rituals and discourses, help to keep these nationalist narratives alive long after the wars have ended. The relationship between war and nationalism not only functions during periods of conflict, but also continues long after the fighting has ended.[2]

One area in which this type of relationship can be clearly seen is in the creation of pantheons of national heroes, most of which in Latin America are peopled by figures linked to warfare.[3] The link between war and the creation of heroes should come as no surprise, particularly as few values are as highly prized as a willingness to die in the defense of the nation. The exaltation of the idea of sacrifice plays a key role in attaining the status of hero, as it provides a dramatic display of the "sanctification of shared values."[4]

What is more, times of war bring with them greater fluidity in social conventions, demand greater social mobilization, and encourage the democratization of the national pantheons. This has a significant impact on the issue examined here: female warriors. Indeed, war has a paradoxical dimension. On the one hand, as Nira Yuval-Davies has written, wars often reinforce the sexual division of labor. Thus, although certain archetypes of warrior women exist—such as the Amazons of the Ancient Greeks—the prevailing discourse maintained that women soldiers were "unnatural," as femininity has traditionally been associated with peace.[5] This led to recurring themes that distance women from the battlefield, limiting their sphere of action to caring for the wounded, feeding the troops, performing charitable work linked to the social consequences of conflict, and so forth.

Nonetheless, themes of sacrifice linked to collective defense of the nation, together with the flexibility of social conventions in times of war, have also formed a route for women to be recognized as heroes. In fact, the archetype of the Amazon enabled the concept of women warriors to enter familiar narratives. The inevitable connection between the Amazons of myth and women involved in armed conflict granted these female soldiers a "mythogenic" background, to use Burke's term,[6] which facilitated their entry into spaces of public recognition.

This very paradox of the role performed by women can be analyzed in historical terms. The wars that Chile and Peru fought in the nineteenth century provide several female heroes whose public acceptance shows the different ways in which this issue was addressed. In Chile, although framework for women's labors during wartime remained traditional, several cases exist of female soldiers—with Candelaria Pérez, the famed "Sergeant Candelaria" as the principal archetype—who challenged these roles and received not only official recognition, but also renown amongst their contemporaries, although this was short-lived. Meanwhile, in Peru, the tendency toward division of roles during wartime remained more rigid than in Chile. Models for the creation of heroes were therefore generally attached to women who followed these roles, either collectively—the *rabonas* and *cantineras* (sutlers)—[7] or individually. Except for a few cases that only came to public recognition after their time, and with tales that took on a more apocryphal style—such as Leonor Ordoñez—female warriors were not widely celebrated amongst their contemporaries. All in all, the presence of women in official recognition of heroes of Peru's nineteenth-century wars seems to have been more successful and long-lasting than in the case of Chile. This later recovery of the memory of female warriors in the Peruvian case has more to do with the need to construct ethnically popular icons according to the official indigenous nationalist agenda than a redefinition of gender roles as a result of the wars themselves. The weight of oral tradition helped to make these heroines moldable characters according to the needs of the populist and nationalist discourses of the 1960s and 1970s, unlike the Chilean case, where these narratives were built in wartime and then forgotten.

I suggest that these differences refer to the ways of conceptualizing war in both countries. In the Chilean case, it was a war of conquest, so the nationalist discourse and its gender roles tended to feminize the adversary and his passivity and submission to the victor army.[8] In this structure, the warlike actions carried out by women, in addition to having historical antecedents, also possessed a positive value for the contemporaries, who exalted the military courage. In the Peruvian case, the late feminine heroization referred more to the particularities of a less martial nationalism than the Chilean, and founded on revenge, collective victimization, and irredentist discourse regarding the conquered territories. Thus, in the Peruvian historical narrative, the action of women tended to be used to show the excesses of the Chilean occupation, focusing on the acts of violence committed by the troops that had a recurring objective in Peruvian women. It was only in the 1960s and 1970s that said discourse, while still important, gave way to the discourses around female heroization, but in the key of nationalist exaltation of popular resistance to the invader. Thus, the differences in female heroic representations have to do with the different ways of conceiving the nationalist discourse in both countries based on their divergent war experience

The "Chilean Amazon": The Sergeant Candelaria

One of the first regional armed conflicts amongst countries that were emerging from their recent independence came in 1836, with the War of the Peru-Bolivian Confederation. In that year, the politician and military Andrés de Santa Cruz led the project of forming a supranational state uniting the republics of Bolivia and Peru. For the Chilean government the establishment of the Peru-Bolivian Confederation implied a threat to the balance of powers in the South Pacific, risking the position of the port of Valparaíso as the center of the region's trade. For that reason, supported by some Peruvian caudillos dissatisfied with the policies of Santa Cruz, he decreed war which lasted for three years. In January of 1839, the decisive Battle of Yungay put an end to the Confederation, gaining Chile new national heroes, chosen from those who played the most outstanding roles in the fighting: General Manuel Bulnes (who went on to take the presidency in 1841); Juan Colipí, whose image evokes the country's indigenous legacy; and the woman, Candelaria Pérez, known as "Sergeant Candelaria."

Pérez came from humble origins and is believed to have worked as a domestic servant in Valparaíso before moving to Peru and finding similar work there. She managed to save enough money to open a small Chilean tavern in the port city of Callao, but the outbreak of hostilities between the two countries forced her to close the business. Having enrolled first as a spy, then as a sutler in the Carampangue Battalion, she finally attained a combat position and fought with outstanding courage in the Battle of Yungay.[9]

The heroine caused great commotion upon returning to Santiago in the victory parade for Chilean troops in December 1839, particularly for the "great emotional power over the imagination of the people whenever something departs from the ordinary progress of human happenings," as the journalist Vicente Reyes noted. This situation could help to explain her popularity following the war, together with a number of other "mythogenic" characteristics, as Reyes spoke of her as a Chilean Joan of Arc, and a "Chilean Amazon."[10] Antonio Barrena, who was himself a veteran, recalled that during those days "everyone would ask after her, everyone sought the honor of being introduced to her so they could shake her hand, and meet the heroine of modern times." Riding a horse that was given to her by one of her admirers, Candelaria Pérez "looked prouder than any Amazon in the gaze of a country in awe of her heroism and eager to see her."[11]

Her popularity among the people of Santiago, together with her performance on the battlefield, led to pressure on the government to award her some official recognition. A short while later, in January 1840, the authorities promoted her to the rank of sergeant—leading to her nickname of "Sergeant Candelaria"—and awarded her a state pension. A few months later, President Prieto himself petitioned Congress to promote her to the rank of sub-lieutenant. Echoing a motion raised by General Bulnes, he justified the promotion claiming that Candelaria Pérez was a "remarkable woman" whose patriotism led her to abandon "her former relations and whatever she possessed" to serve in the army, coping with "the rigors and privations" of the campaign alongside her male counterparts. The crowning moment of her heroism was the Battle of Yungay, when she even "came to rival the most seasoned warriors in the vanguard." This increase in her pension not only covered her material needs, but also comprised an "equitable and fair measure that still affects the honor of the nation."[12] Congress passed the motion a month later, concluding that the measure was "a well-earned award in view of the efforts and patriotism shown by the recipient throughout the Peru campaign, setting an example that perhaps served to encourage the heroic feats of arms that were performed during that war."[13]

However, despite these events, her immediate rush of popularity and renown faded over the course of time, until Pérez was all but forgotten. The only attempts to recover her fame came from literary circles, stemming from a logic of historical curiosity and a yearning for justice. For instance, February of 1849 saw the premiere of the play *La acción de Yungay* by Manuel de Santiago Concha, with the character of "Sergeant Candelaria" as one of the protagonists.[14] The audience included Pérez herself, who was quickly noticed and given an ovation. That was to be the last public recognition that she enjoyed.

The writer Vicente Reyes, fascinated by this remarkable phenomenon of popularity and forgetfulness, interviewed the heroine in 1859. The veteran told the young writer about her adventures with an evident desire to reclaim her place among the country's heroes. Reyes himself was insistent on that point, contrasting the few days of glory when Candelaria entered Santiago after the war, when "leading dignitaries were quick to festoon her with congratulations," with the descent into obscurity suffered by the majority of war veterans. By the late 1850s, the mythical heroine had been reduced "almost to the condition of an invalid," living with a meager pension—increased by 5 pesos in 1852—so Reyes called for her to be granted a further bonus, in recognition of "the uniqueness of her services, and as an encouragement for the future."[15]

In 1869, the prolific historian Benjamín Vicuña Mackenna, like Vicente Reyes, decided to visit Candelaria's modest home to hear about her adventures from her own lips. He wrote that her humble origins—easily perceptible in her "clearly mestizo" physiognomy—had played a role in her public appeal, paving the way for the public's empathy with "modern Chile's most famous Amazon." Vicuña Mackenna was particularly interested in the process whereby Pérez "went from tavern-keeper to heroine," with the war providing the necessary opportunity. Without the war,

Candelaria Pérez would have led an unimportant life. It was only in the context of warfare that her heroic virtues were manifested, in a reflection that transposed the characteristics that the people admired: patriotism, sacrifice, solidarity, and courage. The fame attained by "Sergeant Candelaria" was quick to ebb, as Vicuña Mackenna wrote. When he visited her she was living "on little apart from public charity," and "in the seclusion of a hermit." Without the war that had been her route to heroization, Candelaria Pérez returned to her "meager existence."[16]

One year after that meeting, the newspapers of Santiago announced the death of Candelaria Pérez, on March 28, 1870. Her death reinvigorated the myth of "Sergeant Candelaria." For example, *El Ferrocarril* recalled "her admirable bravery and fearlessness," particularly for a woman.[17] Another newspaper, *El Independiente*, briefly memorialized the "life of this manly woman, who is both one of the celebrities of our army, and an eminently popular heroine," as "her name is familiar to everyone, and nobody is unaware of the feats of Sergeant Candelaria." It reported that her adoption as a folk hero was rooted in oral tradition, "the only thing that has spread word of her, often ascribing incredible actions to her."[18]

The opinion published in *El Independiente* reflects the staying power of popular tradition about the construction of heroic narratives. The tension between neglect of "Sergeant Candelaria" amongst the official authorities and remembrance in oral tradition helps to understand the comment made by writer Ventura Blanco in 1870, who wondered why, as the woman was living out her years under conditions of penury, the "people were still enthusiastically repeating tales of her feats."[19] If Vicente Reyes's goal was to rescue the heroic deeds of Candelaria Pérez in the collective memory, and Vicuña Mackenna was endeavoring to explain the astounding journey that took her from innkeeper to "modern Chile's most famous Amazon," then Ventura Blanco's mission was to mythify the image of "Sergeant Candelaria" in the wake of her death. He too had interviewed her, and went on to present her as the most genuinely popular heroine the country had ever known.

In a nationalist historiographical essay, Blanco called for a "task of justice reparation" for that "exceptional woman," noting the outstanding features of her character and filling the narrative with anecdotes that underscore the bravery of a heroine who nonetheless died in "misery and abandonment." The humbleness of her funeral provides an indicator of this abandonment: only five mourners attended. Blanco closes with a call to buy a burial monument for the heroine, which he described as a "work of holy reparation," and the only way to "snatch remembrance of this extraordinary woman from the obscurity of the grave."[20]

Blanco's proposal bore fruit, as a few years later historian Gonzalo Bulnes mentioned the poem that was carved on the new tomb of "Sergeant Candelaria," perhaps written by Blanco himself:

"Beneath this cross, the key of heaven, lies/An extraordinary, heroic woman,Honor of Chile on Peruvian soil,/So unfortunate, Sergeant Candelaria.Remembering Yungay with holy zeal/The people raise their prayers for her,And in remembrance of her noble story,/Mourn her troubles and honor her name."[21]

The War of the Pacific and Female Heroism in Chile and Peru

A decade after the death of "Sergeant Candelaria," Peru and Chile were once again at war. The War of the Pacific (1879–84), caused by border disputes in the saltpeter zone of the Atacama Desert, was fought between Chile and its Andean neighbors, and had an immense impact on the region not only by redrawing the countries' borders—the incorporation of the Atacama Desert into Chilean territory, leaving Bolivia landlocked—and its repercussions on the power balance in South America, but also for the nationalistic narratives that emerged from the conflict.

The war led to an extensive mobilization of the military forces, the logistical capacities, and the civil society of the combatant nations.[22] Women were certainly not left on the sidelines, not only because the scale of the conflict made their support a necessity, but also because, unlike the previous war, the existence of women's organizations resulted in great visibility for women's participation, in a manner that had not previously been seen during wartime.

Indeed, exhortations in the press in both countries encouraged women to take part. For example, Chilean educator José Bernardo Suárez called on the country's women to come to its defense: "Our virtuous ladies must take on an important role in this war: collecting contributions to provide the government with weapons; organizing dances and gatherings for the same purpose; encouraging mothers not to protest against their sons going off to war; and spreading this message."[23] This type of discourse tends to feature a paradoxical dimension: while encouraging women to take part, it also limits their participation to the domestic roles assigned to them. Women certainly had no place on the battlefield in Suárez's view: "The delicacy of the female sex is at odds with the ideas of destruction that war represents." "Women are born to perpetuate humanity, and not to destroy it," so the female sex had to be "left outside the life of the barracks and the encampment."[24] Such comments were not unique to the situation of Chile. In the words of the Lima newspaper *El Peruano*, "Christian civilization" has excused women of the "duty to take up arms," but does require them "to share with the citizens the exhaustion and sacrifice of war, helping to soften the barbarism of war and its disastrous results."[25]

By taking on these roles, women cooperated actively both on the home front and through several organizations that focused mainly on supplying food and caring for the wounded and children left orphaned by the conflict. Associations like the Society of Saint Vincent de Paul and the Company of Mary, Cruz Blanca in Perú, and the Sociedad Protectora de Santiago y Valparaíso, the Casa de María and the Sociedad del Perpetuo Socorro in Chile provided opportunities for women's participation during the war.[26]

Although the prevailing discourse on women's roles tended to limit them to the home front, a few dissenters called such messages into question. Eva Canel, a Spanish publicist living in Lima, called for an expansion of women's sphere of activity, to "shake off this absurd timorousness" and "sacrifice our wellbeing for the sake of our patriotic love." Some of the "manlier women" could even go to the battlefield "to help those who fall."[27] In fact, in 1882, around twenty women from the city of Arequipa requested permission to create a female military unit, claiming it did not matter to "patriotic love whether it was contained in the body of a woman." The move to the battlefront emerged through the logic of supporting men's efforts in combat, both materially and emotionally: "Our enthusiasm cannot be silenced because we were born women [...] let us mix our blood with that of those who we love most dearly," they concluded.[28]

Some of the most noteworthy women who wanted to take part in the conflict from within the theater of operations included the sutlers, known as *cantineras* amongst Chilean forces, and *rabonas* amongst the Peruvians. During the wars that were fought in the nineteenth century, these were women who brought their peacetime roles to the battle zone: cooking and delivering food, washing soldiers' clothing, and taking care of the sick.[29] Many of them were the wives and partners of soldiers, although sometimes soldiers' mothers also performed these duties. Thus, even though these *cantineras* and *rabonas* accompanied the armies to the battlefield, prevailing discourses called for them to maintain behavior associated with femininity. Indeed, writer Abelardo Gamarra described the Peruvian *rabonas* as the soldiers' "companions in their exhaustion, with no stake in their glories." They were "the type of faithful wife, the model of wives, the angel dressed in sackcloth."[30]

Cantineras and *rabonas* tended to personify the division of roles in wartime, even in combat zones. Nonetheless, the fact that they traveled to the front presented them with a strong opportunity to transition from caring for the wounded to becoming women warriors. However, despite these similarities, there were major differences in social recognition of military action by women between the two countries. In Chile, military action performed by women was quick to be celebrated and lauded with public recognition among contemporaries, and then faded over the course of time. Conversely, in Peru, the model associated with femininity was maintained, although this did not exclude the possibility of military action; and although little recognition was accorded at the time, this did provide women with a route into the pantheon of national heroes, where their deeds were remembered for longer and with more formal success than in Chile.

In "Sergeant Candelaria," Chilean society already possessed a heroic archetype that could be mobilized to recognize further military prowess amongst women, fitting it into a well-established interpretative framework. The Tarapacá Campaign (April–December 1879), the first phase of the land campaign of the war, proved to be the birthing grounds of feminine combat heroism. A case in point would be that of the *cantineras* of the Second Line, Leonor Solar and Rosa Ramírez, martyrs of the Battle of Tarapacá. Solar and Ramírez, who before the outbreak of war had been working as seamstresses, were pulled into the din of battle, pinned down in a farmhouse, wounded, and finally captured by a Peruvian unit. According to a newspaper, the Peruvian troops mercilessly fell into "an orgy of blood and cruelty," mutilating the *cantineras*' breasts and butchering their bodies, before burning down the farmhouse with their bodies inside.[31] Nothing was found amidst the ashes except for the scorched shoe of one of the women, which Patricio Lynch, the commanding officer of Iquique, sent back to Santiago along with other war trophies, as a patriotic relic.[32]

In the same battle, another *cantinera* attained public recognition in a very different way. Dolores Rodríguez—lauded in the press as a "second Sergeant Candelaria"—accompanied her husband to the battlefield as his *cantinera*. When he was killed, she took up his rifle and killed three Peruvian soldiers, before being wounded in the leg.[33] The press highlighted one of her actions as the most praiseworthy: even though the *cantinera* was wounded, she expressed a desire to continue with the campaign instead of returning home, proof of "how much Chilean bravery is capable of, even in the female sex."[34] Dolores Rodríguez was promoted to the rank of sergeant by the Commander in Chief, Erasmo Escala.

The most striking case of a *cantinera* who personified the link between vengeance, warfare, and heroism with remarkable popularity during the war was Irene Morales. Born in 1865 in the district of Chimba, Santiago—the same as Candelaria Pérez— Morales had worked as a seamstress, and entered into her first marriage at the age of twelve. Quickly widowed, she moved north to the city of Antofagasta and married a man named Santiago Pizarro. Pizarro became involved in a quarrel with a Bolivian soldier, was executed in September 1878, and his body abandoned in the desert. The Chilean occupation of Antofagasta in 1879 brought Morales an opportunity for revenge. An eyewitness reported that amongst the crowd that received the incoming Chilean troops, one effusive orator stood out, haranguing the crowd to take up arms and seek out revenge on the Bolivians: that orator was Irene Morales.[35] She dressed as a man in order to join the army, and took part in the landing at Pisagua and the Battle of Dolores, where she fought alongside her male counterparts. On learning of her performance on the battlefield, General Baquedano allowed her to wear a feminine military uniform, and promoted her to the rank of sergeant. She later took part in the capture of Arica—where her unquenchable thirst for vengeance caused her to order the execution of more than sixty Peruvian prisoners—[36] and in the decisive battles of Chorrillos and Miraflores.

Benjamín Vicuña Mackenna saw Morales's story as similar to that of the "widely celebrated Candelaria." And although he never hid his distaste for the "woman warrior, soldierly and masculine," he could not fail to recognize the heroism displayed by Morales in the many battles in which she fought. There can be no doubt that these were exceptional actions and worthy of recognition, but he still urged the heroine to return to the place that society assigned to women in times of peace. It was time, he advised Morales, to swap "her warrior's revolver for her old, honored and beloved sewing machine"[37]—advice that Morales ignored.

Meanwhile, in Peru, the activities of women during wartime fit the established frameworks for feminine behavior under such conditions. Thus, apart from the broad case of the *rabonas*, only two women played a striking role on the front lines, particularly in the irregular combat that resisted the Chilean invasion from 1881 to 1884: Leonor Ordoñez and Antonia Moreno de Cáceres.

While little documentary evidence exists, and significant parts of her life story are known largely from the oral tradition, Leonor Ordoñez seems to have followed a similar path to the Chilean *cantineras*. Born in Huancani in 1837, Ordoñez enlisted as a *rabona* to join her husband in the war; he was then killed in the battles to defend the capital from Chilean forces. Following defeats at Chorrillos and Miraflores, Ordoñez returned to the mountains to take part in the irregular combat led by Andrés A. Cáceres. In the town of Huancani, in the Mantaro Valley, Leonor Ordoñez led a guerrilla band numbering more than forty local farmers who brought sticks, stones, shovels, and

Figure 10.1 Irene Morales, *c.* 1881 (National History Museum).

spears, operating at the orders of Father Buenaventura Mendoza. In the Huaripampa Engagement (April 22, 1882), Ordoñez was captured, tortured, and eventually put to death by troops commanded by Colonel J. Antonio Gutiérrez.[38]

Peruvian feminine heroism, albeit necessarily rooted in a logic based on the society's expectations of women's roles, was perhaps best incarnated by Antonia Moreno. She was not a woman soldier but assumed a remarkable role during the Chilean occupation, although always in the footsteps of her husband, Andrés A. Cáceres, the leader of the mountain resistance groups. Born in Ica in 1848, following the occupation of Lima by Chilean forces, Moreno undertook significant actions as part of the Patriotic Resistance Committee, working to obtain money and intelligence, and to oversee the delivery of weapons and ammunition to the guerrilla fighters in the Andes.[39]

Under the steady gaze of occupying Chilean forces, Moreno decided to take her three daughters and flee to the mountains to meet up with Cáceres. It was her participation in the Breña Campaign that earned her the recognition of the people at large, particularly among indigenous peoples, as she was seen as the leader of the *rabonas* and worked closely with her husband, the so-called "Wizard of the Andes." His legendary charisma certainly played a major role in fostering the heroic narrative upon which much of this recognition was built. Cáceres, who spoke Quechua, had a significant symbolic background for the indigenous people of the Andes, who saw him as a continuation of their Inca heritage. The deployment of this "mythogenic" feature was a key element in his heroization. According to Antonia Moreno's memoirs—an unique historical source about the 1879 women warriors[40]—"the Indians saw Cáceres as a reincarnation of the Inca," calling him "Tayta with so much adoration that they moved him." The same indigenous people of the highlands called her "great mother."[41] It was from that position, repeated time and again in historiography, that Moreno earned public recognition: wife, mother, and selfless companion.

"Amazons" in the Pantheon? Memories, Gender, and Nationalism in the Postwar

In October of 1883, Chile and Peru signed the Treaty of Ancón, putting an end to the war. However, one issue remained unresolved: sovereignty over the regions of Tacna and Arica, that Chile had annexed during the conflict. Disputes over this issue only came to an end in 1929, so the rivalries and shared imageries that arose during the war lasted for decades. Those were prolific years for the creation of *lieux de mémoire* for the war, a time when the hero cult, expressed in public sculptures, the names given to places and streets, and civic rituals, reached a peak of intensity. These dimensions of heroization were limited to a select group of male heroes who stood for patriotic virtues such as bravery and selfless sacrifice.[42]

This process must be understood as something more complex than a simple imposition of nationalism as a state ideology. Indeed, both the nationalistic narrative and the rituals for bringing it into society were often promoted by civil society itself, particularly among veterans. It is relevant to examine this issue considering the concept described by Jay Winter as "collective remembrance," a public recollection that is different from that of the state, but rather founded on a need for public sharing of the experiences linked to something so shocking as warfare, preventing its oblivion.[43]

The postwar years saw the formation of a duality in the remembrance of the warrior women discussed in this essay. In Chile, the country that had seen more cases of women soldiers, the national pantheon of heroes minimized the visibility of women. Even the *Álbum de la gloria de Chile* [Album of the Glory of Chile], which Benjamín

Vicuña Mackenna published during the years after the war, dedicated only a few lines to the part women had played in the conflict. The book, which provides biographical sketches of more than a hundred personalities described as heroes, mentions women only in the broadest category of "the anonymous of the war." They only warranted an evidentiary appendix in his thesis on how Chile had defeated its neighbors thanks to collective national heroism. In this scenario, women's participation in the war should have been described, "be it with the Roman matron's majestic cloak, with the virgin's spotless apron, or with the picturesque suits of the battalion *cantineras*," only mentioning the most outstanding female warfighters: Leonor González, the martyred *cantineras* of the Second Line Regiment, María Quiteria Ramírez, and the "cruel and avenging" Irene Morales.[44]

In general, both the decorations and the pensions were limited to the best-known cases in the public domain. Thus, in 1888 the Chilean government decided to award Irene Morales a life pension of 15 pesos per month, in recognition of her work during the war. A decade later, the same award was made to Josefa del Carmen Herrera, and the following year, 1898, saw the same benefit going to *cantinera* Juana López.[45] This tension between official recognition and oblivion amongst the populous is clear in the cases of Irene Morales and Juana López, both of whom died in obscurity, but this led paradoxically to efforts made among certain circles to restore their place of honor in the collective memory of the war.

Thus, only two years after being awarded her government pension, Irene Morales died of pneumonia on August 25, 1890, in the wards of San Borja Hospital. The poet Rómulo Larrañaga dedicated some of his works to her as an homage, calling her "a shrine of the whole republic": the woman who "cured the wounded" and, when the time came, "herself did battle/with extraordinary valor."[46] Indeed, although Larrañaga had already hatched the idea of a public collection to fund a worthy mausoleum for the heroine, he had to wait until the next century to see it become reality, in November 1903.

The outcome for Juana López followed a very similar pattern of recognition by the authorities, and oblivion and repositioning amongst the people. In January 1904, a few years after being awarded her meager government pension, López died. "Not a soldier, not a musician, nobody was drawn to accompany her remains due to patriotic sentiment. She who gave up four lives for the nation died, abandoned by everyone," wrote a newspaper, together with a call to collect funds to build a mausoleum in her honor.[47] The mausoleum was opened five years later, amidst celebrations of the centenary of Chile's independence, in a patriotic ceremony that was actively attended by both veterans and municipal authorities.

Set against this background, the case of Filomena Valenzuela is truly exceptional. In the north of the country, in lands still disputed with Peru, during the crudest moments of Chilean nationalist propaganda, the female veteran's presence was at the forefront of the commemoration ceremonies supported by patriotic and veterans' groups. During the following years, and until her death in 1924, she resided in the port city of Iquique and played a leading role in festivities associated with the war, turning her into something of a people's icon in the region, a "revered shrine of our army."[48] In fact, the author of a biography written in Iquique during that period—Spanish writer José de la Cruz Vallejo—set out to use her life as an example of values that contemporary society should imitate. If the Chileans were "the manliest of the Hispano-American peoples," Filomena Valenzuela was able to meet that description, being a woman with "a manly disposition." Her participation in the war showed it clearly: "In the years of her youth, when the world shows a woman its most tempting charms, when feminine souls dream of an ideal future, she opted to sacrifice herself for her homeland and suffer the grueling rigours of a military campaign."[49]

In Peru, recognition of women's participation in the war was slightly more significant, although it continued to frame their value in the context of expected wartime gender roles. For example, Antonia Moreno received public recognition following the war, not least due to her position as First Lady, although even this was not without its problems. Indeed, as Iván Millones notes, Andrés A. Cáceres' rise to power in 1886 was as much the consummation of public recognition for his personage—a "living hero"—as the start of his move into political debate, and later displacement. A decade later, once again in power and with every intention of retaining it, he was deposed and exiled after a civil war (1894–95), justified by a strong anti-militarist sentiment sustained by the Civilist Party.[50] There can be no doubt that these events affected the way in which his wife would be remembered. While she was able to enjoy the privileges of power, she was also obliged to suffer its detriment. For example, in 1890, Peruvian intellectual Clorinda Matto de Turner started a series of biographies of significant women in national politics with a profile of Antonia Moreno, as one of the few women who had entered such circles through her own merits. Although the editor of *Perú Ilustrado* saw her husband, Marshal Cáceres, as the main hero, Antonia Moreno had suffered "all of the privations" of the war at his side. Moreno therefore represented a model of the "selfless patriot" and "good Peruvian woman," as she had faced the vicissitudes of war with a combination of "Spartan resolve" and "Christian resignation."[51] But her husband's fall in 1895 besmirched the reputation of Antonia Moreno by association, particularly in Lima's satirical press, which, in the midst of the political polarization brought about by the civil war, set out to vilify her for her humble and provincial origins, calling her a *rabona* or labeling her "Madame Melon." So it was that when Antonia Moreno died, in February 1916, the press barely covered the news, and the obituaries that were published returned to the same entrenched archetypes: a selfless wife and mother.[52] Years later came the official recognition of another female Peruvian war hero, Leonor Ordoñez. In 1918, residents of Huancani sent Congress a request to rename their district after the local heroine, a motion that was passed. A regional law passed in September of 1920 thus approved the creation of Leonor Ordoñez District, an utterly unique form of recognition by both the people and the authorities, for one of the female warfighters considered here. However, paradoxically, this led to the disappearance of any mention of Ordoñez herself until the centennial of the War of the Pacific in 1979–84.

In the postwar memory, the Peruvian women were included in the national narrative, albeit in a systematic manner built on the existing beliefs regarding traditional women's roles. Feminist academic Elvira García dedicated several pages of her 1924 book *La mujer peruana a través de los siglos* [The Peruvian Woman through the Centuries] to examining the role performed by Peruvian women in different wars, but generally most of those women took part in a manner based on the concepts of charity and care for the wounded. Except for characters rooted largely in fiction—like Dolores the *rabona*, an incarnation of the indigenous women in their struggle against the Chilean invaders—or forged in oral tradition like "Marta the cantinera," García continued to see Antonia Moreno as the feminine ideal.[53] A decade later J. Armando Guinet included the same women as the only female individuals to appear in his compendium of heroes *Los grandes del Perú* [The Greats of Peru] albeit with extremely minimal descriptions that once again prolonged the bifurcation of stories about the war's women in Peru: recognition of Antonia Moreno linked to her husband, and the lionization of partly legendary folk heroes, like Dolores the *rabona*.[54]

The most significant process of female hero-making associated with the war took place around the centenary of the War of the Pacific. The years running up to the centennial were marked by a profound change in the country's national narratives, brought about by the left-wing military regime of Juan Velasco Alvarado (1968–75). The central change was the positioning of ordinary people, the indigenous and the mestizo groups—the *cholos*—as core elements of the Peruvian identity.[55] It should therefore come as no surprise that a wide range of heroic icons were selected for veneration, expressing values linked to this new national identity.

The repositioning of a range of heroic discourses around women's participation in the War of the Pacific illustrates this point by expressing the core concept of the time: underscoring the contrast between folk heroism as a collective lionization with no distinction based on race or gender in the resistance against the Chilean invasion, compared with the collusion and clear lack of unity amongst the collaborationist upper classes. Gathering resources, providing supplies, curing the wounded, and accompanying soldiers on the battlefield, Peruvian women asserted their collective heroism, claimed Judith Prieto.[56] Indeed, as a media group stated at the centennial celebration of the Breña Campaign, if "the official history always omitted what the ordinary people went through in its stories of the past," it was time to reclaim the people's contributions during wartime, when women had played a crucial role. To highlight female contribution was also to highlight the contribution made by the people, helping to empower a new narrative about the nation's collective valor, finally giving visibility to the "historic ferociousness and bravery of the Peruvian woman."[57]

Thus, alongside the names that were already a part of the shared consciousness regarding the war, new ones began to appear, although always based on the logic of highlighting resistance to the enemy and appealing to cases that could underscore the regional or racial features of the heroines. This was the work of César Ángeles: academic, journalist, and folklore expert, through novelistic descriptions, filled with patriotic stereotypes and unencumbered by any supporting historical evidence. For example, Hortensia Ceballos, the local heroine of Pisagua, provided a case study in sacrificial resistance to the enemy. When the port city was being invaded, she saw her entire family murdered by Chilean troops and decided to kill herself with her husband's bayonet, thrusting it through her neck. This act showcased her as an "unfading heroine, whose life and deeds comprise the most extraordinary example of heroism and sacrifice."[58] The tale of Catalina Buendía, which was based solely on oral traditions in the Ica region, was even more novelistic and paraded all the elements that the nationalistic rhetoric of the day wanted to hear. An Afro-Peruvian, Buendía had led

Figure 10.2 Leonor Ordoñez Monument, Huancaní, Perú.

resistance against the occupation in the town of San José de los Molinos, where, crying out "Not one step! Long live Peru!" she died, impaled on the staff of the Peruvian flag.[59]

Against this backdrop, it should come as no surprise that the country's leading newspaper, *El Comercio*, launched the "Heroine Antonia Moreno de Cáceres Prize" for teachers and students to prepare a biographical sketch of someone who, in the words of her great-granddaughter, had "used her presence and that of her daughters to foment the spirit of sacrifice and love for the flag" amongst the peoples of the Andes.[60]

The edifying description of Antonia Moreno that was assembled by teacher Grecia Mendoza in a text published for the competition, rescued figures like Moreno from oblivion, after the "oligarchy" had banished their historic narratives.

An examination of the different facets of Antonia Moreno—the epitome of motherly love, marital companionship, and feminine patriotic selflessness, a woman of the people who respected indigenous customs—could be used to provide the youth with a "symbol of inspiration for their moral learnings," an example "to follow, striving tirelessly in the hope that the people would someday receive justice." Most of all, the prevailing narrative during that period saw Antonia Moreno as a model of social integration and national unity: "despite coming from the coast, she learned to love Andean customs, farmers, highlanders, and to live with them in their homeland, taking everything with great responsibility."[61]

Finally, the most decisive milestone that must be mentioned in any discussion of the heroic narrative for women during those years was the incorporation of both Leonor Ordoñez and Antonia Moreno into the official Peruvian pantheon of national heroes, the formal and definitive recognition of their category as heroines. Under Law 1906, the Peruvian government the Crypt of Heroes of the War of the Pacific, a canonical *lieux de mémoire* for the conflict, housing the remains of the nation's most important heroes: Miguel Grau, Francisco Bolognesi, and Andrés A. Cáceres, together with a hundred other figures. It was only in 1988 that the expansion of the mausoleum broke the male monopoly on civic value that had held sway before, with the exceptional burial of two women in that pantheon: Leonor Ordoñez and Antonia Moreno de Cáceres.[62]

Thus, unlike the situation in Chile, where women soldiers were recognized during their lifetime but were then swiftly forgotten, Peru was slower to acknowledge the women who took part in the war and even disregarded their irrefutable contributions on the battlefield but made sure that their recognition was eventually formalized in official *lieux de mémoire*. The move was a success, and to this day has achieved its goal in collective memory. Women won access to the pantheon, albeit not necessarily women combatants.

Notes

1 Miguel Ángel Centeno, *Blood and Debt: War and the Nation-State in Latin America* (University Park: Penn State University Press, 2002).
2 Siniša Malešević, *The Sociology of War and Violence* (Cambridge: Cambridge University Press, 2010).
3 Samuel Brunk and Ben Fallaw, eds., *Heroes and Hero Cults in Latin America* (Austin: University of Texas Press, 2006).
4 Laurence van Ypersele "Héros et héroïsation," in *Questions d'histoire contemporaine: Conflits, mémoires et identités* (Paris: PUF, 2006), 164.
5 Nira Yuval-Davis, *Gender & Nation* (London: Sage, 1997), 93–115.
6 Peter Burke, *Varieties of Cultural History* (Ithaca: Cornell University Press, 1997), 50–52.
7 The *cantineras* and *rabonas*—translated as sutlers in this work—played different roles in times of war, like civilian merchant, camp-follower and nurse functions,

superimposing all these tasks. Many times, they had an army license to provide these services, but not always.

8 See Carmen Mc Evoy "Bella Lima ya tiemblas llorosa del triunfante chileno en poder: Una aproximación a los elementos de género en el discurso nacionalista chileno," in *El hechizo de las imágenes: Estatus social, género y etnicidad en la Historia Peruana*, ed. Narda Henríquez (Lima: Pontificia Universidad Católica del Perú, 2000), 195–222.
9 Ignacio Silva, *La Sarjento Candelaria Pérez* (Santiago: Imprenta Cervantes, 1904), 27–42.
10 Vicente Reyes, "La Sarjenta Candelaria," *La Semana*, Santiago, June 11, 1859, 53.
11 Antonio Barrena, *Vida de un Soldado: Desde la Toma de Valdivia (1820) a la victoria de Yungay (1839)* (Santiago: RIL, 2009), 260.
12 José Varas, *Recopilación de leyes, decretos supremos i circulares concernientes al Ejército, desde abril de 1839 a diciembre de 1858* (Santiago: Imprenta Chilena, 1860), 33.
13 Valentín Letelier, *Sesiones de los Cuerpos Lejislativos de la República de Chile* (Santiago: Imprenta Cervantes, 1903), XXVII, 176.
14 "Avisos," *El Progreso*, Santiago, February 24, 1849.
15 Reyes, "La Sarjenta Candelaria," 56.
16 Benjamín Vicuña Mackenna, "La sarjento Candelaria y la Monja Alférez. Estudio crítico," in *Miscelánea* (Santiago: Imprenta del Mercurio, 1872), 236–242.
17 *El Ferrocarril*, Santiago, March 30, 1870.
18 "Candelaria Pérez," *El Independiente*, Santiago, March 30, 1870.
19 Ventura Blanco, "La Sarjento Candelaria," *La Estrella de Chile*, Santiago, April 17, 1870, 431–435.
20 Blanco, "La Sarjento Candelaria," 431–435.
21 Gonzalo Bulnes, *Historia de la campaña del Perú en 1838* (Santiago: Imprenta de Los Tiempos, 1878), 227.
22 William F. Sater, *Andean Tragedy: Fighting the War of the Pacific, 1879–1884* (Lincoln: University of Nebraska Press, 2007).
23 José Bernardo Suárez, "Más entusiasmo por la patria. Al bello sexo," *El Mercurio*, Valparaíso, April 23, 1879.
24 "Las mujeres soldados," *El Mercurio*, Valparaíso, July 24, 1880.
25 "Actualidad," *El Peruano*, Lima, July 4, 1879.
26 Maritza Villavicencio, "Acción de las mujeres peruanas durante la guerra con Chile," *Debates en Sociología* 10 (1984): 147–158; Juan José Rodríguez, "El bello sexo en guerra: cultura política y género durante la Guerra del Pacífico," *Illapa* 5 (2009): 83–120; and Paz Larraín, *Presencia de la mujer chilena en la Guerra del Pacífico* (Santiago: Centro de Estudios Bicentenario, 2006), 125–167.
27 Eva Canel, "La mujer ante la patria," *La Bolsa*, Arequipa, May 24, 1880.
28 "Las arequipeñas solicitan de la autoridad tomar las armas contra los chilenos," Arequipa, May 31, 1882, in Pascual Ahumada, *Guerra del Pacífico* (Santiago: Andrés Bello, 1982), vols. VII–VIII, 110–111.
29 Nanda Leonardini, "Presencia Femenina durante la Guerra del Pacífico: El Caso de las Rabonas," *Norba* XXXIV (2014): 177–195; and Larraín, *Presencia de la Mujer*.
30 Abelardo Gamarra, "La rabona," *El Peruano*, Lima, June 19, 1879.
31 Rafael Egaña, "Los Mártires del Rancho," *El Nuevo Ferrocarril*, December 11, 1879.
32 "Las Heroínas del 2° de Línea," *El Mercurio*, Valparaíso, April 13, 1882.
33 "Más detalles sobre Tarapacá," *Los Tiempos*, Santiago, December 7, 1879.
34 "Una heroína chilena," *La Patria*, Valparaíso, December 11, 1879.

35 Antonio Urquieta, *Recuerdos de la vida de campaña de la Guerra del Pacífico* (Santiago: Talleres Gratitud Nacional, 1907), vol. I, 50.
36 Nicanor Molinare, *Asalto y toma de Arica, 7 de junio de 1880* (Santiago: El Diario Ilustrado, 1911), 114.
37 Benjamín Vicuña Mackenna, "Las amazonas del Ejército de Chile. La cantinera del 3°, Irene Morales," *El Nuevo Ferrocarril*, Santiago, August 13, 1880.
38 Ahumada, *Guerra del Pacífico*, vols. VII-VIII, 406; Eduardo Mendoza, *Historia de la campaña de La Breña* (Lima: ITAL, 1983), 167ss.
39 Grecia Mendoza, *Antonia Moreno de Cáceres* (Lima: Universidad Nacional Mayor de San Marcos, 1987).
40 The text remained unpublished for years and was published almost a century after the conflict, in 1974. On its contents see Ruth Solarte, "Desplazamientos y resistencia femeninas en la Guerra del Pacífico: las memorias de Antonia Moreno de Cáceres," *Decimonónica* 15, no. 1 (2018): 50-66.
41 Antonia Moreno, *La campaña de la Breña* (Lima: Universidad Alas Peruanas, 2014), 135.
42 Gabriel Cid, "De héroes y mártires. Guerra, modelos heroicos y socialización nacionalista en Chile (1836-1923)," *Mélanges de la Casa de Velázquez* 46, no. 2 (2016): 57-78.
43 Jay Winter and Emmanuel Sivan, eds., *War and Remembrance in the Twentieth Century* (Cambridge: Cambridge University Press, 1999), 6-10.
44 Benjamín Vicuña Mackenna, *El Álbum de la gloria de Chile* (Santiago: Imprenta Cervantes, 1883-1885), vols. II, 588.
45 Ricardo Anguita, *Leyes promulgadas en Chile* (Santiago: Imprenta Barcelona, 1912), vols. III, 82, 392 and 411.
46 Rómulo Larrañaga, "La cantinera Irene Morales," in *Canciones y poesías de la Guerra del Pacífico*, ed. Juan Uribe Echeverría (Valparaíso: Ediciones Universitarias de Valparaíso, 1979), 104.
47 "Juana López," *El Diario Ilustrado*, Santiago, January 27, 1904.
48 "Filomena Valenzuela en la Misa y en el Instituto Comercial," *La Aurora*, Arica, June 9, 1917; see also "Doña Filomena Valenzuela," *La Aurora*, June 6, 1920. "La señora Valenzuela," *La Aurora*, June 7, 1921; "Filomena Valenzuela," *La Aurora*, June 6, 1922; and "La veterana Filomena Valenzuela en la Escuela Superior de Niñas," *La Aurora*, June 11, 1924.
49 José de la Cruz Vallejo, *La cantinera del Atacama, doña Filomena Valenzuela* (Iquique: Imprenta y Encuadernación I División, 1922), 26.
50 Iván Millones, "El mariscal Cáceres: ¿un héroe militar o popular? Reflexiones sobre un héroe patrio peruano," *Íconos* 26 (2006): 53.
51 Clorinda Matto de Turner, "Grabados," *El Perú Ilustrado*, Lima, July 5, 1890.
52 Abel Bedoya, "La señora Antonia M. de Cáceres," *La Prensa*, Lima, February 29, 1916.
53 Elvira García, *La mujer peruana a través de los siglos* (Lima: Imp. Americana, 1924), vols. I, 351-409.
54 J. Armando Guinet, *Los grandes del Perú. Índice heroico* (Lima: Siempre Adelante, 1937), 54, 71.
55 Juan Martínez Sánchez, *La revolución peruana: ideología y práctica política de un gobierno militar, 1968-1975* (Seville: CSIC/Universidad de Sevilla, 2002), 166-174.
56 Judith Prieto, "La mujer peruana y la guerra con Chile," *El Comercio*, Lima, April 5, 1979; and "Resistencia y conspiración femeninas," *El Comercio*, Lima, December 17, 1979.

57 "Mínimo reconocimiento a la mujer peruana," *La República*, Lima, November 15, 1982.
58 César Ángeles, "Pisagua y su heroína," *Expreso*, Lima, February 27, 1979.
59 César Ángeles, "Catalina Buendía de Pecho," *Expreso*, Lima, May 15, 1979.
60 "La esposa del Mariscal Cáceres alentó el espíritu de sacrificio de sus compatriotas, dice bisnieta," *El Comercio*, Lima, April 18, 1979.
61 Mendoza, *Antonia Moreno,* 22, 30.
62 *Cripta de los Héroes de la Guerra de 1879. Guía histórica y biográfica* (Lima: Centro de Estudios Histórico-Militares del Perú, 1999), 46–47; 116.

11

Commemorating China's Wartime Spies: Red Agents Guan Lu and Jiang Zhuyun, and the Problem of Female Fidelity

Louise Edwards

Once the guns go silent, the post-conflict struggle to write and rewrite the history of war commences. The propagandists who sought to inspire popular participation in the war turn their attention to building "correct" memories of it. Postwar reconstruction often involves invoking normative gender roles as governments hoping to build stability and their own legitimacy, be they victors or vanquished, make frequent use of gendered tropes. The political order they seek to consolidate is at once a sexual order. In narrating the histories of the "War of Resistance against Japan" (1937–45) and the three-year Civil War against the Nationalists (GMD) (1946–49), the government of the People's Republic of China (PRC) actively invokes a sexual-political order to legitimize the Chinese Communist Party's (CCP) rule.[1] In PRC wartime commemoration, visions of noble and courageous men defending their staunch and morally upright women abound. Fidelity to the Chinese nation and to the CCP emerge as key attributes of warriors fighting for CCP victory. But what of those people for whom service to the party and the nation requires duplicity—the spies and secret agents gathering intelligence deep behind enemy lines? Their fidelity is not as easily manifest or as readily recreated in postwar commemoration. Their successful wartime service requires lying, cheating, and trickery.

For women spies, commemoration is particularly problematic because postwar normalcy often depends upon the reaffirmation of patriarchal sexual norms, within which female *sexual* fidelity is central. Women agents are regularly assumed to be "honey traps" who use their feminine charms to seduce their targets. A woman's active, cunning deployment of feminine sexuality runs counter to mainstream propaganda in which "our women" are pure. Around the world, wartime governments motivate soldiers with calls to defend the chastity of "women at home" against enemy violation. After war, invocations of female sexual fidelity symbolize the return of "normalcy." In the PRC, idealized visions of "our" women's sexual fidelity are evident in the prosecution of war, the rebuilding of stability after war, and in the re-narration of the war after its cessation. Just as war is a gendered system, so too is its commemoration.[2] The treatment of women spies in commemorative processes exposes the contradictions inherent in the recognition of women's war service presented by the return to a form of patriarchy that depends upon female fidelity.

This chapter focusses on two communist spies from the 1930s to 1940s and shows that for their service to be useful in postwar reconstruction their compromised sexual fidelity must be carefully managed. The renowned author Guan Lu (1907–82) operated in Japanese-controlled Shanghai throughout the War of Resistance using the cover of editor for the leading collaborationist women's magazine—*Nüsheng* (Women's Voice). Her prewar fame as one of the highly desirable Three Female Talents of Shanghai (*Shanghai sange cainü*), along with Eileen Chang (1920–95) and Ding Ling (1904–86), ensured that once she joined the Japanese-sponsored magazine, she was branded a "cultural traitor" (*wenhua hanjian*). Her celebrity status as a desirable beauty, who was supposedly living a glamorous Shanghai life-style consorting with the Japanese military while the rest of the population suffered wartime privations meant that her reputation was irreparably tarnished. The second CCP agent, Jiang Zhuyun (1920–49), worked underground in GMD-controlled Sichuan to mobilize students and workers. Unknown during her short life, Jiang's cover involved living as wife to a male CCP agent despite his already-married status. Their relationship was more than "political" since Jiang soon gave birth to a child. Captured on the eve of the CCP's victory, Jiang was jailed, tortured, and executed. After the war Jiang's story was promoted extensively and she has long been celebrated as a heroic martyr to the nation. Despite her adultery, she is depicted as a revolutionary sister and mother in the patriarchal national family through repeated and often convoluted makeovers that minimize her deviations from the normative sexual-political order and explain away the conjugal irregularity that is at the heart of her war service. Guan Lu, in contrast, became an embarrassment to the CCP and faced extensive public criticism, endured years of imprisonment, and enforced singlehood. She was never recognized during her lifetime for her war service. As this demonstrates, the treatment of the two women spies, one who lived on well beyond the war and the other who died during its course, tells us that sexual fidelity is crucial to the recognition of women's war service.

Guan Lu: A Tarnished Celebrity Trapped by Fame and Longevity

Guan Lu was born Hu Shoumei into the Shanxi family of a prominent local official, the County Magistrate, during the last decade of imperial rule. On the death of her father when she was nine and her mother when she was sixteen, she moved to live with her widowed aunt in Nanjing and eventually relocated downriver to Shanghai where she entered university and joined the city's lively literary scene. Her decision to join the CCP in 1932 at twenty-five years of age was spurred by the Japanese annexation of Manchuria in north-east China. Through the 1930s, she garnered popular fame as a left-wing writer publishing poetry and fiction in a wide range of journals. A celebrity figure, she featured in gossip columns that pondered the glamorous lifestyle of a Shanghai woman writer. Her prominence was confirmed when in 1937, the year the War of Resistance against Japan formally commenced, she penned the lyrics for the theme song of a popular left-wing patriotic film, *Shizi jietou* (Crossroads). Her serialized

Figure 11.1 A glamorous Guan Lu as a celebrity author. Photo credit: Unknown.

autobiography, *Xin jiu shidai* (The era of old and new), appeared through 1938 in the journal *Shanghai funü* (Shanghai women) and narrates the trials and tribulations of a woman of her time.

With Japan's occupation of the city, many left-wing writers quit Shanghai, but the CCP instructed Guan Lu to remain. She undertook a number of small missions until finally in 1939 she was charged with infiltrating Shanghai's collaborationist secret police unit, No. 76. Her goal was to win the trust of one of its key figures, Li Shiqun (1905–43)—himself a former CCP member. The plan drew on good groundwork and Guan Lu had good "cover." Her brother-in-law and elder sister were old acquaintances of Li Shiqun and his wife Ye Jiqing (n.d.) through various political activities. Guan Lu's brother-in-law had been arrested in the early 1930s and met Li by chance during this incident. But the deeper connection came through their wives. In 1932 or 1933, Li Shiqun and Ye Jiqing were again jailed and faced the prospect of a lengthy internal investigation. Guan Lu's sister took care of their child and provided supplies for them while they were in prison. These acts won trust and gratitude.[3] Guan Lu simply had to become friends with her sister's friends to gain access to the highest levels of No. 76. These deep family connections made Guan Lu's cultivation of a friendship with Li

Shiqun and Ye Jiqing credible despite her known leftist sympathies. All the while, she continued to write and gain acclaim as a beautiful, young literary celebrity.

Typical of the celebrity gossip surrounding Guan Lu is a 1937 article in a popular movie magazine which revealed her perceived desirability as a beauty about town. The article commences with the juxtaposition between her "leftist" politics and her "fashionable" lifestyle. "Although Guan Lu is a very progressive poet, she commonly decks herself out in fashionable clothes. When you add that to her really beautiful features, she has no shortage of suitors from amongst the literary world." The article continues reporting that she forgot a previously arranged date with fellow writer Xu Maoyong, went out herself, and returned late at night, stepping over the sleeping suitor on the steps. On encountering him there the next day, in a disheveled state asleep in his clothes, she couldn't bear to look at him and no longer speaks to him.[4] Another article from the same year ran with the racy headline "Are Ba Wei and Guan Lu in Lesbian Love?" Ba Wei (1894–1983) was another woman writer and she and Guan were often compared and critiqued on the basis of literary merit—this article simply sought to sell papers through sexual innuendo.[5]

This frivolous tone would cease once her connections to the collaborationist regime became more public in 1942 when she became editor of *Women's Voice*—a magazine sponsored by the Japanese consulate's Office of Information in Nanjing. The journal's Japanese co-editor Sato Tamura Toshiko (1884–1945) was a feminist writer and journalist with extensive socialist connections—although her socialism was muted during her years in China.[6] Guan was to use her proximity with Tamura to glean information about the Japanese leadership in Shanghai and to deepen the CCP's links with the Japanese socialists.[7] The cost to Guan Lu's reputation from this public role in a collaborationist publication would prove irreparable. Up to this point, the general public simply knew that the author, Guan Lu, was living in Shanghai since they had no knowledge of her "friendship" with leading figures in No. 76. Once she was appointed editor, her status as a collaborator was confirmed. Academic Wu Peichen argues that despite its outward appearance as a collaborationist publication, and unbeknownst to the reading public, most of the contributors to *Women's Voice* "were underground members of the Communist Party."[8] The complexity of the espionage tactics remained unobserved by ordinary people and Guan Lu was branded a "cultural traitor."

As editor, Guan Lu traveled to Tokyo in 1942 to attend the Second Annual Greater East Asia Writers' Meeting, and while there she took messages to Japanese communists from their Chinese counterparts and was able to report back on conditions on the ground in Tokyo.[9] However, the Japanese authorities used these kinds of events to great propaganda effect—publishing photographs of the Chinese delegates to Tokyo and her status as a "leading cultural traitor" was confirmed. Guan Lu's role in strengthening communication between the communist movements in both countries remained hidden to all but the top CCP leadership.[10] Instead, her participation in the meeting became key evidence of her treachery in the immediate postwar period, the Civil War, and after the establishment of the PRC.

After Japan's defeat in 1945, Guan Lu's name appeared on the Nationalist government's list of "traitors." Immediately, the CCP secretly arranged for her evacuation from Shanghai into territory held by their New Fourth Army. Her espionage role was not disclosed in order to make further use of her intelligence—the general public simply thought she had fled in guilt to evade trial. A report by Tang Qing from 1946 in the popular magazine *Shanghai tan* (Shanghai bund) pondered her sudden disappearance saying that she had been spotted in Hong Kong. But, the article continued, given her "shameless" crimes in collaborating with the Japanese there was nowhere safe for her to ever raise her head again. Tang Qing notes that "one report says she is with the New Fourth Army" but uncertainty about the veracity of the report suggests that few readers would suspect or believe she was a spy for the CCP. In Tang Qing's various attacks on this "thoroughly hateful person," he noted that after she returned from the Tokyo conference, she wrote highly positive essays, "totally selling herself out. The fact that she is a woman makes this really scandalous." It was "disgraceful" that a gifted writer should become a traitor—she had no conception of "shame."[11]

The mystery surrounding her whereabouts gave Shanghai's popular press a further opportunity to compare her with Ba Wei. A 1946 article by Ma Xiang describes BaWei as having continued her "progressive" views moving inland during Shanghai's occupation, while Guan Lu instead cast her eyes upon the "Japanese pirates" (*wokou*), lived a very comfortable life, and wrote "peace literature." Ma says that after being listed as a traitor, Guan Lu had fled to Subei to escape arrest. Referring to Ba Wei's return to Shanghai, he wrote, "As one departs, the other arrives." Ba Wei is cited as saying that Guan Lu has "lost her integrity" (*shijie*). Readers are reminded that for women, the term *shijie* connotes the loss of sexual virtue—the article queries whether there is any difference between collaborating with the enemy and sleeping around with men.[12] Later, in 1947, even once her status as a CCP spy was confirmed, a Shanghai paper noted that the intelligence she gathered was worthless because she didn't understand military affairs—her experience revolved around writing love poems and flirting. She "didn't even know how to use a rifle." Confirming her dubious sexual morals and lightweight politics, the author, Ding Ji, noted that she had recently been seen whispering sweet nothings in the street with the branch head of the Xinhua News Agency, Chen Yiqun.[13]

The negative gossip about Guan Lu published in Shanghai partly reflects the anti-CCP sentiment in the city—but it is also revealing of the way women agents are perceived. The sexualized nature of the insults Guan Lu faced, the diminution of her value as an agent, the minimization of the risks she faced, and the amplification of the privileges she secured in her undercover role all reveal the precarity of life as a female spy. The doubt cast upon her sexual fidelity placed her in an invidious position—she became a pathetic figure whose only hope for reputational rehabilitation was death. For centuries sexually compromised women were presented with long histories of chastity martyrs within an ideology that produced the maxim "to starve to death is trivial, to lose one's chastity is serious" (*esi shi jixiao, shijie shi jida*). The party would prove to be no more grateful for her work against the Japanese than the Shanghai media despite its rejection of "feudal traditions" such as chastity martyrs.

While the media speculation about her continued, Guan Lu was trying to resume her pre-espionage life in CCP-controlled areas. She had been separated from her boyfriend, a prominent CCP member, Wang Bingnan (1908–88), during her years as an undercover agent—but now free of that role the couple reconnected. Wang was based in Nanjing serving as CCP "foreign envoy" and requested Party permission to be redeployed nearer to Guan Lu because they planned to get married. Wang's request was denied—Guan Lu's tainted public reputation was regarded as inappropriate for a high-level CCP diplomat.[14] The decision came from the very top echelons of the party with the leading CCP couple, Zhou Enlai and Deng Yingchao, deciding that a marriage between Wang and Guan would run counter to the party's interests.[15] As loyal party members, the couple accepted the ruling and never resumed their relationship. Guan remained single until her death. She struggled to find a meaningful role within the CCP-controlled areas and her mental health became increasingly fragile from this point on. The party's official broadsheet *Xinhua ribao* (New China daily) rejected submissions she made under the name Guan Lu because the name was so tainted that it would generate ire among their patriotic readers.[16] Her inability to resurrect the literary reputation she had secured as Guan Lu caused her great distress. With her literary identity destroyed, and her hopes for marriage thwarted, she shifted around various locations under CCP control, working as a teacher and publishing under pseudonyms.

The years after the formation of the PRC in 1949 would prove to be even more tumultuous. Moving to Beijing, she was assigned work teaching and as an administrator. She continued writing and adopted the Maoist prescriptions for socialist literature—producing a novel *Pingguo yuan* (The apple orchard) in 1951.[17] She secured central government approval for the book to be turned into a movie and commenced the scriptwriting. This creative career success would prove short-lived. In June of 1955, she was accused of being a counter-revolutionary in the so-called "Pan Hannian clique"— Pan (1906–77) had been her "handler."[18] When Mao Zedong accused Pan of being an "internal traitor" (*neijian*) in 1955, his entire team came under suspicion. After a series of trials, Guan Lu was jailed for two years in Beijing's Gongdelin prison. Released in March 1957, Guan soon found herself buffeted by the Anti-Rightist Campaign during which she was repeatedly "criticized." A few years of respite followed but with the advent of the Cultural Revolution, she became an early victim of campaigns to "root out bad elements." On July 1, 1966, Guan was arrested and jailed again—this time in the infamous maximum-security Qincheng prison. Her release came only on May 20, 1975, when authorities resolved that there was no evidence that she was in a counter-revolutionary clique.

During her first incarceration she commenced work on a never-published novel *Dang de nüer, Liu Lishan* (Daughter of the Party, Liu Lishan) that narrated the real-life story of a CCP martyr from the anti-Japanese war. Like Guan Lu, Liu (1921–43) worked underground in cultural affairs, editing newspapers and producing dramatic works for the CCP. She was captured soon after giving birth in Shanxi by the invading Japanese and swiftly executed. Liu's heroic martyrdom no doubt contrasted with Guan's embarrassing existence. Guan Lu's CCP minders confiscated the 300,000-word manuscript in 1967 prior to its completion, according to her sister Hu Xiufeng. When it was returned to her in 1975, the second half was lost.[19]

Her full political "rehabilitation" was granted on March 23, 1982, when the government overturned the 1955 verdict. Pan Hannian was posthumously rehabilitated a few months later in August. Guan Lu committed suicide on December 5, 1982, at age seventy-five, by taking an overdose of sleeping pills. The scholar Barbara

Barnouin said Guan Lu suffered from deep depression.[20] At the commemoration service held at the Babaoshan revolutionary cemetery, Guan was described as leaving four deep impressions. 1. She was intensely loyal to the party and did everything to protect the party. 2. She sacrificed her personal status and reputation without a single word of complaint in serving the interests of the party and the people. 3. She was a warrior who worked in enemy territory to save the country from destruction by the Japanese. 4. "Comrade Guan Lu valued friendship deeply, was steadfast in her beliefs, and her spirit of revolutionary optimism is worthy of our respect."[21] Her one-time love, Wang Bingnan, attended but did not deliver a commemorative speech. Because her death was a suicide, there was no official memorial notice.[22] Other published commentaries also depict her as a noble, self-sacrificing CCP member and "a staunch proletarian warrior" who sacrificed everything for the revolution.[23] Her sister noted her "complete faith in the communist project" and that despite all the hardship of her life she would never complain about the party. The official CCP daily newspaper *Renmin ribao* (People's daily) noted her passing with a brief obituary that acknowledged the injustice of her imprisonment. The article explained that she expressed staunch loyalty to the people and the revolution despite facing years of suspicion and mistreatment by the CCP. The article quoted a poem: "Warhorses love the battlefield, they gallop in service and not in order to win a good reputation. Writing led to calamities that produced injustice, yet still I love the rustle of paper and the fragrance of ink."[24] The commentaries and obituaries all reiterate Guan Lu's political fidelity to the nation and to the party—but nothing could be said to rehabilitate her reputation in sexual terms. The genre of CCP obituaries does not address decades of salacious supposition.

Jiang Zhuyun: Making a "Revolutionary Sister" from an Adulterer

Jiang Zhuyun, aka Zhang Zhijun or Elder Sister Jiang, has been a regular feature of public commemoration of the Civil War in the PRC. A household name in China, she has posters, comics, paintings, poems, plays, opera, fiction, school books, movies, television documentaries, a thirty-episode television series, statues and museums dedicated to her, as well as numerous scholarly histories. Guan Lu enjoyed none of this prestige, despite the posthumous praise at her funeral—with short documentaries of her life only appearing four decades after her 1982 death through suicide. The extent and variety of narratives about Jiang's espionage work stand in dramatic contrast to the relative silence on Guan Lu.

Jiang's comparative anonymity while alive combined with the convenience of her death provided extensive scope for CCP makeovers that would present her as a figure of political and sexual fidelity. Martyrs are perfect fodder for postwar commemorative projects and the more anonymous the better. Their "life-stories" are completely malleable and are uncomplicated by reality. CCP mass propaganda includes copious examples of martyred warriors who achieve celebrity status after death—see for example the posthumous fame of women warriors like Zhao Yiman (1905–36) and Liu Hulan (1932?–47) or the young male soldier, Lei Feng (1940–62).[25] Jiang Zhuyun's sexual and marital history is carefully managed to ensure that as a CCP martyr she emerges as having the fidelity expected of patriots, soldiers, and good women. To make effective use of her memory, her breach of the orthodox sexual-social order needed reconstruction. Ultimately, she would emerge from the multiple recreations of her life as having scrupulous sexual morals and a strong belief in normative patriarchal feminine and maternal values while also being dedicated to the CCP and her comrades.

Joining the CCP in 1939, Jiang Zhuyun was sent to Chongqing's Sichuan University in 1944 to build a secret communist study cell among left-wing students. She and fellow communist Peng Yongwu pretended to be a couple as part of their cover while running this cell, even though Peng was already married. In 1945, within a year of establishing their cell, they were apparently formally married with official narratives describing their "mutual deep affection in work"[26]—a truly revolutionary couple. The divorce that should have preceded the marriage is not mentioned. With the outbreak of the Civil War in 1946, Jiang was instrumental in rallying students to undermine youth support for the Nationalist government through underground newspapers and organizations like the Anti-Riot Movement. In 1947, Peng and Jiang were sent to the town of Xiachuan to foment anti-government activities and Peng was killed during one of the demonstrations. On his death, Jiang resolved to "fight in his place." The couple's young son was left to the care of Peng's "first" wife, Tan Zhenglun, as Jiang moved to a more active and dangerous role in the struggle. Eventually, Jiang was taken captive on June 14, 1948, in Wanxian and held in Zhazidong Prison in Chongqing. During her eighteen months in prison, Jiang faced numerous tortures and privations as the Nationalists sought to root out further communists. Despite the hardship which reportedly included "the tiger bench, the sling, beatings with a steel whip with thorns, electrocution, and having bamboo sticks placed under her ten finger nails" she never capitulated.[27] In holding out, she protected her comrades. Although Mao Zedong declared the formation of the PRC on October 1, 1949, the Nationalists still held pockets of the country—including Chongqing. On November 14, 1949, only days before Communist troops took the city, the Nationalists executed Jiang at age twenty-nine. In the postwar remaking of her as a loyal wife, sacrificing mother, staunch party member, and "elder sister" of the revolution, Jiang's life became a vision of fidelity to family, party, and country suitable for a hero that would help the CCP consolidate the new nation's sexual-political order. Her adultery is either ignored or carefully explained as being part of a grand CCP design.

In 1961, a major novel, *Hongyan* (Red crag), appeared and has been central to building awareness of her heroic sacrifice but also of creating her as a CCP "family member"—a revolutionary elder sister.[28] The authors, Luo Guangbin and Yang Yiyan, were in prison with Jiang which gives their narrative considerable weight. Sister Jiang first features in the book in Chapters 4 and 5 when she learns of her husband's death and requests to replace him in serving the CCP. Peng was described as having told her "If you take the masses to be your father, only then will the masses regard you as their son. Dedicate absolutely all your energies to this task through to your dying days."[29] Her revolutionary fervor increased after her husband's death. In Chapter 15, after she was imprisoned, readers are told of her dedication to her comrades and her unflinching resolve in the face of extremes of torture and denial of food. "Everybody knew that Sister Jiang had endured untold bouts of torture in order to protect the party's secrets and in so doing had won the respect of so many comrades."[30] The novel provides graphic details of her torture, her capacity to endure, and her willingness to sacrifice but it also reveals the family-like care that her fellow prisoners took in nursing her after torture—the reciprocity of revolutionary modeling is central to her story from this point onward. She models excellent revolutionary resolve and others model respect for those qualities in a perpetual cycle of teaching and learning and family support.[31]

The transmission of revolutionary fervor to those around her also has an intergenerational aspect. The prison includes an orphaned infant—collectively cared for by the prisoners—all that they know of her parentage is that she is the child of a CCP member who perished in prison. Sister Jiang reminds her fellow inmates that "The child is ours, we are all her fathers and mothers."[32] With this statement the idea that the CCP becomes the "family" for all Chinese is invoked with her role as Elder Sister reinforcing this familial political bonding. The child becomes a device to remind readers of the new national flag as well. In Chapter 25, word reaches the prisoners that Mao Zedong had declared the formation of the PRC back in October, and jubilant shouts of "The Chinese people have stood up" resound through the cells. The infant's mother, prior to her death, had hidden the PRC's Five Star Flag in her bedding and this is now unfurled. The adults tell the girl, "Child, beloved child! Look at the red flag, it was left by your father and mother."[33] The child, as inheritor of a revolutionary tradition, performs one more significant role before the end of the novel. Sister Jiang's parting words to those around her as she leaves for the execution ground are rousing calls to the imminent success of the revolution with the child called out for special attention. "Child, did you hear Aunty's words? Whether there is roaring thunder or horrendous gales, regardless of the terrifying waves and stormy seas, you must direct the banner of struggle towards the Communist Party!"[34] In 1962, the familial role Elder Sister Jiang plays is one of collective mother to all citizens.

In 1964, following the huge popularity of *Red Crag*, playwright Yan Su composed a seven-act operatic play titled *Jiang Jie* (Sister Jiang) based on the novel. Yan Su's play took Jiang as the protagonist and was a major force in her elevation as an independent heroic figure. The play adopts Maoist prescriptions in its characterization and each of the "positive" characters is as staunch as they can be in their communist dedication. In Act 6 Sister Jiang declares, "I am a revolutionary; and you, you are a counter revolutionary! The only relationship between us is one of revolution or counter revolution and a class struggle to the death!" When her opposite declares that Marx and Lenin are both long dead, Sister Jiang declares, "But there is still Stalin and there is still Mao Zedong!"[35] The collective spirit and intergenerational parenting are continued in his play, but with a slightly different twist. The jailors call for a particular prisoner, and when she comes forward, she is carrying an infant. The mother, leaving the infant as she goes to her death, is asked by Sister Jiang "Comrade, what is your name?," to which she replies: "Communist Party Member!" The aria Sister Jiang sings is replete with exhortations to the abandoned child to never forget the sacrifices, never forget the red flag and to "struggle forward to welcome the people of the new nation."[36]

Critic Ren Jia wrote of the opera, using the new garbled language of revolution:

> Through its political fervour and abundant revolutionary spirit, *Jiang Jie* has produced Sister Jiang as a glorious figure of a proletarian warrior, has extolled Sister Jiang as a representative of communist party members, and as a revolutionary spirit in the times of new revolutionary democracy and the struggle against enemies. [She manifests] the majesty of lofty heroism, courage in defying obstacles, and the noble qualities of being unafraid of sacrifice, and has offered us deep class education and education on revolutionary traditions.[37]

Ren's two-page review is replete with emotive repetitions of "revolutionary glory," "heroic martyrdom," "noble warrior," and "party loyalty." Her personal family circumstances as an adulterer and mother of a semi-legitimate child are ignored just as they were in Yan Su's play. The communist mother is the collective mother and the communist wife is a revolutionary partner in these Maoist years. The importance of the opera was confirmed in the immediate aftermath of the Cultural Revolution when it was released as a movie in 1978 and performed by the Shanghai PLA Arts and Culture unit.[38] Its appearance, despite the formulaic socialist realist script, was hailed as a mark of the liberalizing of the literary and arts scene with the jailing of the ultra-leftists under Mao.[39]

By 1996, the cartoon rendition of her story, also based on Yan Su's operatic play, explains the importance of recognizing the sacrifice of those who fought in the underground but includes the line that they sacrificed themselves in the struggle for "the construction of the modern and strong socialist nation of today."[40] Targeting a youth readership, the images show Sister Jiang as a friend and guide to young teenagers selling papers and helping with communication between communist agents. The final scene includes the sad farewell between Sister Jiang and the infant at the prison doors prior to her execution. Her status as an Elder Sister for all good patriotic revolutionaries is amplified in this version of her story.

From the 1980s onward, the CCP sought to resurrect its reputation as builders of "family values" and family property. In these new times, Sister Jiang's relationship with Peng was revived but it also required some "explanation" by the CCP party historians. In the 1961 novel *Red Crag*, her husband features only in death—he is captured, decapitated, and his head displayed as warning to others—the novel depicts Sister Jiang's horror as she arrived at the town, anticipating a long-awaited reunion with her beloved husband, only to be confronted by the gruesome mutilated body. There was no discussion of their pretense at marriage for the sake of the revolution. Rather, the authors dedicate some pages at the start of their novel explaining that love can coexist with and even spur on revolutionary fervor. Sister Jiang advises her CCP comrades that "Think of the great revolutionary leaders, Marx and Lenin … There were deep and strong bonds of affection between Marx and his wife Jenny, theirs was the love of revolutionaries."[41] This recognition of "bourgeois" sentiments like romance was already pushing the boundaries of acceptability in the Maoist prescriptions of the function of literature and art which dominated in the 1950s through the late 1970s—revealing that

Figure 11.2 Jiang Zhuyun, her husband, and their son. Photo credit: Unknown.

she was actually the "other woman" in an adulterous relationship or a "second wife" in a feudal polygamous marriage would have been impossible.

From the 2000s, the purity of revolutionary martyrs was not as tightly controlled—but Sister Jiang's life with Peng still required some cosmetic surgery. The official narrative became that they were initially living together and "pretending" to be a couple out of dedication to the Party. They only fell in love through their close work on a shared revolutionary goal for the party and they married to confirm this cover. Zhang Nan's 2016 explanation of the situation is revealing. "Peng already had a wife and a son in his hometown. Peng and Jiang's relationship was initially as revolutionary comrades. But engaged in a common pursuit, they gradually developed deep affection for each other, which they tried to repress for moral reasons."[42] Similarly, the 2010 thirty-episode television series *Jiang Jie* dedicates considerable time in the first few episodes to the awkwardness of Peng and Jiang's relationship as it becomes more intimate. But the Party is central to and ultimately authorizes the romantic shift. Peng is advised by his CCP superiors that their cover is threatened by the fact that after a year of "marriage" they have no children. He is advised to *really* become married to Sister Jiang. On discussing the matter, Jiang agrees, and they pick a secret wedding day—a dramatic student demonstration serves as their marriage ceremony. A bond that is marked by a revolutionary protest rather than a public ceremony soon produces a pregnancy, and their cover is confirmed. Another unfortunate twist in the plot sees her separated from Peng for the birth of their child. Significantly, this 2010 rendition of her life includes her decision to be sterilized immediately after the birth—one child is enough for a dedicated patriotic revolutionary.[43] The PRC's One Child Family policy was still operating in 2010 and even wartime spies could be put to use for that contemporary demographic campaign.

These convoluted explanations protect her glorious revolutionary sexual-political fidelity and also bring her back within PRC patriarchal family morality—norms that were not necessarily upheld among the radical youth that joined the CCP in the 1930s and 1940s when radical women and men had multiple partners simultaneously and serially. She and Peng were likely in a *de facto*, adulterous relationship by personal choice. The need to construct a semi-credible moral tale in the Opening up and Reform years is important because the liberalization has meant that scholars and journalists are now undertaking a host of research themselves into the "true story," "unofficial history," and "background facts" for many of China's political and popular celebrities. As publishing has become profit-oriented and more open, popular demand for these narratives is strong. The Party controls the story by presenting a credible, moral narrative that also includes the experience of her son and Peng's first wife. For example, Ding Shaoyi's 2011 book *The Real Family History of Sister Jiang* (Jiang Jie zhenshi jiazu shi) reveals that while Sister Jiang may personally have escaped the harsh treatment in the PRC through her death, her family members and son were subject to considerable pressure during the Cultural Revolution.[44] They suffered despite the impeccable revolutionary credentials of their martyred relative.

Another key mechanism by which the CCP has sought to reinforce her reputation as a good revolutionary mother is through the release of relics—specifically a final prison letter from mother to son. At the Sanxia museum in Chongqing, Jiang's last letter, supposedly written with a bamboo stick just prior to her execution, has been on display since 2007, the 58th anniversary of her death.[45] It runs "Tortures are insignificant hardships for a Communist. Bamboo sticks are only bamboo, but the will of a Communist is forged of iron and steel." Such last letters to sons have been "discovered" for other martyred mothers as well—for example, Zhao Yiman.[46] Museum guides report that the letter and exhibition on Jiang "inspire countless visitors to make contributions to the CPC (CCP)" and that "more than six million people from all over the country each year went to show their respect and condolences to the great member of the CPC."[47] As well as her dedication to the party, these post-Mao renditions of her life emphasize "her deep love for her son as a mother."[48] In the 1961 *Red Crag* version, no mention is made of her final letter to her son and her feelings for him are not mentioned in the final scene where she departs her jailed comrades for her execution. Instead, the final note written speaks only of revolutionary matters: "Never quail before danger, always advance, and meet all trials without flinching, without our hearts beating any faster!"[49] But by 2007, a letter is discovered that becomes a relic proving the dedication of the CCP martyr who died for the cause while sorely missing her son. The separation of mother and child becomes a tool for amplifying her sacrifice as a *woman* secret agent. Her deviation from normative marital practices is smothered by her maternal dedication.

Nostalgia for the Glamour of Wartime Shanghai: A Partial Rehabilitation for Guan Lu

Unlike Sister Jiang there are no hagiographic campaigns for Guan Lu and only in the twenty-first century do historical accounts of her life appear on mass media. A living spy, especially a woman who had been used and abused by the party she had served, was not a reliable subject for commemoration. The woman spy's implicit sexual infidelity could not be remade through narratives of sacrificing motherhood in Guan Lu's case because she had been unable to marry: there was no cleansing patriarchal family role available to her. Moreover, in contrast to women soldiers whose non-normative gender roles in taking up arms alongside "our" (male) troops, Guan Lu's fidelity was not purified by the putative nobility of combat or front-line hardship. While neither was Jiang Zhuyun cleansed by the battlefield, her experience of torture and execution along with her elevated status as a sacrificing mother were central to her utility to the PRC state's commemoration of wartime heroism.

However, four decades after Guan Lu's death, she achieves a partial rehabilitation in a series of television documentaries about her life. In these, the romance of a female undercover spy fighting for the CCP emerges within the burgeoning nostalgia about the glamour of wartime Shanghai—complete with its visions of women in cheongsams visiting opulent dancehalls and restaurants. The "sexiness" and "glamour" of being a celebrity and a woman spy are an underlying thread in all these documentaries. CCTV 10's 2013 two-part documentary, *Kang Ri nü qingbaoyuan Guan Lu* (Guan Lu—female intelligence agent in the War of Resistance against Japan), is typical of the format. It reinforces the extent of her sacrifice as she worked within No. 76 connecting Pan Hanmin to Li Shiqun and repeatedly asserts her unflinching dedication to the CCP. Another documentary screened in 2015 transports viewers back into the romance of 1930s Shanghai frequently featuring her contributions to the successful movie, *Crossroads*. Both documentaries present Guan as a young and beautiful woman celebrity working in extremely dangerous circumstances.[50] Her fidelity to the CCP overrode all personal concerns about reputational damage. A third documentary, CCTV 7's 2013 *Hongse tegong midang* (Secret Archives of the Red Spies) series, includes a forty-minute episode that explains how she isolated herself from her previous left-wing friends in Shanghai once she undertook her undercover role. As a result, she suffered enormous abuse and former friends labeled her a shameless traitor. Viewers are told of her loneliness and hardship, how she facilitated a crucial meeting between Pan Hanmin and Li Shiqun, and how she hoped to be able to join her CCP comrades once her work at No. 76 was over. But instead, she found her undercover role extended through *Women's Voice*. The sacrifice and loneliness that Guan Lu experienced though her life as a result of the work she undertook as a spy are also emphasized. The episode closes with an expert historian interviewee, clad in PLA uniform, explaining that Guan Lu sacrificed her happiness, love, health, and career to build the prosperity and happiness experienced in the PRC today.[51] Unable to be rehabilitated as a mother or wife, Guan Lu's best position is as a blindly-loyal party member who sacrificed her reputation and personal happiness for the CCP. Even after her rehabilitation, Guan Lu emerges as a pathetic figure, rather than a heroic woman warrior, like Jiang.

The contradictions embodied by the commemoration of the wartime female spy are myriad. Women's war service requires a vision of sexual fidelity if these women are to be useful to a patriarchal postwar state that seeks to reassert normative sexual moral boundaries of family values in which good women marry good men. But military states also need courageous women who will sacrifice personal, family happiness when called upon in extraordinary times. So, women spies like Guan Lu and Jiang Zhuyun are depicted as actively choosing to serve the CCP in the full knowledge of the normative sexual moral boundaries they may be called upon to transgress—thus, amplifying their sacrifice. In order to assert the level of fidelity required of a national hero, the party must be the ultimate object of the woman's devotion. Even tainted celebrities like Guan Lu are gathered up—as she was in comments made at her funeral—into a story of the CCP as her supreme devotional object. She was ultimately a pathetic battered wife to the abusive party-husband.

Notes

1. On the postwar sexual/political order, see Louise Edwards, "Women Sex-Spies: Chastity, National Dignity, Legitimate Government and Ding Ling's 'When I Was in Xia Village,'" *China Quarterly* 212 (2012): 1059–1078.
2. Joshua Goldstein, *War and Gender* (Cambridge: Cambridge University Press, 2001).
3. Yu Jizeng, "Hongse tegong chuanqi—daru Riwei moku de nü zuojia Guan Lu" (Tales of Red Spies—the woman author, Guan Lu, who infiltrated the Japanese and puppet regime's den), *Dangshi bocai* (Party history collection) 3 (2013): 50. We have no confirmed dates for Ye, but she is said to have died while serving time for her collaboration. For more detail on Ye and other collaborators, see Yun Xia, *Down with Traitors: Justice and Nationalism in Wartime China* (Seattle: University of Washington Press, 2017).
4. "Xu Maoyong zhuiqiu Guan Lu" (Xu Maoyong chases Guan Lu), *Diansheng* (Light and Sound), 6.20 (1937): 899.
5. 'Ba Wei Guan Lu tongxing lian'ai? (Are Ba Wei and Guan Lu in lesbian love?), *Zhongguo dianying* (Chinese cinema), 1.5 (1937): 6. For discussion of their literary connections, see Amy Dooling, *Women's Literary Feminism in Twentieth-Century China* (New York: Palgrave Macmillan, 2005), 114–115.
6. Tamura's socialist leanings most likely would have been known to the Japanese authorities but her fame as a feminist writer and her ready availability in China possibly enabled her to secure this position. Moreover, she had spent nearly two decades living in North America in the 1920s and 1930s avoiding the censorship and imprisonment of many of her left-wing friends. For detailed discussion about her ambiguous politics during her time in China, see Anne Sokolsky, *From New Woman Writer to Socialist: The Life and Selected Writings of Tamura Toshiko From 1936–1938* (Leiden: EJ Brill, 2015), 44–52.
7. Xiao Yang and Guang Qun, *Yige nü zuojia de zaoyu: ji Guan Lu yisheng* (A woman author's calamity: the life of Guan Lu) (Harbin: Beifang wenyi chubanshe, 1988), 152–153.
8. Peichen Wu, "Satō (Tamura) Toshiko's Shanghai Period (1942–1945) and the Chinese Women's Periodical 'Nü-Sheng,'" *U.S.-Japan Women's Journal* 28 (2005): 109.

9 Xiao and Guang, *Yige nü zuojia de zaoyu*, 152–155.
10 *Hongse tegong midang—Guan Lu* (Secret archives of Red Spies—Guan Lu), broadcast January 4, 2014, http://military.cntv.cn/2014/01/04/VIDE1388830327119695.shtml (accessed August 10, 2018).
11 Tang Qing, "Guan Lu Xianggang baoshang chuxian" (Guan Lu appears in HK papers), *Shanghai tan* (Shanghai Bund) 13 (1946): 11.
12 Ma Xiang, "Ba Wei dama Guan Lu shijie" (Ba Wei curses Guan Lu for having no integrity), *Yi zhou jian* (In one week) 11 (1946): 11.
13 Ding Ji, "Nü zuojia Guan Lu zhuiqiu Chen Yiqun," (Woman author Guan Lu chasing Chen Yiqun), *Shanghai tan* (Shanghai bund), 5 (1947): 4.
14 For Wang's official PRC Ministry of Foreign Affairs biography, see "Wang Bingnan," http://www.fmprc.gov.cn/mfa_eng/ziliao_665539/wjrw_665549/lrfbzjbzzl_665553/t40525.shtml (accessed August 10, 2018).
15 Luo Jiurong, *Ta de panduan: Jindai Zhongguo guozu yu xingbie yiyixia de zhongjian zhibian* (Her trial: loyalty and treachery in the significance of nation and gender in Modern China) (Taipei: Academia Sinica, 2013), 125.
16 Ji Hong, "Jingyu zuican hongse nü tegong: 50 niandai beibu pingfan hou zisha" (Tragic circumstances of a red spy: arrested in the 1950s and commits suicide after her name was cleared), history.ifeng.com (accessed April 28, 2013), http://news.ifeng.com/history/zhongguoxiandaishi/detail_2013_04/28/24768091_0.shtml (accessed October 13, 2019).
17 Guan Lu, *Pingguo yuan* (The Apple Orchard) (Beijing: Gongren chubanshe, 1951, rpt. Beijing: Haitun chubanshe, 2013).
18 A 1940 newspaper report notes that Pan returned to Shanghai from China's wartime capital, Chongqing, with the mission to form "red literary groups" (*chise wenhua zhenying*). But he had barely left his rooms and only his closest literary friends knew he was in town. The "special circumstances in Shanghai" (an allusion to the Japanese occupation) meant he quickly left. This kind of brief visit is typical of a "handler." "Pan Hannian jiang fan yuezhuan Yu" (Pan Hannian plans return to Chongqing), *Shanghai jizhe* (Shanghai reporter), May 25, 1940, 4.
19 Hu Xiufeng, "Huiyi wo de jiejie Guan Lu" (Remembering my sister Guan Lu), in Ding Yanzhao ed., *Guan Lu a Guan Lu* (Guan Lu oh Guan Lu) (Beijing: Renmin wenxue chubanshe, 2001), 17. Guan Lu asked Hu to read an early draft in correspondence they exchanged in the 1960s. See Ding Yanzhao, "Guan Li shuxin xuan" (Excerpts from Guan Lu's correspondence), *Xinwenxue shike* (New literary history studies) 4 (2008): 175.
20 Barbara Barnouin, *Zhou Enlai: A Political Life* (Hong Kong: Chinese University Press, 2006), 86.
21 Xu Xinzhi, "Dao Guan Lu," (Mourning Guan Lu) in Ding Yanzhao ed., *Guan Lu a Guan Lu*, 86–92.
22 Huang Wei, "Hongse jiandie: Guan Lu zhi si" (Red spies: Guan Lu's death), *Chuanqi* (Legends) 1 (2010): 15.
23 Xiao Yang and Guang Qun, *Yi ge nüzuojia de zaoyu: ji Guan Lu yi sheng* (A women writer's calamity: remembering the life of Guan Lu) (Harbin: Beifang wenyi chubanshe, 1988), 1.
24 "Shoudu wenyijie renshi jihui jinian zuoyi nüzuojia Guan Lu" (Leading figures in the capital's literary and art world gather to remember women left-wing writer, Guan Lu), *Renmin ribao* (People's Daily), December 29, 1982, 5. See also Yu, "Hongse tegong chuanqi—daru Riwei moku de nü zuojia Guan Lu," 49–53.

25 See chapters on Zhao Yiman and Liu Hulan in Louise Edwards, *Women Warriors and Wartime Spies of China* (Cambridge: Cambridge University Press, 2016). On Lei Feng, see Louise Edwards, "Military Celebrity in China: The Evolution of 'Heroic and Model Servicemen,'" in *Celebrity in China*, ed. L. Edwards and E. Jeffreys (Hong Kong: Hong Kong University Press, 2010), 21–44.
26 Chen Guozhou, "Sister Jiang: Committed Her Life to Communist Ideals, Died 29," *Women of China*, June 25, 2016, http://www.womenofchina.cn/womenofchina/html1/people/history/1606/2674-1.htm (accessed August 10, 2018).
27 "Jiang Zhujun," http://www.womenofchina.cn/womenofchina/html1/special/13/1038-1.htm (accessed August 10, 2018).
28 Luo Guangbin and Yang Yiyan, *Hongyan* (Red Crag) (1961 rpt. Beijing: Beijing dianzi chubanshe, 2001).
29 Ibid., *Hongyan*, 105–106.
30 Ibid., *Hongyan*, 359–360.
31 Ibid., *Hongyan*, chapter 15. This chapter also includes poems that were written by her fellow prisoners in her honor.
32 Ibid., *Hongyan*, 376–377.
33 Luo and Yang, *Hongyan*, 671.
34 Ibid., *Hongyan*, 671.
35 Yan Su, *Jiang Jie* (Sister Jiang), *Juben* (Play Monthly) 10 (1964): 24.
36 Ibid., *Jiang Jie*, 26.
37 Ren Jia, "Kan geju *Jiang Jie*," *Xiju bao* (Drama news) 10 (1964): 25.
38 *Jiang Jie* (Sister Jiang), dir. Huang Zumo and Fan Lai, Shanghai dianying zhipian chang, 1978.
39 See Wu Ran, "Xuan Jiang Jie" (Selections from Sister Jiang), *Dianying pingjia* (Movie Review) 1 (1979): 13, 19.
40 *Jiang Jie* (Sister Jiang), drawings by Men Qingjiang (Beijing: Renmin meishu chubanshe, 1996), front matter.
41 Lo Kuang-pin and Yan Yi-yen, *Red Crag* (Beijing: Foreign Languages Press, 1978), 62.
42 Zhang Nan, "Jiang Zhujun: A Steel Rose Who 'Bloomed before the Dawn,'" *Women of China*, October 27, 2016, http://www.womenofchina.cn/womenofchina/html1/people/history/1610/3526-1.htm (accessed August 10, 2018).
43 *Jiang Jie* (Sister Jiang), dir. Yan Xiaozhui. Beijing, CCTV, 2010.
44 Ding Shaoyi, *Jiang Jie zhenshi jiazu shi* (The real family history of Sister Jiang) (Wuhan: Wuhan daxue chubanshe, 2011).
45 "Jiang Zhujun's Letter Goes Public," *Women of China*, November 16, 2007, http://www.womenofchina.cn/womenofchina/html1/news/newsmakers/8/9691-1.htm (accessed August 10, 2018).
46 See Edwards, *Women Warriors and Wartime Spies*, 130–135.
47 Chen, "Sister Jiang."
48 "Jiang Zhujun's Letter Goes Public" and Zhang Nan, "Jiang Zhujun."
49 Lo and Yan, *Red Crag*, 527.
50 *Dang'an: Guan Lu—luanshi hongyan de tegong shengya* (Archives: Guan Lu—The career of a beautiful secret agent in troubled times), Beijing TV, 46 mins, 2015.
51 *Hongse tegong midang* (Secret Archives of the Red Spies), CCTV-7, 2013, 25 mins.

12

Vietnam's Martial Women: The Costs of Transgressing Boundaries

Karen Gottschang Turner

> When the enemy comes close to home, even the women must fight.
> (Vietnamese proverb)

"The American pilots never knew that beneath them, our Vietnamese women had woven a fine hairnet of opposition. With their shovels, hoes, and guns, they secured the future of Vietnam."[1] Military historian, Professor Nguyen Quoc Dung, boasted about the contrast between the "well-fed US pilots in their big, heavy planes," and the "simple, modest activities of women, who used their small guns to shoot at airplanes and delicate hands to defuse unexploded bombs." His gendered language describing the work of the 300,000 or more women who joined the Communist armies squares with the official line that a unified population and high morale in Northern Vietnam outweighed US technological superiority.[2] This discourse plays against the insults leveled by American militarists to feminize and thus diminish Vietnam and its leaders, implying that if diminutive Vietnamese women prevailed against the American leviathan, their male counterparts must have presented a far more potent threat.[3]

It was into this sexualized terrain that Vietnamese women entered the war zone and established a new chapter in the nation's history of martial women, who since the first century CE have taken up arms to save the nation from outside invaders. The war the Vietnamese call the "American War" drew far more men than women into the military, but it is the plight of the women who joined the volunteer youth brigades while in their teens, who risked their chance to marry and bear children, that has captured the imagination of cultural interpreters in postwar Vietnam. They were thrust into the war after the United States landed combat troops and began an air war against cities and strategic sites in North Vietnam in 1965, charged with defending the Ho Chi Minh Trail, the 10,000-mile long network of paths and roads running through the jungles sheltered by the Troung Son mountain range. These women's stories are linked with one of the most potent symbols of Vietnam's capacity for endurance. Both the United States and Vietnam viewed the Ho Chi Minh Trail—the "Blood Road"—as the key to the war, the only inland supply route from the northern rear to the Communist armies in the south, and the main target of US airstrikes.[4]

In this paper, I have chosen to focus on Northern Communist women who went into the field after 1965, to build on scholarship on martial women in Vietnam's earlier history.[5] I offer here an analysis that includes my own first-hand encounters with some of the male and female veterans of the youth brigades, placed in context with relevant textual, literary, and cinematic representations. Structured according to the sources, the paper begins with the writings of women veterans, followed by the memoirs and fiction of men who worked with and observed women in the jungles, and concludes with oral histories I collected with Phan Thanh Hao in Hanoi between 1996 and 2000.[6] My goal is to discern why the female volunteers remain absent from Vietnam's pantheon of national heroines. By drawing on a living cultural experience in which the women themselves still have a voice to shape their stories, I hope that these materials will yield clues about the construction of national heroines in other times and places. What I glean from a variety of Vietnamese materials is that women's health problems troubled postwar Vietnamese society less than the transpositions of gender roles and hierarchies of authority that emerged during the war. Women became literally polluted by contact with dirt and blood as they excavated the roads and buried the dead; but more seriously, they crossed the normative gender line between the protectors and the protected. The women who rejected the second-class citizenship that these boundaries were meant to safeguard posed a serious threat to established authority.

Pure and Dangerous

Despite their critically important work on the Trail, value as symbols of the collective energy of a unified nation, and the socialist government's promise to award women full citizenship in exchange for their support, the volunteers have not gained the economic benefits or official recognition merited by their service. As volunteers, they are eligible only for disability benefits but many cannot document wounds treated in field hospitals during the war.[7] The most vulnerable veterans, unmarried women and those too ill to work, live in destitution, even as the nation has prospered since the economic reforms initiated in the late 1980s. Women who cannot or choose not to marry and bear children are marginalized as no longer useful to the patriarchal family and nation state. Those who defy convention to remain single or bear children outside marriage create unease. And many veterans believe that as young people forget the war, their service will be erased, written out of history.

They are right to worry. The women who fought on the Trail are not the heroines who have been officially recognized. It is a militia woman, Ngo Thi Tuyen, famed for guarding a strategic bridge in her hamlet near Thanh Hoa, who was singled out for public honors in the north. Like the Viet Cong guerilla fighters in the south, she carried on under the watch of her family and community. As a war widow who spent years searching for her dead husband's body, remains childless, and lives in poverty, Ngo Thi Tuyen shares experiences with many ordinary women. Her symbolic value is enhanced because she broke her back, not during the war, but when East German filmmakers forced her to re-enact her feat of carrying huge loads of ammunition when she could not call on her revolutionary spirit for energy.[8] The politics of memory operate differently for the volunteers who left home at the time they should have married, to live with men for years in the jungles. True, a few volunteers are honored in their local communities, but as far as I know, no national-level museum exhibitions or public monuments honor them. Male and female volunteers alike were denied graves in the official cemeteries reserved for regular soldiers; instead, their comrades buried them along the jungle paths where they fell.[9]

As I searched for clues to explain these omissions, I recalled a conversation with a retired colonel who had supervised young people on the Trail. Le Trung Tam began our session with a poem he had written to a girlfriend who had been killed there, which in turn sparked a conversation about the dilemma faced by military planners who feared that sending "pure young virgins" to work with men in the dense jungles that sheltered the Trail would compromise their futures as wives and mothers and thus harm the well-being of the nation itself.[10] The French historian, Francoise Guillemot, paints a darker picture, one of heartless exploitations with no concern for women's welfare:

> These teen-soldiers (sometimes even child-soldiers) armed only with simple pickaxes and shovels and with little intellectual grounding, given their few years of elementary school education, often found themselves propelled into the line of fire. They had no military knowledge and so were trained on-the-job.... The shock must have been particularly brutal for girls from urban areas, many of whom only knew how to hold a pen or carry out simple household tasks.... Bodies were struck down by fever, hunger and death. The hostile environment was infamous for its rains, mud, poisonous vines and leeches. Daily life was a series of accidents, bombings, and chemical attacks. The most common feelings were pain, fear, and horror.[11]

Figure 12.1 Young People on the Trail, date unknown. Courtesy of Colonel Le Trung Tam.

Both versions triangulate with other sources. Military records show that Ho Chi Minh himself ordered field commanders to attend to women's health, but because short supplies, especially medications to prevent and treat malaria, went first to the regular soldiers, most volunteers suffered afflictions that persisted long after the war ended. Yet as oral histories testify, although they would not deny the costs of their service, many of the female volunteers do not view themselves as pawns, but rather as active agents who joined the brigades for their own purposes and banked on their wartime sacrifices to press for benefits when they returned home.[12] Their later disappointments did not diminish the pride they earned in the field.

References in a variety of sources to innocent teenagers who entered a zone that violated their bodies and souls resonate with Mary Douglas's work on purity and pollution as a useful framework for sorting through the contradictions and ambivalence that pervade narratives by and about militarized women in Vietnam. The question that guides this chapter, then, is whether conceptions of purity and pollution help to explain the discrimination against women veterans who transgressed boundaries during the war and threatened a fragile postwar culture. As Douglas famously asserts, "Pollution ideas work in the life of society at two levels, one largely instrumental, one expressive....Beliefs reinforce social pressures....I believe that some pollutions are used as analogies for expressing a general view of the social order."[13] She refers to the story of Joan of Arc as one emblematic example of a woman with no place in "proper" society. A woman, peasant, and warrior, she was sacrificed to restore order. Douglas's insights about gendered nationalism are useful for analyzing more contemporary studies of problems generated when the female body becomes identified with the health of the nation-state in times of transition.[14] The impulse to impose boundaries and margins is not limited to "primitive societies," but intensifies in times of disruption, especially the chaos generated by political violence. According to Douglas, "I believe that ideas about separating, purifying, demarcating and punishing transgressions have as their main function to impose system on an inherently untidy experience."[15] Such pressures are rarely directly expressed, but appear in oblique, encoded fragments that suggest dangerous boundary crossings.

Women Frame the War Story

Most Vietnam War stories are shaped by men as is evident from US media and literary productions by male veterans that dominated the 1980s, and more recently, Ken Burn's documentary series, "The Vietnam War."[16] When women are included, they usually serve as accessories, bit players in a masculinized drama. As critic Kali Tal points out: "These fictional relationships with Asian women, whether as mistresses or prostitutes, do not indicate any feeling *on the part of either character or author* that women are human beings worthy of respect."[17] The Vietnamese case is unique, however, because martial women are not an anomaly there and so Vietnamese women can neither be readily silenced nor written out of war. It is true, however, that most of the published sources are authored by male veterans, and in some cases, I agree with Guillemot that they betray male fantasies about the teenagers who seemed so incongruous as they bravely led the trucks and soldiers through the jungles, their white bras and blouses gleaming in the night. But it is precisely these subjectivities that bolster my argument about how perceptions of women as pure, but also sources of disorder, would play a role in postwar culture.[18] As more women become combatants in recent global conflicts, records of first-hand experiences are no longer limited to men, as Eliza Griswold observes: "It's a dubious privilege that a woman can tell war stories as brutal and devastating as a man can, but it's certainly earned."[19]

One of the few women anywhere who has shaped a public war story through film is Nguyen Thi Duc Hoan (1937–2003), whose personal history mirrors that of revolutionary Vietnam. Educated and radicalized in French Catholic schools, she escaped her oppressive Confucian father to join the Viet Minh Armies to teach literacy classes in the countryside.[20] "I was able to use the education in the French Catholic schools but I hated the nuns, who made us feel inferior. They also told us about the French revolution. We wondered why they couldn't apply their principles of liberty and equality to us Vietnamese." In Vietnam's final battle with the French at Dienbienphu in 1954, she served as an artillerywoman. Duc Hoan was one of many women who supported the Indo-Chinese Communist Party's anti-French movement, but she was more fortunate than resistance workers who suffered imprisonment and execution by the colonial powers. These martyrs were often radicalized and organized in French prisons and though their work was essential for the movement, their tough, uncompromising commitment to the cause above family duties troubled some Party men. But as Party General Secretary, Le Duan, reassured his comrades in 1959, if the radical women themselves eschewed motherhood, they must have been raised by good mothers:

> It is in women that we find the essence of our national characteristics. The fine traits of the Vietnamese character are first of all present in Vietnamese women. I noticed in prison that most of our revolutionaries had fine women as mothers.[21]

Duc Hoan managed to juggle a variety of duties to the nation over the course of her life. In her domestic affairs, she continued to resist patriarchal norms, divorcing her first husband, a career army man who could not adjust to civilian life, to marry a man fifteen years her junior. She was also a devoted mother who raised a daughter who followed in her revolutionary path. Professionally, she was above all an artist. After studying acting in Russia, she became one of Vietnam's most beloved actresses by the 1960s.[22] She used the cachet of her background as a privileged, educated artist and veteran to take up problems others feared to raise. In our conversations, she emphasized the themes she cared about: "The old men in charge of war images wanted films that featured bombs and heroic men standing up against the enemy. As for me, I believe that in the end, there are no heroes, for everyone suffers, even the victors."[23]

Duc Hoan circumvented government censors to film the damage that even a righteous war inflicted on individuals and their families, creating sympathetic portraits of individuals who threatened the puritanical norms of the regime: a wife repulsed by her soldier-husband's disfigured face, an officer in the Communist forces who defected to the enemy in a moment of terror and weakness.

Of the female volunteers, she declared: "We took for granted the fact that women always fought when the country was in danger. Our mothers sang lullabies of the heroines of old, the Trung sisters who rebelled against the Chinese in 40 C.E., and the flamboyant Lady Trieu Thi Trinh (d. 248), who rejected domestic duties to fight the Chinese." Like many Vietnamese women, she relished Lady Trieu's famous manifesto against domestic duty: "Why should I imitate others, bow my head, stoop over and be a slave?" No written records verify Lady Trieu's legends, but literary and visual images portray her riding high on her war elephant, her three-foot-long breasts cast aside to better handle her sword. She supposedly died in battle, weakened by the unruly, smelly Chinese soldiers who played on her aversion to dirt and disorder. When a pestilence threatened her home village, the Chinese commander ordered phallic symbols nailed on doors to mute her female power.[24] Duc Hoan rejected links between prowess and masculinity: "Our heroines were never successful in gaining power in the long run, but neither were they crazy like Joan of Arc. Why would a woman dress up as a man when her female beauty is her greatest asset?" She admired the eighteenth-century poetess, Ho Xuan Huong, an early feminist who railed against male authority—mocking in one poem a famous Chinese general: "I see it up there in the corner of my eye; the general's tomb standing all alone. If I could change my fate, become a man of heroic deed, couldn't I do better?"[25] Duc Hoan's heroines were not limited to Vietnam, but like other educated women who supported the socialist revolution and studied in Russia, she admired Soviet women like Alexandra Kollantai. Bolstered by this rich legacy of strong women, Duc Hoan fashioned an exceptional life. But she believed in the collective energies of ordinary people and in a common humanity: "Politics change, but human beings remain the same."[26]

Of women who fought against the United States, Duc Hoan observed: "At no other time in Vietnam's history was the will of the people more necessary for national survival. When even the gentlest Vietnamese woman could be inspired to enter a male world of violence for her country, and when she learned to do the job well, the war became in reality a people's war." Her first film about volunteer youth, *From a Jungle*, presented a somewhat idealized view of their innocence and pure intentions. She was particularly intrigued with youth from Hanoi, so attached to their beloved city that they recreated its street names in their jungle camps. She based her film on her own observations during her time on the Trail, and understood that scenes of young people laughing and playing at one moment, and crying over the dead the next, would trouble viewers: "Old people wallow in their sadness, but the young want to live. I wanted to show young women as full of life, able to laugh, acting as equals with men."[27] She did not overlook the sorrows and dangers as well as the freedom that young people found in the jungles. Strangers who had to survive in one of the most remote and dangerous jungle terrains in the world, the youth had to learn to protect themselves from poisonous plants and wild animals; but they also found beauty in an exotic setting unimaginable to those who never left the lowlands. Hidden, not only from the enemy, but in many cases, from the daily scrutiny of military officials, they formed relationships that at times deviated from normal hierarchies of authority. Her young protagonists embody contradictions: they are pure of heart despite contact with blood and violence, but their self-constructed communities also challenge normal structures of authority.

In her later years, Duc Hoan was not immune to conservative pressures to promote established values. Her final production, aired on television in 1991, features a disabled woman veteran who begs a medicine man, also a veteran, to have intercourse with no attachments so she might have a child. Broken bodies and spirits dominate the production. The female veteran calls herself a "lame duck," and the male calls himself a "drunk who cannot forget the sound of bombs." She envies her sister, a flirt, whose jealous husband beats her. When I questioned Duc Hoan about her sympathetic portrayal of an abusive husband, she replied that he acted out of love for his errant wife, who had failed in her duty to her family. She is best remembered and beloved for her earlier films, and when she died in 2003, "the grand old lady of film" was reduced in official obituaries to her role as a beautiful young actress, whose "sparkling eyes" enchanted audiences.[28]

Writer Le Minh Khue published her semi-autobiographical story, "Distant Stars," when nineteen years old and still stationed on the Ho Chi Minh Trail. Her account of the daily life and dreams of three young girls confined to a cave, charged with detonating bombs and filling craters, plays down their terror and fear, but vividly portrays their response to contact with dirt and destruction.[29] The story is shaped by themes related to purity and pollution. The narrator, an educated urban teenager, observes: "The bombs often buried us. Sometimes, when we came down from the hill we were so buried in dirt that only our gleaming eyes showed through.... At those moments, we called each other the 'Black Eyed Demons.'"[30] Language suggesting that these women remain physically and spiritually pure despite these conditions colors the story. After suffering serious wounds, one girl, looking "light and fresh as a white ice cream bar," refused to go to a hospital in the rear for care.[31] The men in the artillery platoon over the hill, "just a rifle shot away," sentimentalize the girls with offerings of flowers and candies, and one young soldier composes a poem that links their shining courage with clear, bright, distant stars.[32] The female narrator brags about taunting a naïve young soldier: "He was hospitable, didn't smoke and didn't flirt with us girls. As for us, we wouldn't leave him alone." Her message is subversive, for her heroines' disdain for the peasant-worker soldiers they encounter reminds us that powerful class distinctions had not been erased by socialism. This theme of gendered and status displacement is repeated in other writings featuring "Hanoi maidens," members of the bourgeoisie who seemed so out of place to the men as they performed labor usually reserved for their less privileged sisters. Even worse, they disturbed the rigid military order when some of them turned out to be hardier than trained army regulars.

Three decades later, Le Minh Khue's fiction turned dark. "A Small Tragedy," published in 1990, features a reporter who hears an account about young people ordered by a stupid bureaucrat to fill a bomb crater in daylight. "The vanguard youths? I've never seen so many people die like that. They were empty-handed, puny, running like ants.... I tried to dig into one depression to pull out three young girls but they died in such a tight embrace I couldn't even untangle them." In this story, the young girls keep faith with one another, just as they did in "Distant Stars."[33] But in "Small Tragedies" they do not act with brio and instead die for nothing more than to satisfy the whims of a corrupt official with a reputation for other avoidable deaths of the young. Dana Sachs argues that this "lost idealism" became a major theme for writers in the 1980s, when "the hollowness of revolutionary victory brought tragic disillusionment, particularly in light of how much they had sacrificed in order to achieve that victory."[34]

The most famous dissident woman writing in contemporary Vietnam, Duong Thu Huong, deals directly with polluted female bodies in *Novel without a Name*.[35] Like Le Minh Khue, she is a veteran of the volunteer youth brigades, one of three survivors of a group of forty young people. Later as a reporter, she witnessed and described the human impact of Vietnam's border war with China in 1979. Her war story focuses on the mental breakdown of a male soldier who is both admiring and repulsed by the women he encounters in his travels. When he happens upon a woman whose unit abandoned her to deal with corpses in an isolated supply depot, she begs him to have sex. He becomes impotent and angry that he cannot grant her only request. But his angst is not focused on the woman he rejects:

> This woman was born of the war. She belonged to it, had been forged by it. It wasn't that she was so ugly that I had rejected her. I had been afraid to face myself, scared of the truth. I was a coward. Ten years of war had gone by. I had known glory and humiliation, lived through all its sordid games. I had needed to meet her to finally see myself clearly. I had been defeated from the beginning.[36]

This use of a male soldier as a voice to narrate disgust at polluted women's bodies, can be interpreted in multiple ways—as anxiety about the author's own body, or more likely, because Huong is one of the regime's fiercest critics, as an uncompromising stand on the futility and ugliness of war. She shows as well how the physical pollution of war created internal disorder.

Unvarnished by postwar pressures, the wartime diary of a young female doctor represents one of the few sources for discerning the internal world of a woman in the thick of violence. The diary includes themes similar to those in the fiction of Le Minh Khue and Duong Thu Huong—sisterly love and pity for the mangled young solders she treats in her mobile clinic, disdain for righteous, ignorant Party bureaucrats who look down on her for her bourgeois background, and confidence earned by dint of her expertise and courage. Her hatred is directed toward the "American bandits" who kill her people and ravage the land: "Oh! Hatred is bruising my liver, blackening my gut. Why are there such terrible, cruel people who want to use our blood to water their tree of gold?" Dr. Tram was killed in 1970, at the age of twenty-seven. Her diary, brought to light after thirty years, is treasured in Vietnam for its honest idealism and critique of Party ideologues.[37]

These three examples expose some of the reasons that women's war stories can be so potent and dangerous to male visions of war. Le Min Khue, Duong Thu Huong, and Dr. Tram include portraits of fragile, naïve, and at times corrupt men. These women both witness and share in the daily trauma of war with men and do not shy away from describing the bloody business that dangerously reverses gender roles.[38] Ultimately, they suggest how conceptions of purity and pollution bring social disorder to light.

Men Write Women into War

The men who have recorded their impressions of the women they met on the Trail tend generally to take a less jaded view of female bodies than women writers. Reports from the field to Hanoi praised the difference that women made, not only for their courage in standing up and shooting at US airplanes and work on the roads but for raising the morale of the men. Some men did worry about women in harm's way and some patronized the girls, but fewer men registered disgust about their bodies. What is notable is just how much attention Vietnamese men in the field paid to their female comrades.[39] As historian, Hue-Tam Ho Tai points out, in Vietnam, discourse about women's issues was never just about women, but has long served as a means for men to voice their own fears and vulnerabilities.[40]

A case of a woman who dared to challenge accepted norms is embedded in the diary of an artist who complained that he was forced into the war. While in the jungle, he meets a young volunteer, Nguyen Thi Lien, whose male comrades admire for her singing voice as much as her courage under fire. The outsider is shocked when one night she takes the part of a man in a play, "grotesquely dressed in men's clothing." Even worse, when the men laugh, she chides them for their bad manners. But her superiors react with unexpected approval, musing, "I suppose it's her frequent encounters with death that have made her so fearless." When her luck runs out and she is killed, the men mourn her openly. Here, in isolation from normal civilian society, a woman who flaunts gender identities and military hierarchies is accepted because of her fearless dance with death. The story ends with a male soldier's message to the outsider: "If the jungle is a university, as you people call it, for me, it is first of all a school to teach me to love and believe in mankind. In my judgement, this love, and this belief is actually the source of revolutionary heroism."[41] Here, a seasoned soldier, inspired by brave teenager girls, subversively identifies heroism with love.

In a similar story of war as a transformative experience in unexpected ways, urbane young women tease the protagonist, a twenty-eight-year-old, "taciturn peasant," charged with leading "four real Hanoi maidens," referred to as "stars," to their unit.[42] The author, a male veteran, draws on themes familiar from Le Minh Khue's "Distant Stars": smart city girls who outshine a naive countrified soldier. Frustrated when they laugh at his orders and take silly risks, he finds himself in a situation that military training never prepared him to face, "He always spoke carefully, aware of their sharp tongues." Sexual frustration plagues him as well. At one point he almost faints when he comes upon one of the girls asleep in her hammock: "The girl was sleeping soundly, a mass of hair over her angelic face. The buttons of her blouse burst out showing her soft and white throbbing chest. Trong put the lights out and ran, bumping into a tree trunk." As the girls toughen up and keep their spirits high in dangerous situations, he finds his life enhanced by their beauty and courage. He breaks down when one of the girls is wounded and a young widow killed: "He could not repress his feelings. It seemed that the Creator had given women superhuman capacity to suffer things that the men could not." When an older comrade witnesses and criticizes the girl's lighthearted demeanor, Trong defends his charges, to which the older man replies: "This is a war and there are no boys and girls in a war, only soldiers." One of the girls contradicts him: "A soldier is not a fighting machine." The story surely serves as a ploy to voice the male author's own resentments: "Is there any pain greater than this one? Those warmongers must be killed." Trong ends up with a "disease of the heart" that incapacitates him, while the gentlest of the young women marches on to the original destination, now a competent warrior. These inversions of the powerful male and weak female reveal the fragility of gendered hierarchies under duress.[43]

Many men who encountered women in unexpected places maintained traditional attitudes. General Phan Trung Tue, commander of Division 559 in charge of the Trail, praised the female volunteers who built a section of the road through particularly rugged terrain.[44] "It was constructed out of nothing but the determination and spirit of the young volunteers with their primitive tools. It was called by different names: 'Determined to Win,' and 'Road 20' to mark the average age of the volunteers."[45] It was not only women's hard work that intrigued him. When encountering their colorful underwear drying on trees, he likened them to "blossoms of flowers on the mountain slopes." At one point, before an expected attack, he recalled in his memoir: "I decided to keep only the fittest and transfer the girls back to the second line. Girls could be good

Figure 12.2 Veterans of Company 814, Hanoi, 1996. Courtesy of Karen Turner.

at bookkeeping, handling freight, or even manning antiaircraft guns. But they would be no match for Saigon infantry." The women in his unit protested: "As human beings, we are not inferior to anyone. We're members of the Youth Union and we want to know why you have a bias against us." Fearing what he calls an "interminable diatribe," he orders them home, allowing nurses to remain on duty to care for the wounded. What is surprising is not the officer's mixed reactions, but the women's argument that they had stood up to danger like men and deserved to be treated like men. Other stories show how admiration for women was often mixed with paternalism. A truck driver, irritated because the young female guide he depends on to lead him through a secret path giggles at him, tries to intimidate her: "Your limbs are as slender as raw silk threads. It seems as if you just left your schoolwork at home only a few days ago. You'll have a tough job building roads and cutting through the mountains. I'll bet at night you're still weeping inside your blankets and calling out to mama, right?" She has the last laugh: "You get used to it brother."[46] Whether the men who penned these memoirs understood that women were rejecting their paternalism cannot be known, but reading between the lines affords readers useful insights about male responses to women in the field.

Men noticed harm to women's bodies. "Be careful or you'll destroy your sex," a male veteran noted as he encountered a woman pushing a loaded bicycle. He recorded his warning to another brave fighter: "To a girl who had passed her thirtieth year of age, I said: 'We are in wartime. The Party knows your high devotion, but also cares about your future. You must return to the rear, get married and have a family of your own.'"[47] A civilian writer's story about a woman's loss of sexual appeal that resonates with Huong's *Novel without a Name* appeared a decade later than the veteran's memoirs and fiction. In this case, a female veteran of the volunteer corps, so poor that she begs for a living, pleads with a male veteran to impregnate her so that she can at least have a child. He tries manfully, but tells her that her reproductive system is good only for "excretory purposes," that she is no longer a woman.[48] In a story about a veteran couple, both exposed to dangerous chemicals during the war, the wife's "sickly yellow," disease-ridden—polluted—body becomes a public spectacle as their community anxiously awaits the birth of her child. The husband's pain at losing a son when the child is born "horribly deformed" is emphasized more than the mother's loss. In general, men don't worry openly in either their fiction or their memoires about their own sexuality or physical illnesses, despite the fact that more men than women were contaminated by Agent Orange because so many worked closer to the areas where the deadliest defoliants were used.[49] Yet, it is often the women who birthed disabled children or remained sterile who were stigmatized the most.[50]

One of the few fictional accounts to mention rape features a female volunteer pressured by a male officer to run an "inn" in the mountains as a rest stop for truck drivers. Mai, a widow, and one of his best fighters, is put in charge. She fears the men who pass through, and begs the officer to send a man to protect the women. But the newcomer instead of safeguarding them, impregnates one of the girls, who aborts the pregnancy. Desperate, Mai knows that she cannot escape her fate: "She could not tell him about her loneliness and her revolt within herself... she knew that Major Lam, without any constraints, could send her to a vicious spot." Eventually she is killed, sacrificed by a sexist superior who abused his authority. In all of these cases, women's bodies become polluted by the vagaries of war.

In Their Own Words

Published writings by veterans, even diaries and memoirs, often represent self-conscious portraits that take up difficult questions about war and society. By contrast, oral histories and testimonies center on war as a daily struggle against great odds.[51] One male veteran remarked, "Only Vietnamese could do what we did. It was like we were possessed by a spirit." Women expressed pride not just for surviving, but for their courage and contributions. But these people at the bottom of the military hierarchy were not simple storytellers, as I learned when Phan Thanh Hao and I were invited to film Company 814's annual reunion, held at their district Party headquarters in July 2000. There, through song, poetry, and carefully crafted speeches, their designated leaders reminded local Party cadres, and me, their American witness, of their service to the country and the government's failure to award them proper benefits. In general, as I conducted interviews for a book and film with Phan Thanh Hao, we found our informants well aware of their audiences. While the politics of memory and my position as an American woman shaped their accounts, so did group dynamics: we found interviews with individuals struck a different tone than those conducted with the larger group during which members could check or confirm particular stories. In this case, collective memory acted as a force that drew not only from larger national discourses but also from a shared experience that extended from their longstanding local ties. Unlike US veterans who scattered after deployment, the members of Company 814 came of age, went to war, and returned home again to their hamlet of Tu Liem District, just outside Hanoi. They knew one another very well and in general seemed comfortable speaking in a group, though the women often deferred to the men. But these veterans might have been unusual because they had strong female leaders who looked after them and organized gatherings after the war. As one male veteran reminded me about women more generally, "The bitter ones, the sick ones, who put down their guns and returned to their villages, will never talk even to their families and neighbors."[52] The issues raised by subaltern studies—who has the right to speak—in this case did not hinge on gender, or even status. The most disadvantaged woman among her comrades complained about the Party's failure to give her a piece of land as promised, and a small stall keeper who took only one day off a year, to attend the reunion, proudly spoke on film about her pioneering work as a road builder. It was clear that many of these women believed they had earned the right to speak no matter how dismal their postwar existence.

The volunteers left home for many of the reasons that young people go to war anywhere—to follow friends or lovers, hope for better conditions than offered by their impoverished families, or to bow to official pressure. For those who witnessed the destruction of their homes and villages, no persuasion to leave home was needed.

As one woman declared: "When my two brothers were killed by American bombs, I wanted to avenge my family. When my workplace called for volunteers, I was ready to go. I thought if men could fight, so could I."[53] Another woman remembered: "We had to leave home to preserve any hope for a safe place to raise future generations."[54] Some, especially those in the Youth Union, believed it was the duty of young people to save the nation while the elders remained behind. Some women mentioned that they were inspired by earlier heroines who had taken up arms in a time of national crisis. As they retold their stories, these women did not remember themselves as pawns of a propaganda machine but willing participants who linked their futures with that of the nation-state.

Most women enjoyed greater equality in the field than at home. The male volunteers, deemed too unreliable to join the regular forces, had no official authority over the women and in a sense were "feminized" as subordinate to the army regulars and truck drivers. As one male volunteer remembered, "We were told that if a marching soldier needed care, we had to use our clothing to bandage wounds. This was hard on the women."[55] In mixed company with the men, women relished retelling stories of coping with hardship better than men. Nguyen Thi Mau recalled that she and her platoon mates, all women, because the men failed to help, used their bodies to support planks over mountain streams so that the trucks could keep on moving. Another volunteer who lived underground with men for ten years in a mixed communication team remembered that she tolerated boredom and hunger more easily than the men. Women assigned leadership positions in mixed platoons and squads asserted that their capacity for endurance, patience, and diplomacy—traits associated with women—helped them smooth relations with men. One woman remembered: "When they called for volunteers for the 'death defying teams,' my girlfriends and I volunteered."[56] These teams were almost always manned by women, and their testimonies coincide with stories of empowered women in the published literature by male and female authors.

Feminine attributes were considered a strength: "We put flowers in our hair to look nice for a while. And we sang because we believed our songs could drown out the sounds of the bombs."[57] With one notable exception, the story of Miss Lien, I have found no references to women yearning to take on masculine dress or manners. Some women did admit that they fretted when they lost their hair and pale skin, and worried about menstrual problems and malaria. Le Minh Khue remembered in an interview in 1996 that she was terrified when she bumped into a dead body while bathing in a stream and feared dying with her clothes blown off by bomb pressure. She recalled with a shudder that burying the dead, the job of men in peacetime, troubled the women. "Sometimes, the burial trenches were bombed and we had to take the bodies out and rebury them. Some were in pieces, blown apart by bomb pressure. It was bad luck for everyone, but the women didn't complain. We had to do it."[58] She also gained perspective on male authority in the field: "After I held in my arms officers who had once been stern, hearing them cry out for their mothers as they died, I no longer feared them."[59] Le Minh Khue, skilled writer and first-hand witness, clearly signifies that women were polluted by contact with blood and death.

The men of Company 814, like the male writers, never talked about their own physical problems but dwelled on those of the women:

> We men felt sorry for the women because it was harder on them. Sometimes they had to work underwater stripped to their underwear. Those conditions harmed them and they have women's diseases. Some of the women are old now and have no medicine. And some could never marry.[60]

One woman, Nguyen Thi Mau, seemed to generate more comments about her ill health than the others. But as I observed the dynamics of the group, it seemed that the men were uneasy about her because of her strong personality rather than her deafness and malaria-related illnesses. She was better educated than many of them, organized the annual reunion—where she recounted the company's history—and chose to remain single despite offers of marriage. Although she laughed off the men's patronizing remarks in mixed company, Nguyen Thi Mau did not deny her problems in an interview with Phan Thanh Hao and me:

> I got malaria so bad I couldn't even speak or hear. I was sent back to the rear. But there was no place for me. My boyfriend had married someone else and I was too sick to find another man, and I didn't want one anyway. My family didn't want to give my dowry land back to me, and so I went to live for a while in an all-female collective farm with other women veterans with no place to go.[61]

Eventually she recovered enough to become an elementary school teacher. Yet, in a filmed interview in July 2000, her mother lamented, while Mau quietly cried, "Yes, she has many friends. But when she is sick, who will take care of her? Her life is not natural because she has no children, no husband. Her whole life changed because of her sacrifice." Another woman, known for her courage as a platoon leader and member of special teams charged with defusing the most dangerous bombs, also lost her first love after she returned home haggard and sick. Now she boasts that she has admirers and could marry, but chooses not to become entangled: "My life is good now. I have sent my niece to school and she can care for me in my old age. If I got married, I might be a burden to a man. And besides, I would have to take care of him too." Both women admit that they have failed in their filial duty to their families but they also seem content with choices that defy social pressures.[62]

Sociologists in Vietnam have studied women who call on the essentialist belief that motherhood is a function "bestowed by the Creator," to legitimate their decisions to bear children outside marriage.[63] The phenomenon of the single mother, inseminated in many cases by a stranger, has been discussed not only in connection with women veterans, but younger women who simply want to avoid marriage.[64] Some women veterans chose their "seed sowers" carefully: in one instance, a woman offered her inseminator a larger reward if she gave birth to a son.[65] In part to support women veterans who "asked for a child" outside marriage, the Vietnam Women's Union successfully pushed for a law in 1986 that allowed these mothers and their children full legal rights. The women who cannot marry and have children because of health or poverty seem to trouble their communities less than those who deliberately choose to defy patriarchal values. And even more dangerous, the example of female veterans who placed their desire for motherhood over allegiance to the patriarchal family has inspired younger women to follow their unconventional path.

Conclusion

The question that hovers in the background in the written and oral sources—women's sexuality during the war—is one that preoccupies most postwar societies and Vietnam is no exception. A young woman who grew up in the same neighborhood as the survivors of Company 814 told me that as a teenager she realized how different they were: "They walked straight and tall and looked you straight in the eye, unlike most Vietnamese women. Sometimes they wore their old uniforms and medals. We admired them but were a little afraid of them too, because we knew that they had lived with men during the war."[66] Many older women insisted, on the other hand, that "men were better then," and veterans often admonished those who have never known the realities of war that they were too preoccupied with surviving to indulge in romance. One of the officers I interviewed in 1996 admitted that women got pregnant in the field, but assured me that these were romantic lapses, and that the army sent them home with fake marriage certificates to protect them and their offspring. When I asked Professor Dung if accounts of rape in some postwar fiction had any credibility, he angrily replied: "We had no comfort women like the Japanese. People were sick, tired, trying to survive. You must understand that."[67] It is true that even the most skeptical observers of the Vietnamese Communist armies cannot find evidence of institutionalized prostitution. As Gerard Degroot points out in his work on oral histories of British service women during the Second World War: "If war seems erotic it is because some people want it to be so. It is always tempting to make the past more exciting than it was."[68]

Evidence that officialdom worried about women who crossed boundaries appeared in an article on plans to domesticate women volunteers published by the Vietnamese Women's Union in 1975. As an example, it uses a female brigade known for courage and expertise, ironically named for Lady Trieu, the heroine who rejected housework to go out to fight. Now these women are being retrained: "Classes are regularly held for the brigade members, at which they acquire a general education and learn sewing and embroidery. Brigade 609's idea of a good woman is one who works diligently, fights courageously, shows good morals and is likely to become a good wife and mother."[69] Vietnam's celebration of the moral mother over the female warrior follows a path taken by almost all governments after a period of conflict that demands women's service in war but relegates them to the home when national interests shift. In Vietnam, perhaps the problem was not so much that martial women needed to be reminded of their female roles but that men needed to rid their fears that women operated as better soldiers under duress. Moreover, the land itself had been feminized and polluted by the occupation of foreign invaders and their weapons of destruction.

Mary Douglas argues that displaced bodies create confusion after periods that threaten the collective welfare, and Vietnam after 1975 faced a hard road to recovery.

The statistics about survival rates among the volunteers are difficult to calculate, but some estimates of a 10 percent death rate suggest that thousands of women returned from the jungles to find a place in civilian society.[70] There were no historical models for dealing with militarized women in peace time and as governmental agendas shifted, women who could not support the economy or replenish the citizenry drained scarce resources. Few families escaped suffering. Over 3 million people died, and at least 300,000 are missing in action. The land and many of its people continue to be ravaged by chemicals used by the US military. In less materials ways, many people believe that the dead who were never properly buried become ghosts that haunt the living. A revival of conservative Confucian values charges women with participating in and raising families to support a new, global economic order while continuing to shoulder their traditional burdens. Yet, some women manage to follow the path of the volunteers in placing desire over duty.[71] Vietnamese women have been called upon to perform different roles over time and it is not too farfetched to speculate that if the nation should need to call them to arms again, the volunteers might become exemplars, no longer embarrassing reminders of the costs of war, but heroines who sacrificed their lives to save the nation.[72] As historian Hue-tam Ho-tai has so eloquently written, "In Vietnam, memory has no name, but it has many faces, and... they are the faces of women."[73]

Notes

1. Another 60,000 women, often for their technical skills, were recruited to the regular forces of the People's Army of Vietnam. For a history of the shock brigades [Thanh Nien Xung Phong], see Francois Guillemot, "Death and Suffering at First Hand: Youth Shock Brigades during the Vietnam War (1950–1975)," *Journal of Vietnamese Studies* 4 (2009): 17–60.
2. Interview, Hanoi, 1997.
3. See Joshua Goldstein, *War and Gender: How Gender Shapes the War System and Vice Versa* (Cambridge: Cambridge University Press, 2001), 356, for Lyndon Johnson's threats to castrate Ho Chi Minh.
4. See John Prados, *The Blood Road: The Ho Chi Minh Trail and the Vietnam War* (New York: Wiley, 1999).
5. For the female southern guerilla fighters, see Sandra Taylor, *Vietnamese Women at War: Fighting for Ho Chi Minh and the Revolution* (Lawrence: University Press of Kansas, 1999).
6. See Karen Gottschang Turner, *Even the Women Must Fight: Memories of War from North Vietnam*, with Phan Thanh Hao (New York: Wiley, 1998) and "Hidden Warriors: Women on the Ho Chi Minh Trail," produced and directed by Karen Turner with Phan Thanh Hao, Hen Hao Productions, 2004, housed in the Harvard Film Archive. These projects could not have been carried out without the cooperation of Phan Thanh Hao, a Hanoi-based journalist, diplomat, and translator.
7. In 2000, I was informed by members of Company 814 that the state had authorized a one-time payment for anyone without family members, but few qualified.
8. See the chapter on Ngo Thi Tuyen in *Even the Women Must Fight*, 51–69.

9 Among the few named dead are the ten girls crushed to death in a remote cave by US bombs in 1968, their graves tended by women in white.
10 Interview, Hanoi, 1997. My thanks to Phan Thanh Hao and Tri Troung for help with Vietnamese terms. "Pure young virgins" is a rendering of "trinh nu," literally, "girls who have never known love."
11 Guillemot, "Death and Suffering at First Hand," 27, 31.
12 The volunteers signed on for one year terms. Some returned early because of ill health, some stayed in the field as long as a decade, and a few joined the regular forces.
13 Mary Douglas, *Purity and Danger: An Analysis of Concepts of Pollution and Taboo* (London: Routledge, 2001), 3.
14 See, for example, Keely Stauter-Halsted, "Moral Panic and the Prostitute in Partitioned Poland: Middle Class Respectability in Defense of the Modern Nation" 68 *Slavic Review* (2009), 557–581, in which the trope of purity and pollution is used to explain why prostitution threatened the integrity of the nation at a precarious time in its history.
15 Douglas, *Purity and Danger*, 4.
16 Premiered September 2017. Burns includes a few interviews with Vietnamese women, but only one with a US nurse.
17 Kali Tal, "The Mind at War: Images of Women in Vietnam Novels by Combat Veterans," 31 *Contemporary Literature* (1990), 76–96, 82.
18 On the liminal position of these writers, see Mikhael Bakhtin, *The Dialogical Imagination*, trans. Emerson and Holquist (Austin: University of Texas Press, 1981).
19 In "Women, Writing War," *Words without Borders: The Online Magazine for International Literature* (April 2016). For a useful comparison with Vietnamese women soldier's memories, see Svetlana Alexievich, *The Unwomanly Face of War: An Oral History of Women in World War II*, trans. Pevear and Volokhonsky (New York: Random House, 2017).
20 I have written about Duc Hoan as an artist working during the Cold War era: Karen Gottschang Turner, "A Vietnamese Woman Directs the War Story," in *Gender, Sexuality, and the Cold War*, ed. Philip E. Muehlenbeck (Nashville: Vanderbilt University Press, 2017), 204–223.
21 Quoted in Ashley Pettus, *Between Sacrifice and Desire: National Identity and the Governing of Femininity in Vietnam* (New York: Routledge, 2003), 28.
22 For a history of youth and the Viet Minh Forces, see Hy V. Luong, *Revolution in the Village: Tradition and Transformation in North Vietnam, 1925–1988* (Honolulu: University of Hawaii Press, 1992).
23 Interview, Hanoi, 1996.
24 Male heroes earned cult status as well, but minority peoples who have contributed to the national story have been left out. For an analysis of the treatment of stories of national heroines in early Vietnam, see Cong Huyen Ton Nu Nha Trang, "The Makings of the National Heroine," The *Vietnam Review* 1 (1996): 388–433.
25 Hồ Xuân Huong, *Spring Essence: The Poetry of Ho Xuan Huong*, trans. John Balaban (Port Townsend, Washington: Copper Canyon Press, 2000), 94.
26 Interview, Hanoi, 1996.
27 Interview, Hanoi, 1996. The film *From a Jungle* [Tu mot canh rung] was made in 1975 but released in 1978 because the censors objected to her portraits of young love in the midst of war.
28 See Turner, "A Vietnamese Woman Directs the War Story."

29 In Le Minh Khue, *The Stars, the Earth, the River*, trans. Bac Hoai Tran and Dana Sachs, ed. Wayne Karlin (Willimantic, CT: Curbstone Press, 1997), 1–20.
30 "Nhung con quy mat den" is the rendering for "black eyed demons."
31 "Light and fresh" is "mat me" in the original text.
32 "Nhung ngoi sao ruc ro" is the rendering for "bright star."
33 Kkue, *The Stars, the Earth, the River*, 128–217. See Dana Sachs, "Small Tragedies and Distant Stars: Le Minh Khue's Language of Lost Ideals," *Crossroads* 13 (1999): 1–10.
34 This fiction emerged from the mid-1980s to the mid-1990s, a period of relative freedom for writers to express their disappointment in postwar culture. Younger writers are rejecting heroic narratives to explore more quotidian concerns. See Rebeka Linh Collins, "Vietnamese Literature after War and Renovation: The Extraordinary Everyday," *Journal of Vietnamese Studies* 10 (2015): 82–124.
35 Duong Thu Huong, *Novel without a Name*, trans. Phan Huy Duong and Nina McPherson (New York: Penguin Books, 1995), 49. See also Hue-tam Ho Tai, "Duong Thu Huong and the Literature of Disenchantment," *Vietnam Forum* 14 (1994): 82–91.
36 Huong, *Novel without a Name*, 132–133.
37 See Dang Thuy Tram, *Last Night I Dreamed of Peace: The Diary of Dang Thuy Tram*, trans. Andrew X. Pham (New York: Broadway Books, 2007).
38 For a useful essay on the relationship between what is seen and written by combat nurses, see Carol Acton, "Diverting the Gaze: The Unseen Text in Women's War Writing," *College Literature* 31 (2004): 53–79.
39 In contrast, see Gregory Daddis, "Historiographical Essay: Mansplaining Vietnam: Male Veterans and America's Popular Image of the Vietnam War," *Journal of Military History* 82 (2018): 181–207.
40 See chapter 3, Hue-Tam Ho Tai, "Daughters of Annam," in *Radicalism and the Origins of the Vietnamese Revolution* (Cambridge: Harvard University Press, 1992), 88–113.
41 In the memoir of Nguyen Hai Thaoi, included in a collection of memoirs and reports from military men, reporters or other personnel who witnessed activities on the Ho Chi Minh Trail. As the preface states, these were compiled to record the heroic determination of the men and women who worked on the Trail. Most of the authors are not professional writers and it is not clear how much of this material was redacted after the war. I thank Phan Thanh Hao and the Joiner Center for the Study of War and Social Consequences, University of Massachusetts, Boston, for help with an unpaginated manuscript copy in translation from the original publication in "Tap chi van nghe quan doi" [Military and Art Review], Hanoi, 1990. Hereinafter referred to as "Memoirs."
42 Duong Trong Dat, *Maiden Stars* [Nhung ngoi sao con gai]. Translated manuscript copy, courtesy of Phan Thanh Hao. Originally published in Ho Chi Minh City, 1985.
43 While outside the purview of this chapter, Judith Butler's conception of gender as performance is useful for understanding these examples of gendered displacements. Judith Butler, *Gender Trouble: Feminism and the Subversion of Identity* (London: Routledge, 1990).
44 Phan Trung Tue, "Memoirs."
45 General Tue's memoirs are the source for these observations about the volunteers. On his role in building the Trail, see Prados, *The Blood Road*, 112–113.
46 Cao Tien Le, "The Sound of Night," in *Writing between the Lines: An Anthology of War and Its Social Consequences*, ed. Kevin Bowen and Bruce Weigle (Amherst: University of Massachusetts Press, 1997), 6–50.
47 Nguyen Hai Thoi, *Memoirs*.

48 Ngo Ngoc Boi, "The Blanket of Scraps," in *Lac Viet*, 16, ed. and trans. Rosemary Nguyen (1997), 96–123.
49 See Diane Fox, "Book Review: Agent Orange: History, Science, and the Politics of Uncertainty," *Journal of Vietnamese Studies* 8 (2013): 120–125.
50 Nguyen Quang Lap, "The Sound of Harness Bells," in *The Other Side of Heaven: Post-War Fiction by Vietnamese and American Writers*, ed. Wayne Karlin, Le Minh Khue, and Truong Vu (Willlimantic, CT: Curbstone Books, 1995), 287–293.
51 See the analysis of oral histories from ordinary women veterans in Alexievich, *The Unwomanly Face of War.*
52 On the issue of dealing with wounded women veterans, see Rivka Syd Eisner, *Performing Remembering: Women's Memories of War in Vietnam* (Cham: Springer, 2018), 187–258.
53 Interview, Hanoi, 1996.
54 Filmed Interview, Hanoi, July, 2000.
55 Filmed interview, Hanoi, July, 2000.
56 Filmed interview, Hanoi, 2000.
57 Interview, Hanoi, 1996.
58 Interview Hanoi, 1996.
59 Turner, *Even the Women Must Fight*, 135.
60 Filmed interview, Hanoi, 2000.
61 Interview, Hanoi, 1996.
62 Filmed interview, Hanoi, 2000.
63 Interviews in Le Thi, *Single Women in Vietnam* (Hanoi: Gioi Publishers, 2006) reveal the hardship and shunning most of these women endure.
64 See Harriet M. Phinney, "Objects of Affection: Vietnamese Discourses on Law and Emancipation," *positions* 16 (2015): 235–258.
65 Le Thi Nham Tuyet, "Asking for a Child in Anhiep Commune," in *Some Research on Gender, Family and Environment in Development* 1 (Hanoi: 1996), 157–163.
66 Conversation with Van Anh Vo, Worcester, Massachusetts, March 2016.
67 Interview, Hanoi, 1997.
68 Gerard J. DeGroot, "Lipstick on Her Nipples, Cordite in Her Hair: Sex and Romance among British Servicewomen during the Second World War," in *A Soldier and a Woman: Sexual Integration in the Military*, ed. Gerard J. DeGroot and Corrinna Peniston-Bird (New York: Longman, 2000), 100–118.
69 From *Glorious Daughters of Viet Nam* (Hanoi: Viet Nam Women's Union, 1975).
70 Interview with Professor Nguyen Quoc Dung, Hanoi, 1997, in which he commented that records of casualties were not kept as carefully for the volunteers as for the regular soldiers.
71 See *Gender Practices in Contemporary Vietnam*, ed. Lisa Drummond and Helle Rydstrom (Singapore: Singapore University Press, 2004) and Ashley Pettus, *Between Sacrifice and Desire*.
72 Wendy N. Dong warns women's entanglement with warfare and collective suffering is dangerous for the state can call on both to suit its needs. Wendy N. Dong, "Gender Equality and Women's Issues in Vietnam: The Vietnamese Woman—Warrior and Poet," *Pacific Rim Law and Policy Journal* 10 (2001): 194–326.
73 Hue-Tam Ho-Tai, "Faces of Remembrance and Forgetting," in *The Country of Memory: Remembering the Past in Vietnam*, ed. Hue-Tam Ho Tai (Berkeley: University of California Press, 2001), 196–226, 167.

Index

Aberri Eguna 61
activism
   intellectual-political 152
   masculinized 172
   post-First World War 150
   women's 151, 168, 170–1
Africa
   African Other 5
   and colonialism 42, 44, 167
   and Pan-African 44
   and religious practices 46
   and women 13, 44, 49–50
   civilization 10
   diaspora 39, 44, 49
   elite 41
   history of 3, 13, 49–50
   languages 42
   memory 39, 51
   nationalism 43–4
   orature 42
   resistance 40, 45, 50
Album de la gloria de Chile (book) 207–9
al-Durr al-manthur fi tabaqat rabbat al-khudur (biographical dictionary) 155. See also Pearls Scattered
al-Fatat (magazine) 149–50, 153, 172
al-Hilal (magazine) 150, 157–8, 160–2, 165, 167–8
al-Jinan (serial) 166–7
al-Mahrusa (newspaper) 170
al-Muqtataf (magazine) 150
Álava Network 12, 65
Albania 114, 118
Alfonso VII 58
Amazons
   American 168–9
   Ancient Greek 200
   and electoral politics 167
   and other women warriors 171
   and the Amazonian army 166
   as a category 170
   Bohemian 6, 153, 166
   Chilean 201
   Dahomey 12–3
   discourse on 167
   figures of 165
   Guipuzcoan 57
   image of 7
   Lisan al-hal's 169
   modern 151
   phenomenon of 132
   the "two Amazons" 134
Amidai Yui no hama de (play) 78
anti-colonial 8, 93, 96, 99, 150, 164
Antifascist and Antiwar World Conference of Women 120
Apostolou, Electra 120
Arab
   and faith 164
   hero 149, 171
   language 167
   popular epic 149, 164–5
   retelling of Jeanne d'Arc 6, 157
   tradition 132
   travel writers 152
Arabophone 149, 151, 168
Aragón, Agustina de 1, 162
Arana, Sabino 56, 59, 61
Armanusa al-misriyya (novel) 161
Arrigorriaga 58–9
Aryan 5, 27, 29–30
   Aryanism 29
Australia 3, 43
Azurduy, Juana de 1–2, 6

Battle of the Bush (collection of plays and short stories) 103
Bernardo Suárez, José 204
biography 150, 154–168
Black Panther (film) 12

Black Power Movement 44
Black Women's Organization 49
Blood Road. *See also* Ho Chi Minh Trail
Book of the City of Ladies (biography) 163
Boubloulina (film) 117, 119, 124
Bouboulina, Laskarina 8, 115–124
Bouboulina of the Occupation, The (book) 120
Bradley, Hannah 96, 103, 105–7
Breña Campaign 208, 211
Brief History of the United States, A (textbook) 93–4
British South Africa Company 39
Buddhist 77, 85

Canada 131, 134–5, 137–8, 139, 142, 144. *See also* New France, Quebec
Canadian International Development Agency 48
cantineras 200, 205, 209
Castillo Ledón, Amalia de 190–1
Castrejón, Adrián 184, 187
Catholic
  anthropology 66
  Basque Catholicism 55
  Church 140, 182
  conversion 51, 133
  culture 134
  film 141
  French Catholic schools 237
  journal 156
  principles 60
  saint 156
  traditionalism 56
  women 190
Charwe 39–51
Chile
  "Chilean Amazon" 201–3
  and Peru 199–200, 204
  and women soldiers 207–9, 213
  Joan of Arc of in 11
  military action in 205–7, 212
  nationalism 208
China
  War of Resistance 217, 229
Chinese Communist Party 11, 217
civil rights
  for women 190
  movement 44
civilization 81, 105
  "Christian civilization" 204
  and nature 95, 100
civilizing
  and colonialism 10, 100–1
  figures 98, 153
  martial mothers 95
  the wilderness 99, 106
class
  and women 83–4, 124, 167
  conflict 113, 225, 239
  education 226
  middle in Mexico 182
  popular in Mexico 188
  upper in Chile 211
colonial(ism). *See also* postcolonialism
  colonial America 10
  liberation 8
  administrators 27
  policy 27, 95
  New France 133
  French colonial rule 167
  and womanhood 96, 98, 101
  European 49
  and gender 28
  theory 29
  and censorship 29
  mission schools 43
  and memory 46, 93, 95
  as a legal system 50
  struggle 8, 26, 40
  British 10, 30, 40
  archive 44–5
  settler colonialism 45–6, 50–1
  narratives and discourses of 24–5, 95
  New England 97–8, 105, 107
  and woman warrior 5, 9
Colossus of Maroussi, The (travelogue) 118
Chūshingura (play) 75. *See also The Treasury of Loyal Retainers*
commemoration
  of Greek Revolution 117
  of heroines 114
  process of 134, 185
  public 223
  wartime 217
"Company 814" 242, 244, 246–7

Confucian 76, 84, 237, 248
Cult
 hero and heroine 3, 9, 55, 107, 199, 208
 warrior 97
Cultural Revolution 222, 226, 228
culture
 Afro-Brazilian 6
 Anglo-American 98
 Czech 6
 female 42, 44
 French-Catholic 134
 Japanese 75
 oral 8
 patriarchal 47
 popular 32
 postwar 236–7
Cumann na mBan 60
Cut Nyak Dhien 8
Crossroads (movie) 218, 229

Dalai Lama 40
Democratic Army 121–4
Dang de nuer, Liu Lishan (novel) 222
dictatorship
 Greek 113
 Spanish 65
Distant Stars (novel) 239–41
Doctrine of Lapse 25, 27, 35
Dolores, Battle of 206, 208
Domesticity. *See also* separate spheres
 and motherhood 95
 and representations of women 99
 cult of 100
 discourse of 161, 167
 Indigenous 102
 trained 153
Dustan, Hannah 96–7, 100–8
Dutch Hivos 48

EAM (National Liberation Front) 120
East India Company 24–6, 31
Edo (Tokyo) 77
Egypt
 "the Egyptian Jeanne d'Arc" 11, 150, 155
 and renaissance 150
 and women 149, 153, 162, 172
 conquest of 161, 167
El Comercio (newspaper) 212

El Coyote: Corrido de la Revolución (book) 188
El Ferrocarril (newspaper) 203
El Independiente (newspaper) 203
El Peruano (newspaper) 204
ELAS (Greek People's Liberation Army) 120
Emakume Abertzale Batza 60
emotion
 and femininity 61
 and self-restraint 65–6
 and sexuality 186, 189
 patriotic 66, 143
Empire. *See also* colonialism
 British 2, 23, 39, 40–9, 51
 Ottoman 113, 115, 172
 Spanish 185
EPON 120. *See also* United Panhellenic Organization of Youth
equality
 gender 13, 114, 169, 192, 245
 women advocate for 13, 120
Erauso, Catalina de 58, 61, 185
Espinosa, María de la Luz 186
espionage
 and the Chinese Communist Party 221
 and women 220
 Second World War 65
 narratives about 223
 tactics 220
ETA. *See* Euskadi Ta Askatasuna 65
Euskadi Ta Askatasuna, ETA 65
euskera 55
Euzkadi (Basque Country) 61
exhibition
 and women 153, 228
 museum 235
 photo 120
 public 77

Faber, Enriqueta 186
Fatat al-sharq (magazine) 164, 171–2
faith
 "faithful wife" 103, 205
 and nation 171
 fighting for 164
feminine
 and masculinity 28, 172, 179
 and nationalism 173, 185, 190

civilizers 94
discourses on 153, 185
iconography 151
identity 11, 187–9
organizations 60
political authority 149
power 26
qualities 10, 153, 168, 189–91
roles 11
solidarity 154
sphere 102
subjects 57
feminism
   1930s 193
   1970s 67
   modern forms of 190
feminization
   of Amelio/a Robles 185, 188
Feso (book) 41, 43
Fiction. *See also* individual novel titles
   and Indigenous people 104
   and men 234
   and women 29, 31, 85, 100, 154, 236, 243
   heroines 66
   Victorian 27
   writers of 95
Fin-de-siècle 150–1
First Peoples 47. *See also* Indigenous
First War of Independence 5, 24
Flow Red the Ganges (novel) 29
Flashman in the Great Game (novel) 28
Flow Red the Ganges (novel) 29
folk
   heroines 81, 203, 211
   and Indigenous 30
   songs 23, 86
Francoism 65
French Canada 131–144. *See also* New France, Quebec
From a Jungle (film) 238
Fukushima 77

Gandhi, Mahatma 26
Gandhi, Sonia 32–3
Gender
   fluidity 151, 163, 169, 171
   invisibility 171–3
   politics 166, 169
   roles 31, 103, 132, 138, 157, 200, 229, 234, 239, 241

Gentleman's Journal (magazine) 166–7, 170
Getsudō kenmonshū (reportage) 81
Glorious Revolution 97, 228
Go Taiheiki Shiroishi banashi (play) 78
Granja, José Luis de la 55
Greece
   emancipation of 117
   emancipation for Greek women 121, 124–5
   Greek War of Independence 114–6
   revolution 115
   warrior women 76, 113–4
Grey, Lord 138–9, 144
Guan Lu 217–8
Guerrero (state government) 179, 181–88
guerrilla 179, 181–7, 206
Guipúzcoa 57, 64

Haru 76
Haudenosaunee 131–3, 136, 141–3
Heroines of the Greek Revolution, The (book) 116, 120
hierarchy
   and women 123
   church 141
   military 244
   of family 76
   traditional 187
Hindu Widows 27
Hua Mulan 1–2, 12–3
Hitler, Adolf 26, 118
Ho Chi Minh 233, 236, 239
Ho Chi Minh Trail 233, 239. *See also* Blood Road
Hongyan (novel) 224. *See also* Red Crag
Hongse tegong midang (documentary) 229
Hundred Years War 1
Hungary 123

identity
   gender 150, 179, 182, 184, 186, 192
   national 183, 188–9, 211
   transgender 193
Inca 107, 240
India
   and sexuality 28
   cultural mores 27
   leaders 23
   nationalism 10, 23–4, 26, 29–30
   Queen (Rani) 29

Remarriage Act of 1856 27
womanhood 31
Indian National Army 26
Indigenous. *See also* individual nations by name
   and memory 93, 101
   and nationalism 200
   and the female body 94
   colonialism 42, 47, 100
   female warrior 9, 12, 98, 101, 108, 151, 165, 211
   hero 3, 104
   representations of 103
   resistance 30
Indo-Chinese Communist Party 237
industrialization 55
Ireland 13, 60, 103
Islam(ic)
   and piety 165
   and women 162
   tradition 155
   warrior women 149, 151
Israel 1
Italy 118

Japan
   "Japanese pirates" 221
   communists 220
   kokka (state-family) 76
   War of Resistance against 217, 229
Jeanne d'Arc. *See also* Joan of Arc
   French 1
   and women's rights 7
   and gender fluidity 11
   Indian 29
   film 141, 143
   canonization 156
   Chinese views of 171
   Turkish 172
   Egyptian 150
   Arab 4, 6, 149, 151–2
   heroine 2, 4
   warrior 6, 7
Jhansi 5–8, 23–34
Jhansi Ki Rani (film) 29, 30
Jhansi Ki Rani (novel) 29, 33
Jhansi Ki Rani (television series) 31
Jiang Jie (play) 226
Jiang Jie (television series) 226

Jiang Zhuyun 227–8
Joan of Arc. *See also* Jeanne d'Arc
   Aryan 29
   Chilean 201
   episodes devoted to 13
   French 139
   of Canada 144

Kang Ri nü qingbaoyuan Guan Lu (documentary) 229
King Philip 100, 103
King Philip (play) 105–6
Kōkoku nijūshikō (print) 78
Korea 84
kwaHarare 40

La Coronela (dance performance) 12, 184–5, 189, 191
La mujer peruana a través de los siglos (book) 211
La Negra Angustias (novel) 191
language
   African 42
   Arabic 167
   body 192
   English 43–4
   euskera (Basque) 55
   gendered 233
   of gender 50
   of revolution 226
   racialist 166
Larramendi, Manuel de 57
Last Night of a Nation, The (drama) 104
Latin America
   heroes 199
   history of 185, 193
La accion de Yungay (play) 202
Le Mercure Galant (court publication) 133
Le Minh Khue 143, 239–41
Lebanon 155–7, 162
LGTBI 192
Libe (character) 10, 56, 58–9, 61, 64, 66
Libe (play) 59
Libe: Melodrama histórico (book) 63
Lisan al-hal (newspaper) 168, 170–1
Los grandes del Perú (compendium) 211
López, Juana 209
Le Japon, Histoire et Civilization (book) 81

Macedonia 123
Manikarnika (film) 12, 31–4
Manikarnika 25
Manto Mavrogenous (film) 115–7, 119
Maoist 222, 225–6
Mares, Encarnación "La Chona" 186
martyr
   Chinese Communist Party 222–4, 226–9, 237
   female 11, 61
   depicted as 122
   heroic 114, 218, 224
   male 164
   warriors 223
Lachmi Bai of Jhansi: The Jeanne D'Arc of India (novel) 29
Ma'rid al-hasna' fi tarajim mashahir al-nisa' (dictionary) 153
Madeleine Vercheres (film) 131
Madeleine Vercheres (poem) 135
Mao Zedong 222–5
masculinity
   British 28
   constructed 183
   expressions of 179
   and feminine, femininity 28, 32, 167
   perceptions of 182
Mavrogenous, Manto 115–7, 119
Mazowe 39, 43, 51
medieval
   anecdotes 151
   France 150
   queen 162
   stories 61
   writings 168
Meiji 78, 81
Mekatilili wa Menza 1, 3
Meletzis, Spiros 120–2
memory
   politics of 234
   collective 114, 203, 209, 213, 244
   policies 199
   popular 11, 39–41, 45, 47, 49–50
mestizo
   groups 211
   women 2, 202
México
   and women 191
   City 188

history of 193
leader 184
rural 182
State 181
Mexican Revolution 179
microhistory 45, 181
militiawomen 64
misogyny 55–9, 66, 169
Miyagino 10, 12, 77–8, 80–9
modernity. *See also* modernizing
   and colonialism 173
   and gender 150, 167
   and politics 56
   and women 172
   indigenous 151
modernizing. *See also* modernity
   and nationalism 150
   intellectuals 164
   societies 150
Monja Alférez, La 185
Morales, Irene 206, 208–9
Moreno de Cáceres, Antonia 206, 211–3
motherhood
   "republican motherhood" 98
   and domesticity 95
   and women 235, 246
   desire for 145
   ideals 100
   martial 103, 105
   narratives of 229
   themes of 101
Munguía
   battle of 10, 59
Musume katakiuchi kokyō no nishiki (comic) 78

National Confederation of Veterans of the Revolution (Confederación Nacional de Veteranos) 187
Nazism 65
Nehanda
   and warrior woman 44–5, 50
   children 44
   legend of 43–4, 49–50
   name of 42
   narrative of 50
   of popular memory 47, 49
   organization 49
   spirit of 46

Nehanda (book) 47–8
New England
    "New England Kitchen Scene" 93–4, 97
    colonial 96, 98–9
    women 102
New France
    and colonialism 131–3
    heroine of 140
New Zealand 43
Nguyen Thi Duc Hoan 235
Nguyen Thi Lien 241
Nguyen Thi Mau 245–6
Njinga, Queen of Angola 8, 13
Norway 1
Novel without a Name (novel) 240, 243
Nüsheng (magazine) 218

Obregón, Álvaro 184
Official Mutiny Narratives (book) 28
oral tradition 9, 39, 200, 203, 206, 211
Ordoñez, Leonor 200, 206, 210, 212–3

Panagiotidou, Eleni (Titika) 210, 121
Pantheon
    Amazons in the 199
    of national heroes 205, 208, 211
Pardo Bazán, Emilia 57
Phil-o-rum's Canoe (poem) 135
Psarogianni, Olga Mastora 123
patriarchy
    and female fidelity 218
    and racism 49
    categories of 24
    in male-dominated societies 32
Pearls Scattered 157, 162. *See also* al-Durr al-manthur fi tabaqat rabbat al-khudur (biographical dictionary)
Pérez, Candelaria 9, 200–3, 206
Perú
    and Chile 199–200, 204, 205
    and feminine heroism 207
    and hero-making 9
    and women 201, 207–9
    historical narrative 201
    national heroes 211
Peru-Bolivian Confederation 201
Perú Ilustrado (magazine) 209
Pingguo yuan (novel) 222

Popular History of the United States, A (book) 93, 95
postcolonial. *See also* colonial
    conceptualizations of gender 24
    literature 47
    nation 30
    rule 26
Psichi Vathia (film) 122

Qatf al-zuhur fi ta'rikh al-duhur (book) 152
Quebec 7, 134, 136, 138, 140. *See also* Canada, New France
Queen Nanny (movie) 8
Queen's Desire, The (novel) 28
Queen Victoria 23, 26, 29, 39
Qur'an 157, 168

rabonas 200, 205–7
race
    and femininity 57
    and sexuality 27, 29–30
    degeneration of 55, 58
    mixed-raced 2, 9
Rane: A Legend of the Indian Mutiny, The (novel) 28
Rani Lakshmi Bai 5, 10, 23–34
Rani of Jhansi (character) 28–9, 31–2
Rani of Jhansi (novel) 28–9
Rani of Jhansi (comic) 31
Rawdat almadaris al-Misriyya (school magazine) 166
Red Crag (novel) 224–6, 228. *See also Hongyan*
Real Family History of Sister Jiang, The (book) 228
religion
    and missionaries 153
    and spiritual practices 46
    and language 134
    politics of 32
religiosity 56, 150, 157
Remarriage Act of 1856. *See* India
Renmin ribao (daily) 223
Republican Motherhood 98. *See* motherhood
Rhodesia
    Revolt in Southern Rhodesia (book) 45, 51
    Rhodesia Literature Bureau 43

rights
  civil 44
  gay 2, 192
  land 8
  language of 173
  political 11, 55, 151, 167–8
  women's 6, 7, 149, 150, 168–9, 170, 190
  worker's 182
Robles, Amelio/a
  biography of 181
  feminization of 188–192
  masculinization of 186–8
  perception of 179
Rodríguez, Dolores 205–6
Romancero Alabés (poem) 58
Rose, Sir Hugh 25, 28

samurai 75–78, 82–6
Santa Cruz, Andrés de 201
Saudi Arabia 13
Semiramis 152, 162–5
Sendai 76–8, 81–3
settler colonialism 40, 45. *See also* colonial
sexual
  and politics 224, 228
  fidelity 11, 84, 217, 223, 230
  order 217
  virtue 221
sexualized 8, 58, 163, 221, 233
Shiga Danshichi 76–7, 85
Shinobu 10, 12, 77–8, 81, 83–9
"Shiroishi-banashi Miyagino, Shinobu Monogatari" (drama) 86
Shiroishi Castle 77, 86
Shizi jietou (film) 218
Small Tragedy, A (novel) 239
social Darwinism 55
Socialist Party of Guerrero (Partido Socialista de Guerrero) 187
Soga monogatari (tale) 78
Sota, Manuel de la 61, 63–4
Souli 114–24
Sovereignty [sic] and Goodness of God, The (narrative) 93
South Africa
  and apartheid 48
  history 50
  peoples of 40
  police 40
  Union of 43
Soviet Union 123
Spain
  Civil War 64
  Monarchy 199
separate spheres 97–8. *See also* womanhood

Tabaqat (genre) 155
Taiheiki kikusui no maki (play) 78
Taka 76
Tale of Shiroishi and the Taihei Chronicles, The (play) 78
Times, The (newspaper) 166–7
Tokugawa period 75, 81, 84
Treasury of Loyal Retainers, The (play) 75. *See also* Chūshingura
Treaty of Ancón 208
Trieu Thi Trinh 238
Trung sisters 236
Turkey 172

United Panhellenic Organization of Youth 120. *See also* EPON
US Global Fund for Women 48

Valenzuela, Filomena 209
Varona, La 58
Verchères, Madeleine de 11–2, 131, 134–6, 138–9
Vietnam
  War 236
  Vietnamese Women's Union 247
Vietnam War, The (documentary series) 236
Vizcaya 58

Weetamoo 9, 93–8, 101–3
Widowed Queen, The (novel) 29
With the Partisans in the Mountains (photo collection) 121
Woman warrior
  Bolivian 6
  colonial 5, 9
  cult 97
  global 9
  imagined 151

    of French Canada 131, 143
    of the revolution 115, 124
    representations of a 23
    figure of 14, 60, 115, 144, 151, 164–5
    term 10
    tradition 131, 133–4, 144
    womanhood of 7
womanhood
    and femininity 44
    and woman warrior 7
    heroic 5, 29
    manifestation of 95
    martial 94, 98, 104, 107–8
    nationalist 171
Women Leaders in African History (book) 49
Women's Voice (magazine) 218, 220
World Congress of the Communist
    International Youth 120

Xin jiu shidai (serialized autobiography) 219
Xinhua ribao (daily) 222

Xochipala 179, 181, 184–91

Yoshiwara 78
Yotarō 76, 83, 85
Youth Union (Vietnam) 243, 245
Yugoslavia 118, 123
Yui Shōsetsu 77, 81–2
Yungay, Battle of 201–3

Zaloggo
    dance 119, 125
    legend 124
Zaloggo; the Castle of Freedom (film)
    117
ZANU-PF 47–8, 50–1
Zapata, Emiliano 181, 181–3
Zapatista 9, 179, 193
Zimbabwe
    National Museums and Monuments of
        50–1
Zuria, Juan 59

www.ingramcontent.com/pod-product-compliance
Lightning Source LLC
Chambersburg PA
CBHW070024010526
44117CB00011B/1706